ELECTRICAL ENGINEERING DEVELOPMENTS

LIGHT-EMITTING DIODES AND OPTOELECTRONICS

NEW RESEARCH

ELECTRICAL ENGINEERING DEVELOPMENTS

Additional books in this series can be found on Nova's website
under the Series tab.

Additional e-books in this series can be found on Nova's website
under the e-book tab.

ELECTRICAL ENGINEERING DEVELOPMENTS

LIGHT-EMITTING DIODES AND OPTOELECTRONICS

NEW RESEARCH

JOSHUA T. HALL

AND

ANTON O. KOSKINEN

EDITORS

Nova Science Publishers, Inc.

New York

NOTICE TO THE READER

Additional color graphics may be available in the e-book version of this book.

LIBRARY OF CONGRESS CATALOGING-IN-PUBLICATION DATA

Light-emitting diodes and optoelectronics : new research / editors, Joshua T. Hall and Anton O. Koskinen.
p. cm.
Includes bibliographical references and index.
ISBN 978-1-62100-448-6 (hardcover)
1. Light emitting diodes. 2. Optoelectronic devices. I. Hall, Joshua T. II. Koskinen, Anton O.
TK7871.89.L53L527 2011
621.3815'22--dc23
2011035487

Published by Nova Science Publishers, Inc. † New York

CONTENTS

Preface   **vii**

Chapter 1   Approaches for Fabricating High Efficiency Organic
                Light Emitting Diode for Flat Panel Display
                and Solid State Lighting: An Overview   **1**
                Jow-Huei Jou, Sudhir Kumar, Yung-Cheng Jou
                and Shih-Ming Shen

Chapter 2   Reliability Estimation from the Junction
                to the Packaging of LED   **43**
                Yannick Deshayes

Chapter 3   The Next-Generation Intelligent and Green Energy LED
                Backlighting 3D Display Technology for the Naked Eye   **83**
                Jian-Chiun Liou

Chapter 4   Intersubband Transition in CdS/ZnSe Quantum Wells for
                Emission and Infrared Photo Detection   **117**
                S. Abdi-Ben Nasrallah, N. Sfina, S. Mnasri and N. Zeiri

Chapter 5   Inorganic-Organic Hybrid Emitting Material
                Fabricated by Solvothermal Synthesis   **151**
                Takeshi Fukuda

Chapter 6   Enhanced Efficiency of ZnTe-Based Green
                Light-Emitting Diodes   **173**
                Tooru Tanaka, Katsuhiko Saito, Qixin Guo
                and Mitsuhiro Nishio

Chapter 7   Self-Introduced Lattice Distortion, Invisible Cavity and Hidden
                Collective Behavior of a Polymeric Nanofiber Laser   **195**
                Sheng Li Wei-Feng Jiang and Thomas F. George

Chapter 8   Photonic Bandgap Defect Structure on iv-vi Semiconductor:
                Resonating Cavity without Cleaving   **211**
                Shaibal Mukherjee

Chapter 9   Modelling of Widebandgap Light Emitting Diodes:
                  From Heterostructure to LED Lamp   **231**
                  K. A. Bulashevich, I. Yu. Evstratov, S. Yu. Karpov,
                  O. V. Khohlev and A. I. Zhmakin

Index   **287**

PREFACE

This book presents topical research in the field of light-emitting diodes and the systems, uses and efficiency of optoelectronics. Topics discussed include fabricating high efficiency organic light-emitting diodes for flat panel displays and solid-state lighting; reliability estimation from the junction to the packaging of LED; next-generation intelligent and green energy LED backlighting 3D display; inorganic-organic hybrid emitting material fabricated by solvothermal synthesis and photonic bandgap defect structure based on IV-VI semiconductors.

Chapter 1 – At present time organic light emitting diodes (OLEDs) are a promising technology for solid state lighting and flat illuminated displays. OLED devices reduce the energy consumption due to its low driving voltage and high energy efficiency. Low driving voltage to some extent may be regarded as less heat production, which in turn shall be most favourable for the next generation illumination since green house effect has become a critical issue. The other features even characterize OLED devices; they are, e.g., colour tunable, colour-temperature tunable, very high colour rendering index, flexible, transparent, fully dimmable, sustainable materials and environmentally friendly etc. Its exquisite and well-designed traits meet with the trend that "Things are getting smaller". Multifarious diversity in its structure design creates unlimited possibilities for future development. Parallel light-emitting beam like sunlight, which does not dazzle the eyes while study and wide-view angle for application in display are another two splendiferous characteristics of OLEDs. In last few years the OLED device efficiency has significantly improved by the modification in device structure and/or geometry. The authors are describing an assortment of techniques, like thin device layer thickness, effective carrier confinement, high carrier transporting ability, low carrier injection barrier, high electroluminescent (EL) active molecules, effective host guest energy transfer, exciton generation on host, balance carrier injection and wider recombination zone and wider recombination zone and improve the light out coupling efficiency, using various out-coupling techniques like substrate modification, texturing mesh surface, aperiodic dielectric stacks, microlens arrays, low index grid, effect of micro cavity, surface plasmon and photonic crystal, to improve the device efficiency.

Chapter 2 – Selection tests and life testing can be used to satisfy the requirements of early and long term failures. But besides the fact they are costly and time consuming. Generally they do not give efficient information about failure mechanisms and defect nature in particular for optical modules. In this context, much effort has been conducted to clarify degradation mechanisms and eliminate the causes. Many studies have demonstrated the sensitivity of the LED to ageing tests leading to packaging or semiconductor degradations.

Chapter 3 – Autostereoscopic displays provide 3D perception without the need for special glasses or other headgear. Drawing upon three basic technologies, developers can make two different types of autostereoscopic displays: a 2D/3D switchable display, head-tracked display for single-viewer systems and a time- multiplexed multiview display that supports multiple viewers. This study investigated the method of using an autostereoscopic display with a synchro-signal LED scanning backlight module to reduce the crosstalk of right eye and left eye images, enhancing data transfer bandwidth while maintaining image resolution. In the following they introduce how LED backlight and optical film elements are used in autostereoscopic 3D display designs including two-view and multi-view designs.

Glasses-free 3D Display Technology, two ways of manufacturing a two-view spatially multiplexed autostereoscopic display. (a) Lenticular: An array of cylindrical lenslets is placed in front of the pixel raster, directing the light from adjacent pixel columns to different viewing slots at the ideal viewing distance so that each of the viewer's eyes sees light from only every second pixel column. (b) Parallax barrier: A barrier mask is placed in front of the pixel raster so that each eye sees light from only every second pixel column.

Lenticular lens technology chiefly involves adding periodic arrays of lenticular lens to the outside of a general display. The design must be optimized in accordance with viewpoints and pixel size in order to produce images that can then be transmitted in varying directions. Since each sub-pixel occupies a different position, its off-axis distance to the lenticular lens will vary. With proper design, parts of the sub-pixels can appear in distinct viewpoints. 3D display can then be realized by employing pictures or images taken at these different viewpoints. Since this technology creates parallax through lens optics, which provides a very high optical efficiency, the only brightness loss comes from interface reflection, lens material transmittance, and light scattering.

A parallax barrier is a device placed in front of an image source, such as a liquid crystal display, to allow it to show a stereoscopic image without the need for the viewer to wear 3D glasses. Placed in front of the normal LCD, it consists of a layer of material with a series of precision slits, allowing each eye to see a different set of pixels, so creating a sense of depth through parallax in an effect similar to what lenticular printing produces for printed products. A disadvantage of the technology is that the viewer must be positioned in a well-defined spot to experience the 3D effect. Another disadvantage is that the effective horizontal pixel count viewable for each eye is reduced by one half; however, there is research attempting to improve these limitations.

A novel time-multiplexed autostereoscopic multi-view full resolution 3D display is proposed. This capability is important in applications such as cockpit displays or mobile, portable, or laptop systems where brightness must be maximized but power conserved as much as possible. The effects are achieved through the creation of light line illumination, by means of which autostereoscopic images are produced, and by simultaneously concentrating the light emitted by the display toward the area the viewer's head is. By turning different illumination sources on and off, it is possible to aim both the concentration area and the 3D viewing area at the observer's head as the observer moves. A variation on the system allows two or more persons to be tracked independently. Cross talk (ghosting) can be reduced to the point of imperceptibility can be achieved.

Chapter 4 – Theoretical study can make a significant contribution to experimental studies and may have profound consequences as regards practical applications of devices.

A great attention has been devoted to the intersubband transition and semiconductor based quantum wells or quantum dots are promising candidates for high-speed intersubband optical devices relying on the quantum confinement of carriers. In fact, the current microelectronics technology does not satisfy the demands of future communication for faster data transfer and a new breakthrough is needed. With the advent of modern epitaxial growth techniques, it has become possible to grow II-VI low dimensional structures and tailor the band to achieve the desired properties for device applications. The band gap engineering allows the design of new optoelectronic devices operating in a specific spectral range. In particular, efforts to improve the optical performance of the CdS/ZnSe SCs has been a major subject of research in the field of II-VI optoelectronic due to the promising applications of this family based light emitters and detectors.

With this motivation, the authors have proceeded, since a few years, to expand their studies in the areas of compact tunable lasers, infrared photodetectors, and applications in non-linear optics based on CdS/ZnSe system. Various devices are proposed and investigated.

Chapter 5 – Eu-complex is one of the most interesting lanthanide with the organic ligand owing to its high photoluminescence (PL) quantum efficiency by the ultraviolet (UV)-excitation. Therefore, Eu-complex is expected as the red-emitting phosphor by the UV-light irradiation for the white light-emitting diode. However, the important problem for the practical application is that the poor long-term stability. Nowadays, the athors demonstrated the improved long-term stability against the UV-light irradiation by encapsulating the sol-gel derived glass network around Eu-complex. The vapor and oxygen are protected by the silica glass layer to react Eu-complex, resulting in the high long-term stability. Especially, the long-term stability was drastically improved by annealing in the autoclave container with high pressure, called as the solvothermal process. By optimizing the silane alkoxide and the solvent in the sol-gel starting solution, the authors successfully achieved the particle structure of the silica glass coated Eu-complex with the diameter of less than 100 nm. Among the tested combinations of the sol-gel process, the combination of TEOS and ethanol is the best condition to realize the high PL quantum efficiency and the long-term stability. Therefore, this organic-inorganic emitting material will be applied as phosphors for practical applications.

Chapter 6 – ZnTe is expected to use in a variety of optoelectronic devices such as pure green light-emitting diodes (LEDs) because of its direct transition band gap of 2.26eV at room temperature (RT). Although p-type ZnTe was obtained easily, the growth of an n-type material was difficult due to a self-compensation effect and an incorporation of residual impurities. Among several attempts to realize the n-type conductivity in ZnTe, a thermal diffusion technique showed high potential for obtaining n-type ZnTe, resulting in a development of ZnTe-based green LEDs. In this chapter, the authors review their recent research results on the development of ZnTe LEDs with a special emphasis on the thermal diffusion technique through the oxide layer, which plays an important role as a diffusion limiting layer. Using the Al-diffused layer with a controlled Al-concentration, the output power of ZnTe LED has been increased significantly.

Chapter 7 – A polymeric laser generally undergoes two unconventional processes: (1) microscopic lattice and electronic evolution and (2) macroscopic localization of light emission. After an external pulsed laser beam is applied to pump a conjugated polymer comprised of parallel polymer fibers, such as polythiophene, the external excitation immediately destroys the periodic structure of the polymer chain, self-introducing localized lattice distortion along the polymer fiber chain. Along with the continuous optical pumping, the electron populations of the exciton in a single polymer fiber are reversed. With multiple light scattering in the fibers, the external gain then counteracts the leakage that causes non-coherent light emission, and finally localizes the light in the middle of the fiber bunch. Concurrent with the localization of the polymer fiber laser, an associated hidden collective behavior is also uncovered. It is revealed that the multiple scattering, instead of phase tuning in the traditional resonator, results in the cavity of the fiber laser to be comprised of only randomly-distributed polymer fibers, and accordingly, to be "invisible."

Chapter 8 – For a few decades, lead chalcogenide diodes have been one of the most popular and commercially available semiconductor mid-infrared (MIR) lasers. However, due to the low operating temperatures and low external efficiencies, the performances of lead salt lasers could not reach the desired level. Even at low cryogenic heat-sink temperature, lasing output powers for single mode operation are less than few mWs. Moreover, an affinity towards multimodal operation and mode hopping is quite familiar. The threshold level of laser operation for IV-VI materials is enhanced due to the existence of four-fold L-valley degeneracy near band extrema. The degeneracy does not get lifted off for orientation which is the most common growth orientation for QW laser structures. This, in turn, limits exploiting the supreme advantages of lead chalcogenide materials for high temperature and long wavelength operation i.e., reduced threshold level originating from a low nonradiative recombination rate. Also growth orientation leads to inferior epitaxial material quality because it does not allow dislocation gliding phenomenon.

Chapter 9 – Light-emitting diodes (LEDs) made from widebandgap materials, first of all, from group III nitrides, are among the most perspective optoelectronic devices serving both as sources of green, blue and ultraviolet light and as the basis of white LEDs for solid state lighting. Nowadays, growth technology of group-III nitride semiconductors and engineering of III-nitride advanced heterostructure devices has leaved far behind a detailed understanding of their operation. This is largely because of non-ordinary physical properties of III-nitrides like spontaneous electric polarization, strong piezoeffect, extremely low acceptor activation efficiency, and a high dislocation density inherent in epitaxial materials. All said, in combination with a complex multi-layer heterostructure and carrier degeneration typical for high power LEDs, hampers an intuitive designing of device based on analogies with conventional III-V compounds. Numerical simulation provides a deeper insight into physical mechanisms underlining the device operation and allows reduction in both time and cost of LED development and optimization. Difficulties of numerical simulation of widebandgap LEDs stem from the high anisotropy of the die, stiffness of the governing equations, extremely low concentrations of intrinsic carriers, a presence of strong build-in electrical fields, and large band offsets at the heterointerfaces.

The chapter describes approaches to III-nitride and II-oxide LED modelling including hybrid (1D/3D) ones and coupling of electrical and thermal processes and light generation/propagation. Results of modelling at different levels—a heterostructure, a die, a die with encapsulant, a LED lamp — are presented and used to analyze factors affecting internal quantum efficiency, high-current efficiency droop, current crowding, efficiency of light conversion in white LEDs, light extraction efficiency.

In: Light-Emitting Diodes and Optoelectronics: New Research   ISBN: 978-1-62100-448-6
Editors: Joshua T. Hall and Anton O. Koskinen   © 2012 Nova Science Publishers, Inc.

Chapter 1

APPROACHES FOR FABRICATING HIGH EFFICIENCY ORGANIC LIGHT EMITTING DIODE FOR FLAT PANEL DISPLAY AND SOLID STATE LIGHTING: AN OVERVIEW

Jow-Huei Jou[*], *Sudhir Kumar, Yung-Cheng Jou*
and Shih-Ming Shen
Department of Materials Science and Engineering, National Tsing Hua University,
Hsin-Chu, Taiwan, Republic of China

ABSTRACT

At present time organic light emitting diodes (OLEDs) are a promising technology for solid state lighting and flat illuminated displays. OLED devices reduce the energy consumption due to its low driving voltage and high energy efficiency. Low driving voltage to some extent may be regarded as less heat production, which in turn shall be most favourable for the next generation illumination since green house effect has become a critical issue. The other features even characterize OLED devices; they are, e.g., colour tunable, colour-temperature tunable, very high colour rendering index, flexible, transparent, fully dimmable, sustainable materials and environmentally friendly etc. Its exquisite and well-designed traits meet with the trend that "Things are getting smaller". Multifarious diversity in its structure design creates unlimited possibilities for future development. Parallel light-emitting beam like sunlight, which does not dazzle the eyes while study and wide-view angle for application in display are another two splendiferous characteristics of OLEDs. In last few years the OLED device efficiency has significantly improved by the modification in device structure and/or geometry. We are describing an assortment of techniques, like thin device layer thickness, effective carrier confinement, high carrier transporting ability, low carrier injection barrier, high electroluminescent (EL) active molecules, effective host guest energy transfer, exciton generation on host, balance carrier injection and wider recombination zone and wider recombination zone and improve the light out coupling efficiency, using various out-coupling techniques like

---

[*] Author whom correspondence should be addressed: jjou@mx.nthu.edu.tw.

substrate modification, texturing mesh surface, aperiodic dielectric stacks, microlens arrays, low index grid, effect of micro cavity, surface plasmon and photonic crystal, to improve the device efficiency.

Keywords: Organic light emitting diode, high efficiency, solid state lighting, flat panel display

1. INTRODUCTION

An efficient, single hetero-junction structure, organic light emitting diode (OLED) was first demonstrated by the Tang and VanSlyke in 1987.[1] OLEDs have aroused extensive research attention from both academia and industry in the last two decades.[2-3] OLEDs are increasingly attracting interest because of their great potential as high-quality flat panel displays, liquid-crystal-display back-lighting, automotive and solid state lighting.[4-8] In these days, OLED technology has already been used for small displays in numerous commercial products such as MP3 players, cameras, mobile phones and i-pads etc. Generally, for this category of small displays, passive matrix (PM) driving technology featuring a somewhat simple scheme is used.

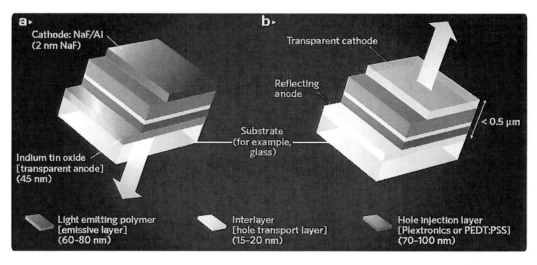

Figure 1. Device structure of (a.) bottom-surface emitting OLEDs architecture and (b.) top-surface emitting OLEDs architecture [9].

Figure 1 shows schematically the bottom-surface emitting and top-surface emitting OLED architectures. [9] However, to produce high-resolution and high-information content large-area displays, the active matrix (AM) driving mode has to be used, which implements a driving unit underneath each pixel and, hence, requires much more complicated addressing circuits than those of PM displays. Furthermore, to achieve stable and uniform luminance, an even more complicated addressing scheme is needed. In either PM or AM-OLED display or illumination, high device efficiency is always highly desired in order to prolong lifespan and standby time, as well as to be energy-saving. Here we will discuss several device structures approaches and light out-coupling techniques to fabricate efficiency enhanced OLEDs for flat panel displays and solid state lighting.

2. APPROACHES FOR HIGHLY EFFICIENT DEVICES

2.1. Device Structure

Over the past two decades, tremendous advances in the area of OLEDs have been achieved mainly through the synthesis of efficient lumophores and the development of improved device structures. [10–13] Tang et al. have reported that by doping the luminescent layer with highly fluorescent molecules, EL efficiency has been improved by about a factor of two in comparison with the undoped cell. [5] In this doped system, exciton formation in a guest dopant molecule would result from either energy transfer from an exciton formed on the host or sequential trapping of a hole and an electron directly on the guest molecule. High-brightness OLED with a double-quantum-well structure has been demonstrated by Huang et al. [14]

2.1.1. Low Carrier Injection Barrier

By lowering the injection barrier for either holes or electrons to enter the host, the driving voltage of the device can be effectively reduced and thus yield a higher efficacy. The improvement on lowering the device driving voltage could be easily seen and evidenced by investigating the following device with a layer of co-host composed of different ratios of TCTA, hole transporting material (HTL), to TPBi, electron transporting material (ETL), shown in Figure 2. Each device is doped with an invariant doping concentration of 3wt% of Ir(2-phq)$_3$ but the TCTA and TPBi ratios were varies as revealed in Fig2 (b).

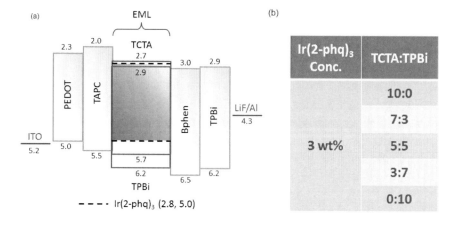

Figure 2. (a) Device structure with an orange emissive layer, Ir(2-phq)$_3$, doped in a layer of co-host with different ratios of TCTA to TPBi. (b) The exact ratios of TCTA to TPBi varying from 10:0 to 0:10 at the same doping concentration of a whole. [17]

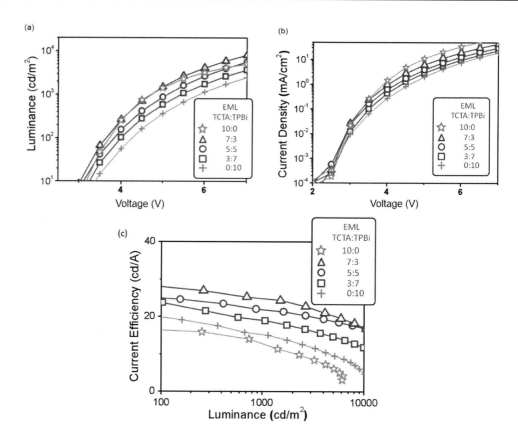

Figure 3. Electroluminescent characteristics of the device with various ratios of TCTA to TPBi which are represented by (a) Luminance (cd/m^2) versus voltage (V) (b) Current density (mA/cm^2) versus voltage (V) (c) Current Efficiency (cd/A) versus Luminance (cd/m^2). [17]

Figure 3(a) illustrates the effect of different ratios of TCTA to TPBi on driving voltages at 10 nits. Since the difference of HOMO level of TAPC and TPBi is 0.7 eV, which is higher than that of TAPC and TCTA, it is expected that increasing the percentage of TPBi will build up a higher barrier for holes, which are dominant carriers in OLEDs [15, 16], and render the transportation of holes more ineffective, thus yield a higher driving voltage as a result. However, the device with a mixing ratio at 7:3, rather than 10:0, has the lowest driving voltage, indicating that the dependence of driving voltage on the ratio of mixing host is not linear.

Nevertheless, it is of interest to see that the two devices with ratios of TPBi in the mixing host beyond 50%, i.e. 0:10 and 3:7 of TCTA: TPBi, have the highest and the second highest driving voltages respectively among the all, suggesting that higher barriers for holes do increase the difficulty for holes in entering the EML. [17] The disorder of the device with the ratio at 10:0 can be partly explained by introducing the concept of carrier injection balance and will be further elucidated in Sec. 2.1.5.

2.1.2. Effective Carrier Confinement

Aside from the importance of low carrier injection barrier on device efficiency, the device capability of effectively confining carriers in the recombination zone of the EML is another pivotal issue when one is concerning how to obtain a well-designed architecture yielding high performance. Lowering the injection barrier is one thing, confining carriers effectively is quite another. Low injection barrier promises a relatively low driving voltage, yet gives no guarantee of blocking carriers in the interface of EML and adjacent layers.

To prove that both the hole-confining layer (HCL) and the electron-confining layer (ECL) have positive effect, i.e. carriers are more likely to be confined in the EML, on device performance when both carriers are properly used. Figure 4 exhibits three devices where both Device II and III contain an additional ECL between the layer of PEDOT:PSS and the emissive layer, while Device I was made without any ECL as a comparison counterpart. From the schematic routes of electrons in three Devices, it's clear that electrons in Device II and III encounter an higher barrier when they are trying to escape from the route of dopant to PEDOT:PSS. As a result, more electrons are constrained in the EML and thus enhance the recombination probability which would finally lead to a higher efficiency.

Figure 5(a) exhibits a plot of current density versus voltage which shows that devices with an ECL a relatively high current density at the same applying voltage. This suggests that more electrons will flow through the EML so that a higher recombination probability could be expected. Figure 5(b) illustrates this fact that Device II and III have a relatively higher current efficiency at the same current density. Table I further illustrates what benefit of the employment of a proper ECL can bring. It's clear that Devices II and III with an ECL have better performance in all respects including power efficiency, current efficiency, and external quantum efficiency (EQE).

Another similar example is shown in Figure 6, but it was fabricated with a HCL, Bphen, as a testing material to see if the HCL have the same positive effect on device performance. Structures shown in Figure 6(a) and 6(b) have the same cascading property such that both devices have relatively low driving voltage. Device without the HCL exhibits fair good performance, however, as the HCL was deposited device performance was raised to an extent of exceeding the 20% of theoretical limit of EQE to 21.1% at 100 nits.

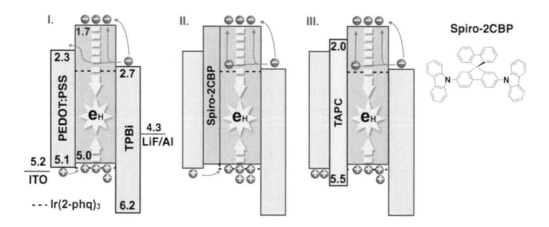

Figure 4. The energy diagram of the studied devices using Spiro-2CBP as host in all three Devices, among which a layer of Spiro-2CBP in Device II is deposited as a hole blocking layer (HBL) and a layer of TAPC in Device III as an another HBL. Also shown are plausible distribution of holes and electrons in the studied devices and the skeletal formula of the organic compound, Spiro2-CBP. [18]

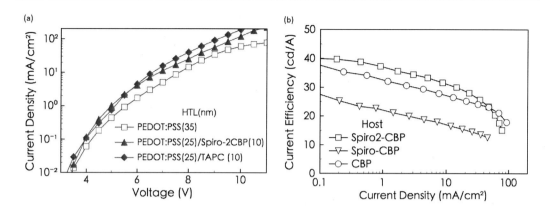

Figure 5. Electroluminescent characteristics of the three devices with the total thickness fixed at 35 nm. (a) A plot of current density versus voltage is given to show that devices with an ECL have higher current density at the same voltage. With higher current density in Device II and III, the recombination probability of electrons with holes could be expected to increase which reflect on (b) the plot of current efficiency versus current density. [18 (a)]

By properly confining the holes from escaping, power efficiency, current efficiency, and EQE in this case are raised by 51%/24%, 24%/22%, and 32%/28% at 100/1000 nits respectively, details of which are listed in Table II. It is noteworthy that the driving voltage of the device in Figure 6(a) is lower than the one without the ECL, indicating that the role of the HCL here may not only contribute to the performance improvement but make electrons easier to enter the EML when comparing the device in Figure 6(b).

Table 1. Electroluminescent characteristics of the Spiro-2CBP composing devices, comparing with those of the Spiro-CBP- and CBP composing counterparts [18(a)]

Devices	Hosts	Driving voltage at 10 cd/m^2 (V)	Current efficiency (cd/A)	Power efficiency (lm/W)	EQE (%)
			At 100/1000 cd/m^2		
I	Spiro2-CBP	3.7	39.5/34.5	23.5/16.8	16.9/15.2
II	Spiro2-CBP	4.5	23.3/18.4	12.2/7.1	11.0/8.7
III	CBP	4.5	35.1/30.0	19.8.13.0	15.4/13.2

Table 2. Comparison of the power efficiency, current efficiency, external quantum efficiency (EQE), and driving voltage (defined as the drive voltage at 10 cd/m^2) of the orange OLED devices with different ETLs [18 (b)]

ETL	Power efficiency (lm/W)	Current efficiency (cd/A)	EQE (%)	Driving Voltage (V)
	At 100/1000 cd/m^2			
Bphen/TPBi	47.2/31.7	47.4/41.7	21.1/18.6	2.7
TPBi	31.2/21.5	36.0/32.6	16.0/14.5	3.0

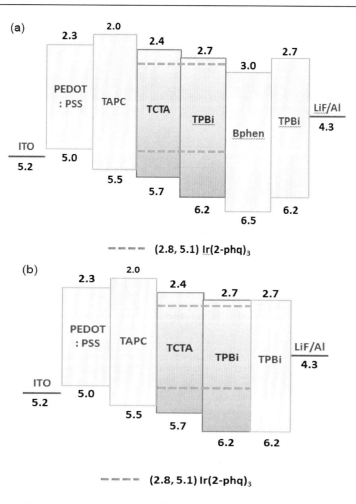

Figure 6. Comparison between two devices, one of which shown in the inset of **(a)** is fabricated with a hole-confining layer (HCL) and the other, (b), without. [18 (b)]

A vigilantly chosen proper HCL or a suitable ECL are moderately enhanced the device efficacy. By the introducing of confining layer in the devices, it may increase the carriers (electrons and holes) recombination probability, but in the absence of a suitable confining layer right next to the EML, carriers would less likely to be blocked at the EML interface and thus lead to a less recombination probability. In other hands it is also played a key role to reduce the device driving voltage, because an appropriate confining layer balanced the device architecture and yields a more stepwise one. So the device power efficiency would be proficient to increase, on the basis of a specific extent depending on the slump in the driving voltage, by the doping of an appropriate confining layer. [18]

2.1.3. Effective Excitons Generation on Host

In order to obtain high efficiency in OLED devices, phosphorescent emitters are recently more favorable for researchers than the fluorescent emitters since they allow full harvesting of all electrically generated singlet and triplet excitons in OLED devices. [8,19-21] However, in most cases of phosphor-composed devices, one of the main obstacles researchers have to confront in order to reach high efficiency is how to avoid the occurrence of triplet-triplet annihilation. [22-24] Such avoidance of triplet-triplet annihilation is necessary and vital for phosphor-based devices due to its devastating impact on the device efficiency should it occur. Hence, whether exciton is generated effectively on the host becomes critical to tackle when it comes to a phosphor-composed device.

2.1.4. Effective Host-Guest Energy Transfer

It is well known that device efficiency depends not only on the energy level distribution among host and guest but on energy transferring capacity from host to guest. Figure 7 depicts the energy level diagrams of three different green or greenish-blue dopants in the same host, CBP

Although energy level distribution of emitters in Device II and III shows scant variation, their performance shows huge difference. In Figure 8 (b), current efficiency of Device II is all the time higher than that of Device III at various current densities. This fact has clarified the fact that host-guest energy transfer plays a significant role in determining device efficiency.

Table 3. Current efficiency of the CF$_3$BNO-composing devices fabricated via wet and dry process, compared with BNO and Ir(ppy)$_3$ composing counterparts [22]

Guest	Current Efficiency (cd/A) At 100cd/m^2 by	
	Wet process	Dry Process
CF$_3$BNO	89.1	35.3
BNO	61.3	41.1
Ir(ppy)$_3$	35.2	49.5

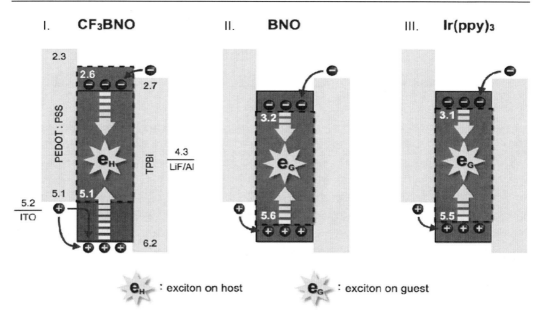

Figure 7. Energy-level diagram of the CF$_3$BNO-composing blue-green device with a CBP host. For comparison, also shown are the energy levels of the two dopants, BNO and Ir(ppy)$_3$. [22].

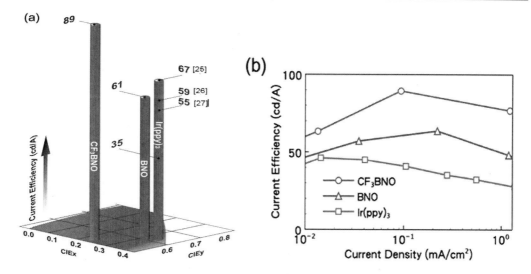

Figure 8. Current efficiency results of the three studied OLEDs and previously reported green ones via wet process on the (a) CIE diagram and (b) current density. [22]

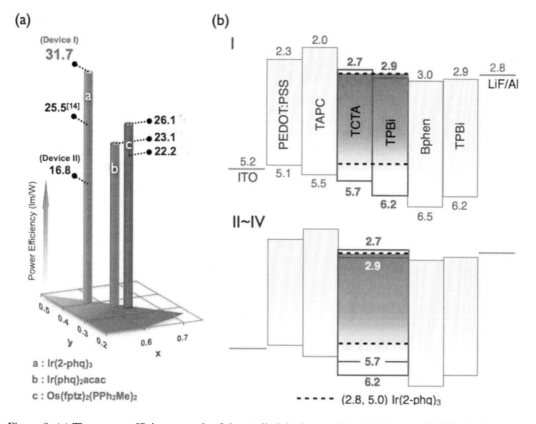

Figure 9. (a) The power efficiency result of the studied devices under a brightness of 1 000 cd m2, compared with those of previous reports in the CIE diagram. (b) Energy diagrams of the studied devices. Device I employs a double EML architecture, while Device II to Device IV use a single EML mixed with different ratios of TCTA to TPBi, which are 7:3, 5:5, and 3:7 corresponding to Device II, III, and IV, respectively. [17]

Extraordinary high efficiency of Device I may be attributed to its high quantum yield, 95%, than that of BNO, 42%, and that of Ir(ppy)$_3$, 38%. Other reasons contribute to high efficiency of Device I and II may result from their shorter excited-state lifetime, 0.30µs for CF$_3$BNO-composing one, 0.32 µs for BNO-composing one, and 1.2 µs for Ir(ppy)$_3$-composing one. [22]

2.1.5. Balanced Carrier Injection

Since exiton is generated at the cost of a pair of hole and electron, balanced carrier injection in the EML turns out to be another major factor influencing the device efficiency. Figure 9 shows how carrier injection distribution affects device performance. In Figure 9(a), EML consists of two transporting layers, TCTA as a hole transporting layer (HTL) and TPBi as an electron transporting layer (ETL) while EML of Devices in Figure 9(b) is a mixed host of TCTA and TPBi.

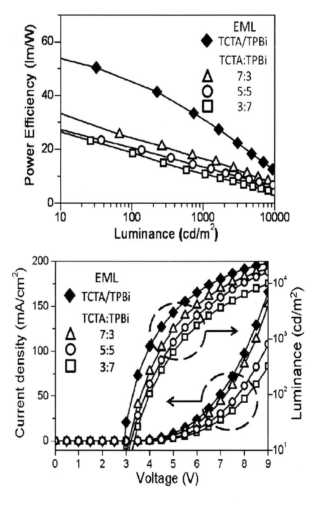

Figure 10. Effects of EML architecture on the resultant current density and power efficiency at varying luminance. [17]

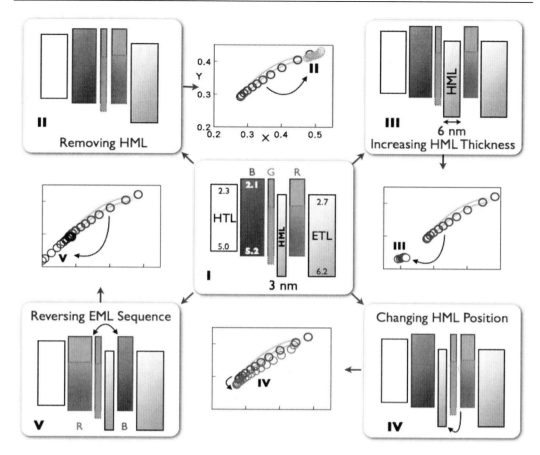

Figure 11. Effects of the hole-modulating-layer, TPBi, and sequence of the emissive layers on the color temperature and color temperature span of the sunlight-style color-temperature tunable OLED, as indicated by the emission track, against the daylight locus, on the CIE chromaticity space . [28]

In Figure 10, it is clear that Device I exhibits better power efficiency at various luminance and higher current density at various voltage. Several different ratios of TCTA to TPBi were applied in mixed host layer of Device II~IV, their performance shows relatively less difference. However, the performance does rise rapidly when a single EML with two host materials is separated into two EMLs with one host material deposited onto the other. The variation of performance among Device I and Devices II~IV becomes more conspicuous.

Figure 10 shows power efficiencies of Device II~IV, in the sequence of Device II, III, and IV, are 19.9, 21.7, and 25.0 lm/W at 100 nits and those at 1000 nits are 11.8, 13.7, and 17.2 lm/W at 1000 nits, respectively. It' clear that the power efficiency of Device I, 47.2 and 31.7 lm/W at 100 and 1000 nits, is nearly two times those of Device II~IV, indicating carriers are much more balanced in Device I. Apart from the balanced carrier injection, another possible mechanism that leads to high efficiency may due to the utilization of double-EML device with a stepwise energy-level as well as its comparatively low injection-barriers for both holes and electrons. In addition, there is a 0.5 eV gap for holes at the TCTA/ TPBi interface in the double-EML device, and a 0.2 eV gap for electrons at TPBi/ TCTA, which effectively confines charges and leads to an efficient excitons generation within the emissive zones. [17]

2.1.6. Carrier Modulation Layer

As discussed in Sec. 2.1.5, OLED devices performance is greatly depend on the balanced carrier injection in EML. In case of single emissive layer devices, the EML thickness can be easily optimized to obtain the highest device performance while in multi-EMLs OLED devices, the optimization of each layer thickness becomes difficult because each EML with proper thickness in its own device may not yield the best outcome when combined with other EMLs to form a multi-EML device. Instead of adjusting each layer thickness in multi-EML devices, simply using a carrier modulation layer makes ease the optimization to achieve the best device performance.

Figure 11 exhibits five kinds of architectures with the center one (Device I) as a standard device. Device I is composed of three EMLs together with a hole modulating layer (HML). Its emission with varying applied voltage is capable of fully covering a wide span of daylight chromaticity with color temperature changing from 2300 K to 8200 K, which nearly contains the whole range of daylight at different times and regions. Such a sunlight-style OLED device with a thin HML elucidates how a HML distribute carriers in each EML. Device II~V clearly depict that the thickness, insertion position, and the employment of a modulation layer are all dominating factors that could seriously affect their performance.

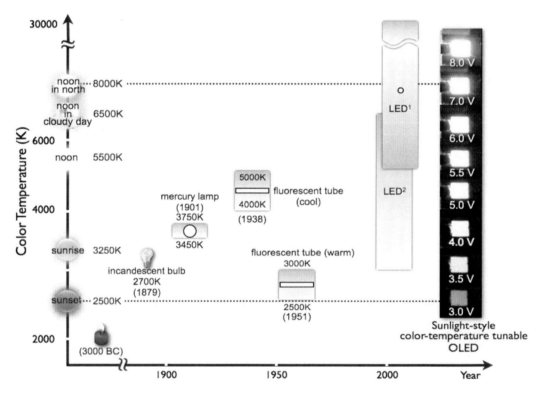

Figure 12. Sunlight-style color-temperature tunable OLED with color-temperature span from 2300 to 9700 K, compared against the color temperature and/or color-temperature span exhibited by candles, as well as typical electricity-driven lighting devices, including the point-type incandescent bulb, mercury lamp, and the latest LED, and line-type cool-white and warm-white fluorescent tubes. LED, using a multi-facet quantum wells microstructure design, shows a color-temperature ranging between 5000 and 20 000 K, and LED2, using an arrayed LED clusters of multi-color combination, between 3000 and 6500 K. [28]

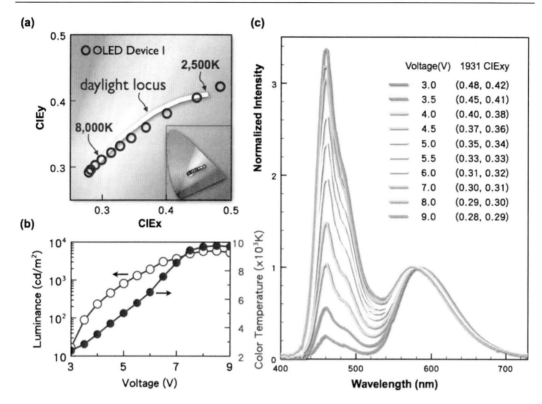

Figure 13. Chromaticity and color-temperature characteristics of the sunlight-style OLED-device I. **(a)** Its emission track on CIE 1931 chromaticity space matches closely to the daylight locus between 2500 K (sunset) and 8000 K (sunlight in northern country). **(b)** Its color-temperature changes from 2300 to 9700 K and brightness from 20 to 5900 cd/m^2 as the voltage increased from 3 to 9 V. **(c)** Its electroluminescence spectra at various applied voltages. Three peaks, appearing at 460, 495, and 590 nm, corresponding to the blue DPASN, green BPTAPA, and red DCJTB, respectively. [28]

The triplet energy state and carrier mobility of a modulation layer are two major factors needed to take into account before select a proper modulation layer. If the device consists of phosphors, consideration on the difference of triplet energy states between the HML and its adjacent layers will be put in the first place since phosphorescence comes from triplet excitons that constitute three- quarters of bound electron-hole pairs that generate during charge injection. Triplet energy transfer is common in phosphorescence (PH) OLEDs and greatly affects, either positively or negatively, the device performance. The higher triplet energy state of the modulation layer employed, the tougher the triplet excitons of phosphors can transfer to the either side of the modulation layer. Figure 12 depicts the history and evolution of the artificial lighting sources. Rarely few can cover a wider range of color temperature than Device I.

Figure 13(a) shows the track of the shifting CIE coordinates, at various applied voltage from 3 to 9V, which nearly overlaps that of the daylight locus. Figure 13(b) shows the variation of luminance and color temperature versus voltage. Figure 13(c) shows the electroluminescence spectra of Device I. That the blue emission gets stronger indicates the recombination zone is shifted to the left side of the HML at higher applied voltage. [28]

2.1.7. Polymeric Nanodots (PNDs)

The OLEDs performance can further enhance by incorporation of high surface-charge polymeric nanodots (PNDs) in the HTL, PEDOT:PSS. [29, 30] In contrast to electrically neutral quantum dots and nanodots used in dry processes, [31] polymeric nanodots can be synthesized with a precisely controlled size. Figure 14 shows the mechanism of different effect of nanodots on the width of recombination zone. It is found that the quantity of charge and charge property of the incorporated nanodots will result in recombination zones with different width.

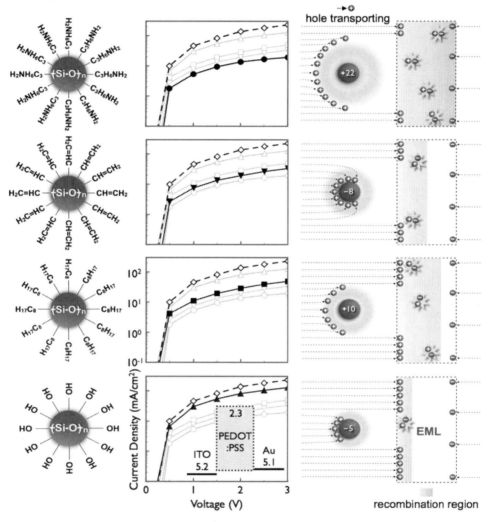

Figure 14. (Left) Schematic illustration of the structures of the four different PNDs studied: Am-PND, V-PND, Al-PND, and H-PND. (Middle) Incorporation effects of different PNDs at 0.7 wt % on the current density of a hole-transporting-only device consisting of a hole-injection layer of PEDOT:PSS sandwiched by a high work-function electrode pair. Devices with (●) Am-PND, (▼) V-PND, (□) Al-PND, (△) H-PND, and (◇) without any PND. The Am-PND device showed the lowest current density, implying that its highest positive charge was capable of most effectively repelling the injection of holes due to a strong repulsive effect and, consequently, small hole-transport flux. (Right) Schematic illustration of the plausible effects of charge intensity of the PNDs on the hole-transport flux, width of recombination region, and exciton population in the emissive layer. [29]

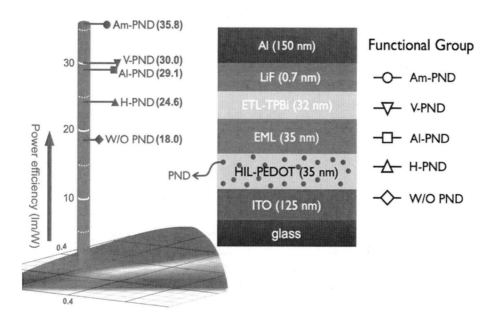

Figure 15. Schematic illustration of device architecture with (O) Am-PND, (▽) V-PND, (□) Al-PND, (△) H-PND, and (◇) without PND. Note that the best power efficiency (24.6 lm/W), found in the H-PND added device in panel a, was obtained at its optimized concentration of 7 wt %; its efficiency was 20.1 lm/W at 0.7 wt %. [29]

For example due to the high repulsive effect a positively charge PND (Am-PND) repelled the hole injection and its small hole transport flux, balanced the hole injection against the electron and maximized the carrier recombination probability. While the negatively charged PND (V-PND) doped device revealed high hole modulation functionality.

It is extensively capable to trap the significant amount of holes and prevented excess of holes in the emissive layer to balance the carrier injection. Again the carrier recombination probability is extensively increased and similarly enhances the device performance. In Figure 15, the enhancement of incorporation of different charge properties of nanodots on the device efficiency composed of a blue EML. It's noteworthy that the device efficiency is improved up to twice that of the one without incorporation of nanodots. [29]

2.2. Light Outcoupling

Light out-coupling efficiency, expressed as the ratio of surface emission to all emitted light, has been assumed to be around 20% due to the critical angle of total refection within devices. [32-34]As presented in Figure 16 a large fraction of the optical energy generated in an OLED is wasted since it couples into modes that are substrate trapped (ST) by total internal reflection, wave guided (WG) modes, [34] or surface plasmon polariton (SPP) modes at the metal cathode/organic interface. [35-38] The light generate into the emission layer needs to be coupled out from the devices passing through transparent or semitransparent electrodes.

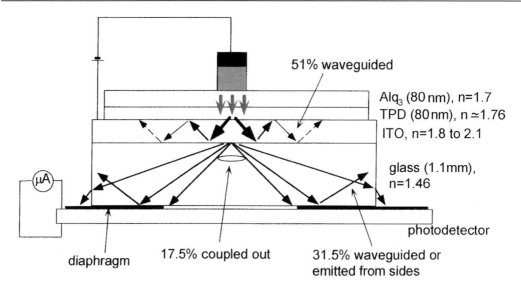

Figure 16. Schematic illustration of loss of light (82.5%) due to waveguide modes and light coupled out (17.5%) from the devices. [34]

Since the light is out coupled from a dense medium to air, part of the light generated within devices suffers total internal reflection (TIR) is trapped and finally absorbed in the devices. The basic reason for light being trapped into device is the high refractive index of light emitting, cladding and substrate layers. [39, 40] Improved light extraction is a key factor for increasing the OLED efficacy. Many approaches have been used to improve the out-coupling efficiency OLEDs.

2.2.1. Substrate Surface Modifications

Substrate surface modification is an important technique to improve the device performance, by enhancing the outcoupling efficiency. [41, 42] It was found that the ITO/organic modes may be concealed because thinness of the ITO/organic layers and ~ 50% of the total generated light is emitted externally in some structures. [40] Numerous substrate modification techniques have been implemented for enhancing the external out coupling efficiency of OLEDs. [5, 34, 43] By the using of rough surface to extract the substrate wave-guided mode is one of the simple techniques. [43] Figure 17 presented the sand-blasting and OLED fabricated at the opposite sand-blasted side. [44] Due to the roughness of substrate the wave-guided modes at glass–air boundary are coupled out into air and coupling efficiency increases on increasing the roughness of substrate. But in this case ITO/organic wave-guided modes remain unaffected as shows in Figure 17 (a). The structure of electroluminescence emission pattern of OLEDs through surface and edge of the glass was characterized by Kim et al. [45] It was found that in the absence of micro-cavity the surface emission is nearly Lambertian while the edge emission comprises discrete substrate reflection and leaky waveguide modes. Lambertian angular emission profile is desired to minimize eye strain and fatigue for displays applications. [46] A combined classical and quantum mechanical model was used to calculate the distribution of light emission from a planar OLED. [47, 48]

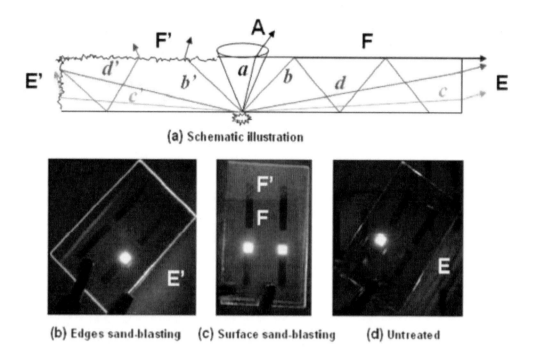

Figure 17. (a) Schematic illustration of light propagation inside the devices, photos of devices fabricated on (b) edges sand-blasting substrates (c) partial surface sand-blasting substrates (d) untreated substrate. [44]

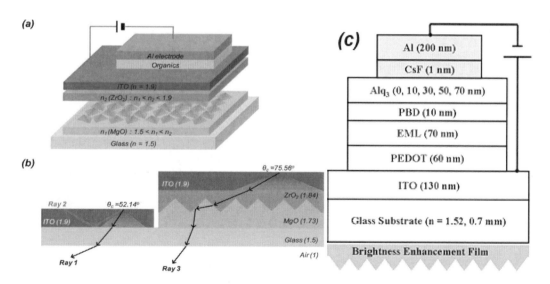

Figure 18. (a) Schematic diagram of bottom-emitting OLEDs with the embedded RIML at the glass/ITO interface. The RIML is composited of electron-beam-evaporated MgO and MgO/ZrO$_2$ (b) Schematic illustration of the mechanism for improving device out-coupling efficiency with a microfacet-structured RIML [52] (c) Device structure with brightness enhancement film, external mode the substrate waveguided mode, and the anode/ organic waveguided mode in an organic EL device. [53]

A novel substrate structure which is increased the external outcoupling efficiency of OLED devices. [34] Riedel and co-workers presented an ITO free OLED design with Bragg grating structure in a high refractive index layer acting as light extraction elements for substrate modes and wave guide mode and reported that the emission is enhance upto 300% for 77nm grating depth. [49] Introducing the scattering effects at the substrate surface by means of techniques such as applying a transparent coating on the substrate with embedded small particles. [50], texturing the substrate surface by sand blasting. [44, 51] Figure 18 shows the refractive index modulation layers (RIMLs) with nanofacet structure to modified the substrate surface and improve the light out-coupling efficiency of devices.

Power efficiency of OLEDs device were improved 34.7% (at 10 mA/cm^{-2}) and luminance value at 220 mA cm^{-2} was increased from 28750 to 34200 cd m^{-2}. [52] Krummacher et al. proposed the substrate modification by the use of brightness enhancement film (BEF) as shown in Figure 18 (c). However, the effect of light out-coupling enhancement due to BEF film is increased and shows a strong dependence on the device structure. As increased the Alq3 layer thickness the fraction of substrate waveguided light intensity increases compared to the fraction of light out-coupled intensity. [53] Tsutsumi et al. reported that the high-refractive-index substrates can enhance the light extraction upto 80%. [54, 55]More recently, using patterned surface of high-refractive-index substrates or a refractive index matched half spheres, can further extract light from the glass mode, thus the white OLEDs efficiencies improve to 90 lm/W and 124 lm/W, respectively, have also been reported by Reineke et al.. [56] A pattern of pyramids (with period 0.5 mm) by cutting 90° grooves into a high index glass, similar to microlens arrays, to couple out more light. [56, 57] Biomimetic AR silica cone array was directly etched on the opposite side of the ITO coated fused silica substrate. The AR surfaces were dramatically suppressed the reflection loss and increase light out-coupling efficiency of phosphorescent white OLEDs. It was also improved the luminance efficiency of devices by a factor of 1.4 compared to non modified substrate. [57]

2.2.2. Texturing Mesh Surfaces

Texturing mesh structured on the substrate is one of the potential and cost effective technique to highly improve the light out-coupling efficiency of OLEDs. Substrates with texturing meshed surface has been enhanced the external quantum efficiency of OLEDs. [58, 59] Cheng et al. demonstrated a textured meshed surface as (Figure 19) was fabricated using poly-dimethyl-siloxane (PDMS) and porous anodic aluminium oxide (AAO) as templates. [58] The original nanowire array turns into disordered meshed structures during fabrication. PDMS meshed structure was fabricated by the use of self organized nano porous AAO. Textured mesh structure is coupled out the TIR light at glass–air boundary. As the incident angle is larger than the critical angle of TIR so the meshed surface is much brighter than flat surface.

Quantitative analysis of out coupling efficiency has been carried out using the varied incident angle of laser beam and measured the transmitted power emitting from the meshed surface. The out-coupling enhancement factor (g) [58] is defined as

$$g = \frac{(L_m - L_p)}{L_p}\%$$

where L_m and L_p are the total luminescences of meshed-surface and flat-glass OLEDs The out-coupling efficiency of OLED with the textured meshed structure was enhanced upto 46% and was found to be insensitive to optical wavelength and hence is useful to preserve the color spectrum of OLED reported by Cheng et al. Solution processes base planarized photonic crystal (PC) substrates were used to enhance the out-coupling efficiency of OLEDs. [58] Light extraction increased 38% when OLEDs fabricated on the solution processed PCs have also been demonstrated by Cho et al. [60] Ziebarth et al. reported the effect of two-dimensional grating on the backside of glass substrates for the enhancement of light out-coupling efficiency. Effect of two-dimensional Braggs grating enhance the directed emission by 70% and 25 % external quantum efficiency of polymer light emitting diodes. [59]

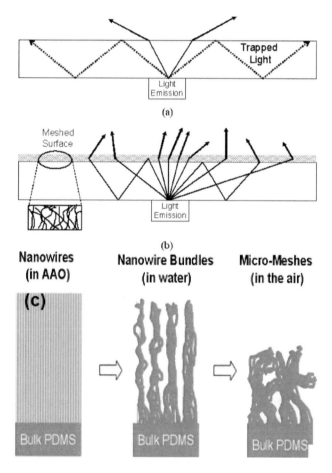

Figure 19. Schematic illustration of optical ray trajectories in a thin film (a) without and (b) with a mesh on the surface. **(c)** Transformation progress of PDMS nanowires in a cross-sectional view. [58]

2.2.3. Multilayer Cathode Structures

Light outcoupling efficiency of OLED depends on the various types of transparent and semitransparent cathodes structure and their methods of deposition. Indium-tin-oxide (ITO), one of the transparent conducting oxides (TCOs), is the most popular cathode material in OLEDs for its high transparency and relatively low resistance. ITO films are usually deposited by DC magnetron or radio frequency (RF) sputtering and those bombardments cause defects and conduction paths to form, leading to degradation of structural, optical, and electrical properties of organic films, and furthermore, the occurrence of a large leakage current ($10{-}1{-}10{-}2$ mA/cm^2 at reverse bias of -6 V). [61] Even though the addition of some buffer layers (CuPc/Al, MgAg, etc) can reduce the plasma damage, In a conventional sputtering techniques very difficult to entirely avoid the plasma damage because organic films are highly sensitive to radiation, heating and charging. [38, 62] MSTS is effective in reducing plasma damage and so in improving the performance of OLEDs reported by Kim et al. [63] Plasma damage-free deposition of ITO top cathode layers for TOLEDs by means of a mirror shape target sputtering (MSTS) technique. [38] Chen et al demonstrated a significant enhancement in light outcoupling efficiency of a top emitting OLEDs improved by using semitransparent multilayer cathode structure [64].

2.2.4. Aperiodic Dielectric Stacks

Aperiodic dielectric multilayer stacks may be designed that would allow normally incident photons or photons within a certain angle to normal incidence, to pass while other photons. The aperiodic stacks may also be used as encapsulation structures for sealing an optoelectronic device. Out coupling efficiency of polymer and small molecular weight OLEDs with randomly oriented emitting dipole is approximately 30% and 20% respectively, since in that case the emissive dipole tends to align in the substrate plane during film deposition. [38] As shown in Figure 20 (a) the TEM images of 9-layer SiO$_2$ /SiNx dielectric stack was deposited by plasma enhanced chemical vapor deposition on glass substrates prior to ITO deposition. The spectral power density at different azimuthal angles (Figure 20 (b)) for an OLED with a 9-layer SiO$_2$/SiN$_x$ stacked substrate structure. [65, 66] Agrawal et al. reported that the SiO$_2$/SiN$_x$ aperiodic dielectric stacks can significantly improved the outcoupling efficiencies in OLEDs. They achieved the index contrast of $n=2.6/1.08$ and stacks can increase the brightness of a typical OLED by a factor of 2.7, while maintaining a Lambertian emission profile within a 60° viewing cone. The aperiodic dielectric stacks do not appreciably affect the total radiative lifetime of the exciton. [66]

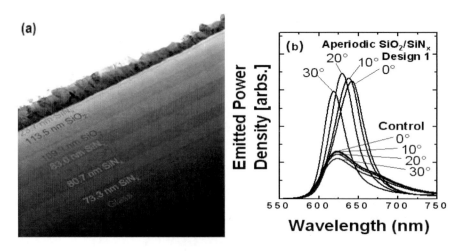

Figure 20. (a) TEM micrograph of the 9-layer aperiodic dielectric stack optimized for the red OLED considered in the text for 60^0 viewing cone. [Adopted from reference 65] (b) Calculated spectral power density at different azimuthal angles for an OLED with a 9-layer SiO$_2$/SiN$_x$ stack optimized for a 60^0 viewing cone. For comparison, the response of the control OLED structure is also shown. [66]

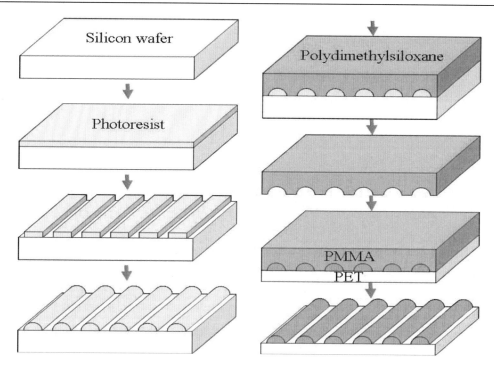

Figure 21. A simple schematic layout of microlens array fabrication process. [74]

Aperiodic stacks increase the exciton coupling to selected UB modes and reduce the coupling for all other modes by redistributing the local photon density of states [67, 68]. The extraction efficiency of organic light emitting devices was significantly enhanced by depositing metallic nano-wires on the glass surface and indium tin oxide (ITO) anode. A non-periodic nano-porous alumina film was used by Peng et al. [69] and outcoupling of light from OLED. The major advantage of such aperiodic nano structures is that the emission color and intensity are independent of viewing angle.

2.2.5. Microlens Arrays

Microlens arrays (MLAs) can strongly enhance the device efficiency, but the devices are too bulky and they violate the development trend of modern electro-optical devices, which is significantly improved the outcoupling efficiency and minimize the light loss due to the TIR and waveguiding. [37, 70-73] As revealed in Figure 21 the fabrication of MLA is simple [71, 74], and it can also be easily applied to large area substrates [37].

This method also has the merits that the MLA is external to the OLED device and, therefore, does not affect its operation. In order to optimize the coupling enhancement factor it is important to study the details concerning the impact of geometrical structure of the microlens. However, little study has been undertaken on analysis of the involved basic physical process and optimization of microlens pattern [37, 70] A hexagon-based MLAs with different fill factors were used to increase the outcoupling efficiency of the planar green OLEDs. [75] Use of ordered and disordered MLAs on the backside of substrate surface have been demonstrated. [37, 70] Though the spherical ordered arrays of microlenses are increased the external quantum efficiency (EQE) as a factor of 1.5 and power efficiency upto 85% theoretically and 70% practically, MLAs can be used for displays if the substrate and microlens layer thickness are optimized. Möller and Forrest were reported the MLAs thickness influences the image blur problem. [37] The image blurring is completely eliminated even for smallest fonts when microlens layer is directly applied on the surface of top emitting phosphorescence OLEDs. [37, 74, 76] Further, it can be controlled by the thin MLAs structure. Figure 22 shows SEM micrograph of low height MLAs attached to an OLED. [77]

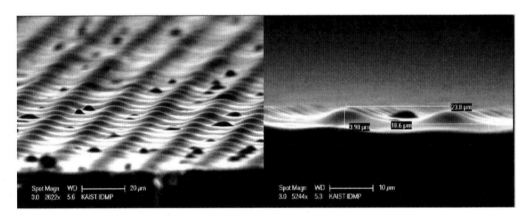

Figure 22. SEM images for the microlens array. The diameter and height of the microlens array is 20 and 3 μm, respectively. The distance between lenses is 25 μm. [77]

Lee et al. reported that the vertical and parallel-aligned cylindrical MLAs, the parallel-aligned one can not only enhance the current efficiency and power efficiency but also have better image quality of the displays. Parallel-aligned cylindrical MLAs are good choice for the large sized OLED display panels. [74] Out-coupling efficiency of self organized MLAs based OLEDs were, demonstrated by the Huang et al., increased 24.5% and particularly enhanced the large viewing angle without any obvious color change. [78]

Figure 23 (a) − (d) showing the schematic diagram of microlens attached OLED and SEM images of various elliptical shapes microlenses. [79] Figure 23 (e) represented the schematic picture of outcoupling enhancement by the use of microlens array on high refractive index glass substrate. [80] Peng et al. fabricated refractive micro-lens arrays and showed enhancement in light out-coupling efficiency over 65% using high refractive index glass substrate. [71, 80] The performance of planar OLEDs for display application was significantly affected by the MLAs [72] and it was found that MLAs not only improve the efficacy but also maintains the CIE index and color invariance for wide viewing angles.

Recently, Kim and Choi approached the imprint lithography technique for MLAs fabrication. Only two photoresists were used to form a spherical imprint mold without additional fabrication and this mold was used for MALs fabrication during imprint lithography. It is an economical and also suitable for large area MLAs fabrication. The improvement in power efficiency and current efficiency 69% and 42% respectively at 900 cd/A^2. The image blurring due to the microlenses was controlled, while luminance enhanced (43%), by the use of low height MLAs. [77]

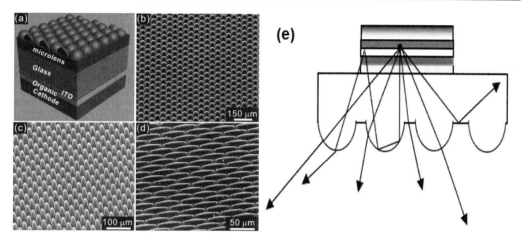

Figure 23. (a) Schematic illustration of an OLED attached with microlens array. (b)–(d) A 45° tilted view of SEM images of microlens arrays with (b) hemispherical (D50), (c) short-elliptical (D50-20), and (d) long-elliptical (D50-10) microlenses [79] (e) Schematic explanation of outcoupling enhancement by the use of microlens array on high refractive index glass substrate. [80]

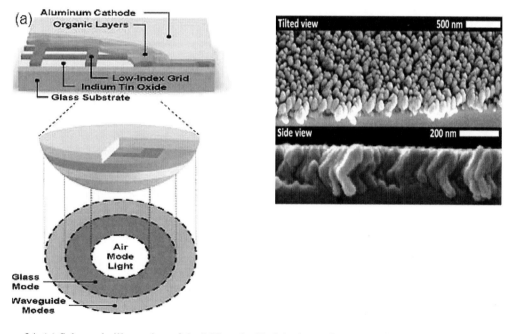

Figure 24. (a) Schematic illustration of the LIG embedded device and an approximate correspondence of sections of the modelled hemisphere to each set of modes [82] (b) The surface (top) and cross-sectional (bottom) scanning electron microscope images of obliquely deposited porous SiO_2 used for the Ultra LIG. [83]

2.2.6. Low Index Grids

Insertion of low index grids (LIG) or ultra LIG between the indium–tin–oxide (ITO) layer and organic layer as shown in Figure 24 (a), is significantly improved the OLED efficiency. [81, 82] When LIG combined with microlenses the light outcoupling efficiency increased the 2.3 folds experimentally and simulations predict that the enhancement factor can be improved to 3.4 times, compared to conventional OLED devices. [81] Sun et al. reveal a wavelength independent technique for extracting waveguided light into the air and glass modes by embedding LIG patterned at the ITO anode and active organic interface of OLEDs, while microlens array was attached outside the glass substrate to extract the substrate waveguiding mode [81]. Simulations show that the light extraction and EQE can be improved upto 50% from the high refractive index region using the ultralow index (1.03) grid compare to conventional OLEDs.

Highly porous 100-nm-thick SiO$_2$ films were deposited on a commercial pre-patterned ITO coated glass at an oblique angle using an electron beam (e-beam) evaporator. In Figure 24 (b), the SEM image revealed the obliquely deposited porous SiO$_2$ based ultra LIG. [83] Deposited ultra LIG was also used to extract the trapped light into the device. [83]

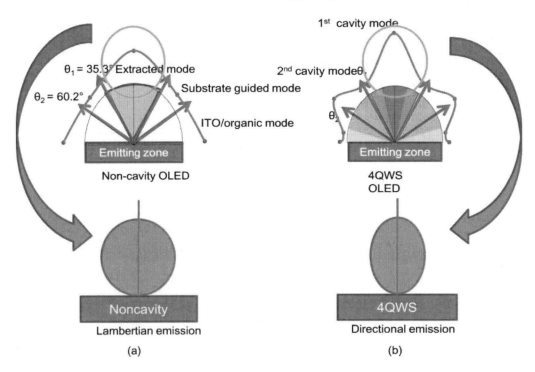

Figure 25. Schematic diagrams of the angular mode distribution for a) non-cavity and b) microcavity devices. [85]

2.2.7. Effect of Micro-cavities

The light extraction to air is strongly influenced by the micro-cavity formed between glass and cathode. [84] Figure 25 shows the schematic illustration of angular mode distribution for non cavity and microcavity based OLED devices. [85] OLED can be admired as 1-D micro-cavity because the total thickness of organic films in the device is of the order of wavelength of light. Weak and strong micro-cavities are two classes of micro-cavities in the OLEDs [86]. Weak Microcavity is formed with the conventional OLED structure due to the metal cathode and high refractive index anode (ITO) [87] while a strong micro-cavity OLED structure usually consists of a metal mirror on one side and a highly reflective dielectric multilayer structure on the other side [88]. Effect of micro-cavity on the light extraction efficiency [87], photoluminescence and electroluminescence spectra of OLEDs [89-93] and the emission characteristics including efficiency enhancement of both small molecule [94-96] and conjugated polymer [97, 98] based electroluminescent devices. Nakayama et al. [99] and Takada et al. [89] demonstrated the control of spontaneous characteristics of OLED with a planar micro-cavity structure. A strong micro-cavity determines the electric field mode distribution, which in turn modify exciton spontaneous emission life-time and the quantum efficiency and hence electroluminescence spectrum and intensity is enhanced [97, 100-102] With the help of strong Microcavity the intensity enhancement of the order of 1.5–2.5 is observed as compared to conventional OLED device.

Lee et al. reported the improved efficiency based phosphorescent (PH) OLEDs in a microcavity structure as shown in Figure 26 (a) Microcavity is used to improve the down conversion efficiency and effectively enhanced the substrate mode by the redistribute of optical modes in device. Light out-coupling efficiency of Flrpic based PHOLEDs were significantly improved, due to the light scattering from phosphor film. The maximum luminous efficacy of white OLEDs was reported 99 lm/W by the combination of microcavity OLED with down conversion phosphor and microlens. [85]

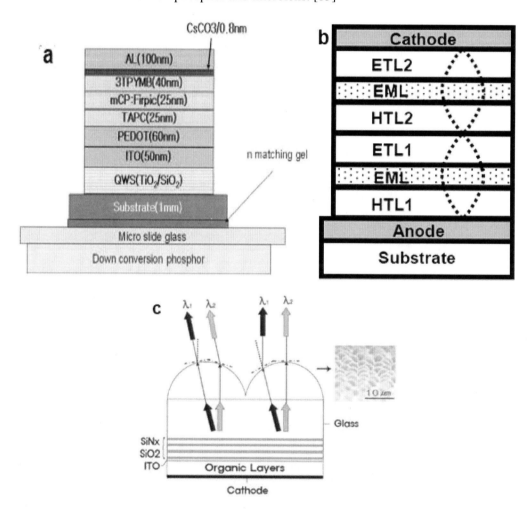

Figure 26. Schematic illustration of (a) A cross-section view of down conversion white OLED with microcavity structure [85] (b) two-unit tandem device structure with microcavity [103] (c) the microcavity OLED structure with irregular microlens arrays (left) and a scanning electron micrograph of the microlens arrays (right). [104]

A fivefold enhancement in OLED luminance was achieved by Cho et al. [103] using micro-cavity two-unit tandem OLED (Figure 26 (b)). Significant enhancement in light out-coupling efficiency and current efficiency can be obtained by means of using top-emitting micro-cavity OLEDs with micro-lenses. Figure 26 (c) shows the structure of a micro-cavity top-emitting OLED with micro-lenses [104]. Peng et al. reported an effective anode modification technique along with a careful design of micro-cavity in which the emission spectrum is independent of viewing angle. With this modified design and optimized micro-cavity top-emitting OLED a current efficiency enhancement of 65% and total out-coupling efficiency of 35% was achieved as compared to the conventional OLED. [102]

2.2.8. Surface Plasmons

Surface plasmon (SP) resonance is another mechanism which can effectively couple the light in organic mode [105]. SPs are playing great role in merging photonics and electronics at nano-scale dimensions [106]. In this decade, the role of SPs in the enhancement of light intensity in LEDs and OLEDs [107-110] has been studied extensively. For out-coupling efficiency enhancement from OLEDs using SP modes requires special methods and structure to convert the non-radiative SP modes into radiative modes.

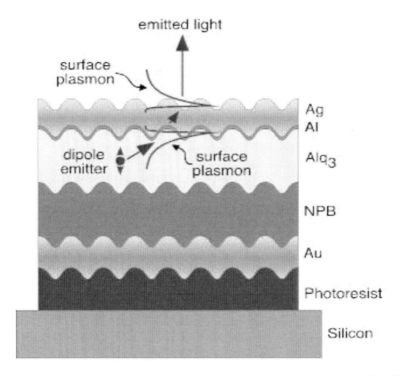

Figure 27. Schematic illustration of device structure with surface-plasmon cross coupling (SPCC) mechanism. The grating period was 550 nm and the peak-to-valley amplitude was about 60 nm. [111]

The formation of diffraction grating on the surface of an OLED for example a metal–insulator–metal structure for white OLED have been used for extracting the trapped light. Extraordinary transmission of organic photoluminescence (PL) was observed by Gifford and Hall using SP cross-coupling from silver-coated films of Alq3 deposited on a corrugated substrate (Figure 27). [111] Top-emitting OLEDs significant enhancements in PL and electroluminescence have been observed. [33, 112] The enhancement is produced by the coupling of SP polarization using micro-structured dielectric grating fabricated on the top of the metal film. However, calculations show that up to 40% light that would have been lost due to excitons quenching by SP modes was coupled out using micro-structured grating. [38] The photoluminescence emission from a structure containing a micro-structured thin metal film [113, 114] that support coupled SP polarization is over 50 times greater than the planer structure. [113]

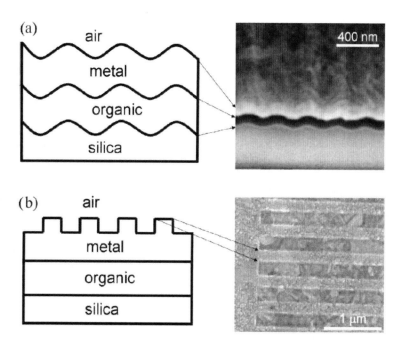

Figure 28. Schematic illustration of structures used in this study and its SEM images (a) A metallic film with a grating profile on top and bottom surfaces and SEM image of its cross section taken at an angle of 52^0 (b) Textured top metal surface and SEM image of its plane view. [114]

Giannattasio et al. presented two non conformal metal geometries in Figure 28 and reported that the OLED device efficiency simply increased using non conformal geometry [114] for different metallic interfaces. Yates et al. reported the efficiency improvement of dendrimer based OLED mediated via SPP mode. Non radiative SPP modes are coupled to light by a wavelength scale periodic microstructure and solvent assist micro-molding technique is used to fabricate the microstructure. [115]

2.2.9. Photonic Crystal

Inclusion of photonic crystals (PCs) in an OLED is one of the simple and effective techniques to enhance the light out-coupling efficiency of the OLEDs devices. Use of submicron size PCs with OLEDs structure is suppressed the light loss in waveguide mode and by the 50% improvement in the out-coupling efficiency of devices. [33, 116-120] The nanoimprint lithography (NIL) process has been applied to direct fabrication of nanoscale PCs [121] in a glass substrate at high pressure of about 20 bars and high temperature of about 300°C so it is also called the hot embossing technique. [116] Ultraviolet (UV) NIL is widely used because it is a cost-effective, simple, and low-temperature process. [120, 122] Cho et al.

reported the effective solution process to fabricate PC substrates by using the UV NIL and sol-gel processes. The PC structure was fabricated using NIL using a UV-curable acrylate and was planarized by using a ZnO layer formed by the sol-gel process because it resulted in a smooth surface with rms roughness of 13nm as shown in Figure 29.

Out-coupling efficiency of planarized PC layer integrated OLEDs (Figure 29 (c)) was improved 38% compared to conventional OLED devices. As revealed in Figure 30 (a) the emission spectra are almost same between the two devices, the intensity ratio at the peak wavelength corresponds to the increment of the extraction efficiency of the emitted light. [122]

The total efficacy improvement is much larger than the efficiency enhancement in the vertical direction shown in Figure 30 (b), because of the strong diffraction at higher viewing angles shown in SPPC OLED. [122] Jeon et al. reported the 50% enhancement in the device efficiency by the use of polymeric PCs structures OLED substrate, which was prepared by an etchless UV-NIL process. The maximum luminance was also increased approximately 30% compared to conventional devices. [120]

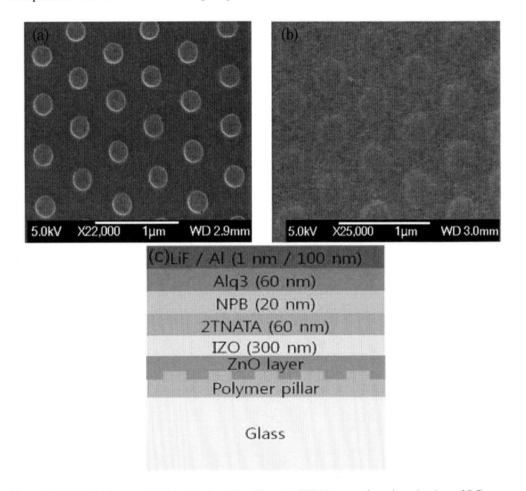

Figure 29. (a) SEM image of PC substrate with pillars (b) SEM image after planarization of PC substrate by the solution process (c) Device structure of OLEDs fabricated on the solution processed PC substrate. [122]

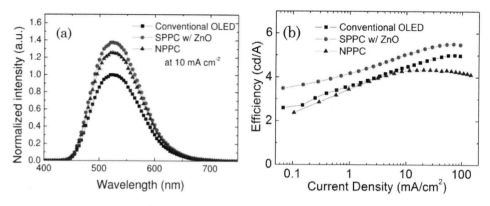

Figure 30. (a) Integrated emission spectra with relative emission intensities at 10mA/cm^2 and **(b)** luminance efficiency of various devices in the normal direction: a conventional OLED (square), a SPPC OLED (circle), and a PC OLED before planarization (triangle), respectively. [122]

2.2.10. Effect of Nano Porous Films, Nano Wires, Nano Particles and Nano Pillars

Effect of nano-structured films both metallic and magnetic nano-particles on the enhancement of extraction efficiency of OLEDs has been demonstrated. [123-125] Both periodic and aperiodic nano-patterns have been used for extracting the light of OLEDs. Lee et al. were reported the significant improvement in light extraction efficiency of periodic nano-patterned OLEDs. [119] Periodic nano-patterned structure converts the guided waves into external leaky waves (in high-refractive index ITO glass/organic layers). The light extraction efficiency of OLED is enhanced 80% theoretically and more than 50% was carried out experimentally by the use of PC nano pattern. [124] The light out coupling efficiency of OLEDs were improved by the use of non-periodic nano-porous alumina film. [126] Figure 31 (a) and (b) shows the top and side view, after pore extension in 5wt% H$_3$PO$_4$ solution, of the SEM micrograph of nano porous alumina film. [126] The experimental out-coupling efficiency was increased 50% without influencing the electrical properties of OLED. Metallic nano wires deposited ITO coated glass revealed the enhancement in the light out-coupling efficiency of OLEDs. [127]

Extraction efficiency enhancement is depended on the scattered light from the trapped photons. The large scattering efficiency and high transmittance of the periodic metallic nano-structures have the possibility for extracting light from the OLEDs. Price et al. was demonstrated a large area nano-OLED array. [128] It has been shown that reducing the size of an OLED can produce higher device densities per unit area which can be used for high resolution displays. The OLED was grown on 2-D PC polymer nano-structure the light extraction is enhanced either the multiple scattering of photons by lattice of periodically varying refractive indices in the PCs, light generated in the band gap region can couple only to radiation modes and is radiated outward, or the refractive index periodicity creates a cut-off frequency for guided modes. In the presence of nano-structured PC the OLEDs light out-coupling efficiency enhanced about 50% compared to conventional OLEDs. [120, 129]

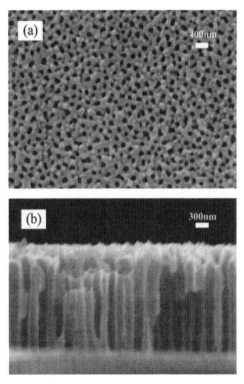

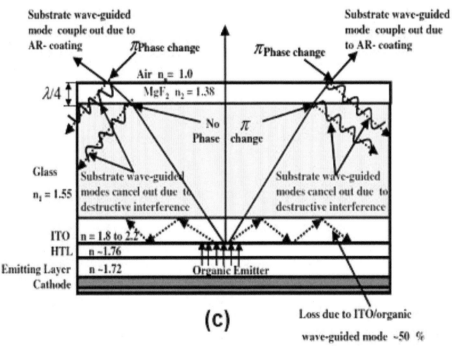

Figure 31. SEM micrograph of the porous AAO films. (a) Top view of the sample before pore extension and (b) side-view of the sample after pore extension in a 5 wt% H_3PO_4 solution. [126] (c) Schematic diagram of the phenomenon of anti-reflection (AR) coating using single-layer MgF_2 for the extraction of substrate-waveguided modes [125].

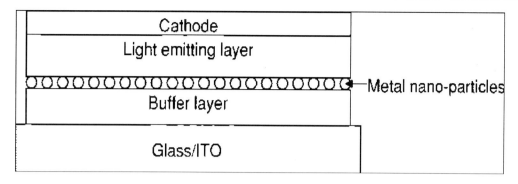

Figure 32. The device structure incorporating metal nano particles in the interface between buffer layer and the phosphorescence based light emitting layer. [130]

In recent times, influence of nano-particles on the performance of OLEDs has been the topic of extensive research. More interestingly, Jou et al. demonstrated that the power efficiency of both phosphorescent and fluorescent OLEDs has been outstandingly improved by the incorporation of highly negatively and positively charged polymeric nano dots (PNDs) in the hole transporting layer [29-31]. It is found that the efficiency enhancement strongly depended on the charge density of PNDs [29] because it caused the strong holes repulsion effect. Consequently, this process resulted the moderately balance holes transfer and enhanced the probability of exciton formation by the charge recombination in the proper recombination zone. [29, 30]

Fluorescent OLEDs, incorporation of metallic nano-particles within the organic host layer can be used to quench the triplet non-radiative excited states and hence suppresses the photo-oxidation and enhances the luminous stability. When gold nano-particles were incorporated into the phosphorescent OLEDs between alternative buffer layer and emission layer, as shown in Figure 32, then luminance efficiency was enhanced approximately 33% reported by Choulis et al. [130] Reported improvement in luminance efficiency is endorsed to the interaction between phosphorescent OLEDs and gold nano-particle surface plasmons which accelerates the spontaneous emission rate and enhances the device performance. Sun et al. demonstrated the EQE [131] of OLEDs has been enhanced by the use of doping magnetic nano particles. The average transmittance was increased from 85.5% to 95.9% for 0° incidence. The improvement of transmittance at large incident angle is more important for reducing TIR and enhancing the EQE. From 0° to 50° incident angles, there was a flat transmission with efficiencies higher than 90%. AR nano pillar imprinted substrate transmission was greatly improved, comparatively unpatterened substrate, from 52.8% to 89.1% for 60° incidence. [131]

Figure 33 (a) represented the flat PC substrate and substrate surface covered with nano pillars. While the ANSI contrast ratio (intensity of white side divided by black side) was 9.81 for the pattern film and 5.17 for the flat PC film. Due to the wide AR, the image contrast ratio was also greatly improved as revealed in Figure 33(a). In Figure 33 (b), AFM images of anti reflective nano pillars which were used for the substrate film patterning. Ho and co-workers were reported the 89% enhancement in ANSI contrast ratio and efficacy was improved up to ~70% as compared to the flat substrate because nanopillars are acted as an AR layer. [132]

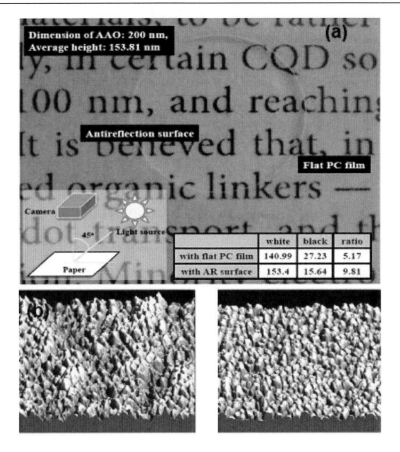

Figure 33. (a) Picture of a paper covered with the polycarbonate film under bright light source and the diagram of the experimental setup (b) AFM images of AR nano pillars made at 150 ^{0}C, 20 kgW/cm^2 and 160 ^{0}C, 25kgW/cm^2. [132]

CONCLUSION

OLEDs are unique technology for ultra thin, light weight full color display and solid state lighting sources in the future. By the use of various device structure approaches, like low carrier injection barrier, effective carrier confinement, effective excitons generation on the emission layer, fair host guest energy transfer, low carrier modulation layer and introduced the highly charged polymeric nano dots (PNDs) as embedded particles in the hole transporting layer, the OLEDs device efficacy are effectively enhanced. In other hands maximum light (~80%) is trapped in the organic/ITO layers, and in the substrate (glass) due to the TIR and wave-guiding of light. Trapped light is eventually absorbed and wasted as heat and minimize the OLEDs efficacy and life time. To extract the maximum possible amount of coupled light we have also discussed the various light out-coupling techniques, like substrate sand blasting, texturing mesh surfaces, multilayer modified cathodes, aperiodic dielectric stacks, LIGs, MLAs, SPPs, porous nano particles, nano wires, nano pillars micro and nano cavity effects, which are used for couple out the trapped light to enhance the out-coupling efficiency and improved the performance of OLEDs. Recently, Reineke et al. presented an improved WOLED structure which reaches the fluorescent tube efficacy by combining the proper device structure with high refractive index substrates and using a periodic out-coupling structure (90 lm/W at 1000 cd/m^2). The WOLEDs power efficiency has the potential to be raised to 124 lm/W, if light out-coupling can be further improved. [56] The future of OLEDs is bright because of its high efficiency, high brightness, design flexibility and eco-friendly technology.

ACKNOWLEDGMENTS

Authors are cordially thankful to National Science Council, National Tsing Hua University and Taiwan Semiconductor Manufacturing Company in support of funding for the presented work.

REFERENCES

[1]   C. W. Tang and S. A. VanSlyke, (1987) Organic electroluminescent diodes, *Appl. Phys. Lett., 51,* 913 - 915.

[2]   C. W. Tang, S. A. VanSlyke, C. H. Chen, (1989) Electroluminescence of doped organic thin films, *J. Appl. Phys., 65,* 3610 - 3616.

[3]   L. S. Hung, C. H. Chen, (2002) Recent progress of molecular organic electroluminescent materials and devices, *Mater. Sci. Eng., R. 39,* 143 - 222.

[4]   J. Kido, M. Kimura, K. Nagai, (1995) *Science, 267,* 1332.

[5]   S. R. Forrest, (2003) The road to high efficiency organic light emitting devices , *Org. Electron., 4,* 45 - 48.

[6]   J. H. Jou, W. B. Wang, S. Z. Chen, J. J. Shyue, M. F. Hsu, C. W. Lin, S. M. Shen, C. J. Wang, C. P. Liu, C. T. Chen, M. F. Wu, S. W. Liu. (2010) High-efficiency blue organic light-emitting diodes using a 3,5-di(9H-carbazol- 9-yl)tetraphenylsilane host via a solution-process, *J. Mater. Chem., 20,* 8411- 8416.

[7]   J. H. Jou, W. B. Wang, M.F. Hsu, C. P. Liu, C. C. Chen, C. J. Wang, Y. C. Tsai, J. J. Shyue, S. C. Hu, C. C. Chiang, H. Wang, (2008) Small nano-dot incorporated high-efficiency phosphorescent blue organic light-emitting diode, *PIERS, 4,* 351- 355.

[8]   W. D'Andrade, S. R. Forrest, (2004), White organic light emitting devices for solid state lighting, *Adv. Mater., 16,* 1585 - 1595.

[9]   D. Fyfe, (2009) Organic displays come of age, *Nature Photonics, 3,* 453 - 455.

[10]   R. H. Friend, R. W. Gymer, A. B. Holmes, J. H. Burroughes, R. N. Marks, C. Taliani, D. D. C. Bradley, D. A. Dos Santos, J. L. Bredas, M. Logdlund, W. R. Salaneck,(1999) Electroluminescence in conjugated polymers, *Nature, 397,* 121 - 128.

[11]   P. K. H. Ho, J. S. Kim, J. H. Burroughes, H. Becker, S. F. Y. Li, T. M. Brown, F. Cacialli, R. H. Friend, Molecular-scale interface engineering for polymer light-emitting diodes, (2000) *Nature, 404,* 481 - 484.

[12]   C. W. Tang, (1996) Organic electroluminescent materials and devices, *Inf. Disp., 12,* 16 - 19.

[13] C. Adachi, M. A. Baldo, M. E. Thompson, S. R. Forrest, (2001) Nearly 100% internal phosphorescence efficiency in an organic light emitting device, *J. Appl. Phys. 90*, 5048 - 5051.

[14] J. Huang, K. Yang, S. Liu, H. Jiang, (2000) High-brightness organic double-quantum-well electroluminescent devices, *Appl. Phys. Lett. 77*, 1750 - 1752.

[15] T. Fuhrmann, J. Salbeck, (2003), Organic Material for Photonic Crystals, *MRS Bull. 28*, 354 - 359.

[16] F. So, J. Kido, P. Burrows, Organic light-emitting devices for solid state lighting, (2008) *MRS Bull. 33*, 663 - 669.

[17] J. H. Jou, P. H. Wu, C. H. Lin, M. H. Wu, Y. C. Chou, H. C. Wang, S. M. Shen, (2010), Highly efficient orange-red organic light-emitting diode using double emissive layers with stepwise energy-level architecture, *J. Mater. Chem. 20*, 8464 - 8466.

[18] (a) J. H. Jou, S. M. Shen, S. H. Chen, M. H. Wu, W. B. Wang, H. C. Wang, C. R. Lin, Y. C. Chou, P. H. Wu, J. J. Shyue (2010) Highly efficient orange-red phosphorescent organic light-emitting diode using 2,7-bis(carbazo-9-yl)-9,9-ditolyfluorene as the host, *App. Phys. Lett., 96*, 143306 - 143309. (b) J. H. Jou, P. H. Wu, C. H. Lin (2010) High efficient orange organic light emitting diode with double emissive layer, *EM-NANO 2010*, Japan.

[19] M. A. Baldo, D. F. O'Brien, Y. You, A. Shoustikov, S. Sibley, M. E. Thompson, S. R. Forrest, (1998) Highly efficient phosphorescent emission from organic electroluminescent devices, *Nature, 395*, 151 - 154.

[20] Y. Kawamura, K. Goushi, J. Brooks, J. J. Brown, H. Sasabe, C. Adachi, (2005) 100% phosphorescence quantum efficiency of Ir(III) complexes in organic semiconductor films, *Appl. Phys. Lett., 86*, 071104 - 071106.

[21] C. Adachi, M. A. Baldo, M. E. Thompson, S. R. Forrest, (2001) Nearly 100% internal phosphorescence efficiency in an organic light emitting device, *J. Appl. Phys., 90*, 5048 - 5051.

[22] J. H. Jou, M. F. Hsu, W. B. Wang, C. L. Chin, Y. C. Chung, C.T. Chen, J. J. Shyue, S. M. Shen, M. H. Wu, W. C. Chang, C. P. Liu, S. Z. Chen, H. Y. Chen, (2009) Solution-processable, high-molecule-based trifluoromethyl-iridium complex for extraordinarily high efficiency blue-green organic light-emitting diode, *Chem. Mater., 21*, 2565 - 2567.

[23] J. H. Jou, C. J. Wang, Y. P. Lin, Y. C. Chung, P. H. Chiang, M. H. Wu, C. P. Wang, C. L. Lai, C. Chang, (2008) Color-stable, efficient fluorescent pure-white organic light-emitting diodes with device architecture preventing excessive exciton formation on guest, *Appl. Phys. Lett., 92*, 223504 - 223506.

[24] J. H. Jou, Y. P. Lin, M. F. Hsu, M. H. Wu, P. Lu, (2008) High efficiency deep-blue organic light-emitting diode with a blue dye in low-polarity host, *Appl. Phys. Lett., 92*, 193314 - 193316.

[25] X. Yang, D. C. Müller, D. Neher, K. Meerholz, (2006) Highly efficient polymeric electrophosphorescent diodes, *Adv. Mater. 18*, 948 - 954.

[26] N. Rehmann, D. Hertel, K. Meerholz, H. Becker, S. Heun, (2007) Highly efficient solution-processed phosphorescent multilayer organic light-emitting diodes based on small-molecule hosts, *Appl. Phys. Lett., 91*, 103507.1 - 103507.3.

[27] S. C. Lo, N. A. H. Male, J. P. J. Markham, S. W. Magennis, P. L. Burn, O. V. Salata, I. D. W. Samuel, (2002) Green phosphorescent dendrimer for light-emitting diodes, *Adv. Mater., 14*, 975 - 979.

[28] J. H. Jou, M H. Wu, S. M. Shen, H. C. Wang, S. Z. Chen, S. H. Chen, C. R. Lin, Y. L. Hsieh (2009) Sunlight-style color-temperature tunable organic light-emitting diode, *Appl. Phys. Lett.*, *95*, 013307 - 013309.

[29] J. H. Jou, W. B. Wang, M. F. Hsu, J. J. Shyue, C. H. Chiu, I-M. Lai, S. Z. Chen, P. H. Wu, C. C. Chen, C. P. Liu, S. M. Shen, (2010) Extraordinary high efficiency improvement for OLEDs with high surface-charge polymeric nanodots, *ACS Nano*, *4*, 4054 – 4060.

[30] J. H. Jou, C. C. Chen, Y. C. Chung, M. F. Hsu, C. H. Wu, S. M. Shen, M. H. Wu, W. B. Wang, Y. C. Tsai, C. P. Wang, J. J. Shyue; (2008) Nanodot-enhanced high efficiency pure-white organic light emitting diodes with mixed-host structure, *Adv. Func. Mat.*, *18*, 121 - 127.

[31] J. H. Jou, M. F. Hsu, W. B. Wang, C. P. Liu, Z. C. Wong, J. J. Shyue, C. C. Chiang; (2008) Small polymeric nano-dot enhanced pure-white organic light-emitting diode, *Org. Electron*, *9*, 291 - 295.

[32] T. Tsutsui, E. Aminaka, C. P. Lin, D. U. Kim, (1997) Extended molecular design concept of molecular materials for electroluminescence: sublimed-dye films, molecularly doped polymers and polymers with chromophores, *Philos. Trans. R. Soc. London A*, *355*, 801 - 813.

[33] Y. J. Lee, S. H. Kim, G. H. Kim, Y. H. Lee, S. H. Cho, Y. W. Song, Y. C. Kim, Y. R. Do, (2005) Far-field radiation of photonic crystal organic light-emitting diode, *Opt. Exp. 13*, 5864 - 5870.

[34] G. Gu, D. Z. Garbuzov, P. E. Burrows, S. Venkatesh, S. R. Forrest, M. E. Thompson, (1997) High-external-quantum-efficiency organic light-emitting devices, *Opt. Lett. 22*, 396 - 398.

[35] C. F. Madigan, M. H. Lu, J. C. Sturm, (2000) Improvement of output coupling efficiency of organic light-emitting diodes by backside substrate modification, *Appl. Phys. Lett. 76*, 1650 - 1652.

[36] P. T. Worthing, W. L. Barnes, (2001) Efficient coupling of surface plasmon polaritons to radiation using a bi-grating, *Appl. Phys. Lett. 79*, 3035 - 3037.

[37] S. Möller, S. R. Forrest, (2002) Improved light out-coupling in organic light emitting diodes employing ordered microlens arrays, *J. Appl. Phys. 91*, 3324 - 3327.

[38] L. H. Smith, J. A. Wasey, W. L. Barnes, (2004) Light outcoupling efficiency of top-emitting organic light-emitting Diodes, *Appl. Phys. Lett. 84*, 2986 - 2988.

[39] W. L. Barnes, (1999) Electromagnetic crystals for surface plasmon polaritons and the extraction of light from emissive devices, *J. Lightwave Technol.*, *17*, 2170 - 2182.

[40] M. H. Lu, J.C. Sturm, (2001) External coupling efficiency in planar organic light-emitting devices, *Appl. Phys. Lett.*, *78*, 1927 - 1929.

[41] M. Scheffel, A. Hunze, J. Birnstock, J. Blässing, W. Rogler, G. Wittmann, A. Winnacker, (2001) Enhanced light extraction by substrate modification of organic electroluminescent devices, *Proceedings of Eur. Conf. Org. Elect. Related Phenomena*, *01*, 158.

[42] R. Windisch, P. Heremans, A. Knobloch, P. Kiesel, G. H. Dohler, B. Dutta, G. Borghs, (1999) Light-emitting diodes with 31% external quantum efficiency by outcoupling of lateral waveguide modes, *Appl. Phys. Lett. 74*, 2256 - 2258.

[43] N. K. Patel, S. Cinà, J. H. Burroughes, High-efficiency organic light-emitting diodes, (2002) *IEEE J. Select. Top. Quant. Electron. 8*, 346 - 361.

[44] S. Chen, H. S. Kwok (2010) Light extraction from organic light-emitting diodes for lighting applications by sand-blasting substrates, *Opt. Exp.*, *18*, 37 - 42.

[45] J. S. Kim, P. K. H. Ho, N. C. Greenham, R. H. Friend, (2000) Electroluminescence emission pattern of organic light-emitting diodes: Implications for device efficiency calculations, *J. Appl. Phys. 88*, 1073 - 1081.

[46] M. Agrawal, P. Peumans, (2007) Design of non-periodic dielectric stacks for tailoring the emission of organic lighting emitting diodes, *Opt. Exp.*, *15*, 9715 - 9721.

[47] Y. Xu, J. S. Vuckovic, R. K. Lee, O. J. Painter, A. Scherer, and A Yariv, (1999) Finite-difference time-domain calculation of spontaneous emission lifetime in a microcavity, *J. Opt. Soc. Am. B, 16*, 465 - 474.

[48] A. V. Tikhonravov, (1993) Some theoretical aspects of thin-film optics and their applications, *Appl. Opt.*, *32*, 5417 - 5426.

[49] B. Riedel, J. Hauss, U. Geyer, J. Guetlein, U. Lemmer, M. Gerken, (2010) Enhancing outcoupling efficiency of indium-tin-oxide-free organic light-emitting diodes via nano structured high index layers, *Appl. Phys. Lett. 96*, 243302.1 - 243302.3.

[50] J. Shiang, T. Faircloth, and A. Duggal, (2004) Application of radiative transport theory to light extraction from organic light emitting diodes, *J. Appl. Phys.*, *95*, 2880 - 2888.

[51] B. Riedel, I. Kaiser, J. Hauss, U. Lemmer, and M. Gerken, (2010) Improving the outcoupling efficiency of indium-tin-oxide-free organic light-emitting diodes via rough internal interfaces, *Opt. Exp.*, *18*, A631 - A639.

[52] K. Hong, H. K. Yu, I. Lee, K. Kim, S. Kim, J. L. Lee, (2010) Enhanced light out-coupling of organic light-emitting diodes: spontaneously formed nanofacet-structured MgO as a refractive index modulation layer, *Adv. Mater.*, *22*, 4890 - 4894.

[53] B. C. Krummacher, M. Mathai, F. So, S. Choulis, and V. E. Choong, (2007) Light extraction from solution-based processable electrophosphorescent organic light-emitting diodes, *J. of Disp. Tech.*, *3*, 200 - 210.

[54] T. Nakamura, N. Tsutsumi, N. Juni, and H. Fujii, (2005) Thin-film waveguiding mode light extraction in organic electroluminescent device using high refractive index substrate, *J. Appl. Phys. 97*, 054505.1 - 054505.6.

[55] T. Nakamura, H. Fujii, N. Juni and N. Tsutsumi, (2006) Enhanced coupling of light from organic electroluminescent device using diffusive particle dispersed high refractive index resin substrate, *Opt. Rev.*, *13*, 104 - 110.

[56] S. Reineke , F. Lindner , G. Schwartz , N. Seidler , K. Walzer , B. Lussem , K. Leo , (2009) White organic light-emitting diodes with fluorescent tube efficiency, *Nature, 459*, 234 - 238.

[57] Y. Li, F. Li, J. Zhang, C. Wang, S. Zhu, H. Yu, Z. Wang, B. Yang, (2010) Improved light extraction efficiency of white organic light emitting devices by biomimetic antireflective surfaces, *Appl. Phys. Lett.*, *96*, 153305.1 - 153305.3.

[58] Y. H. Cheng, J. L. Wu, C. H. Cheng, (2007) Enhanced light outcoupling in a thin film by texturing meshed surfaces, *Appl. Phys. Lett. 90*, 091102.1 - 091102.3.

[59] J. M. Ziebarth, A. K. Saafir, S. Fan, M.D. McGehee, (2004) Extracting light from polymer light emitting diodes using stamped Bragg grating, *Adv. Funct. Mater. 14*, 451 - 456.

[60] H. H. Cho, B. Park, H. J. Kim, S. Jeon, J. H. Jeong, and J. J. Kim,(2010) Solution-processed photonic crystals to enhance the light outcoupling efficiency of organic light-emitting diodes, *Appl. Opt., 49*, 4024 - 4028.

[61] H. K. Kim, D. G. Kim, K. S. Lee, M. S. Huh, S. H. Jeong, K. I. Kim, and T.Y. Seong, (2005) Plasma damage-free sputtering of indium tin oxide cathode layers for top-emitting organic light-emitting diodes, *Appl. Phys. Lett. 86*,183503.1 - 183503.3.

[62] H. W. Choi, S. Y. Kim, K. B. Kim, Y. H. Tak, and J. L. Lee, (2005) Enhancement of hole injection using O_2 plasma-treated Ag anode for top-emitting organic light-emitting diodes, *Appl. Phys. Lett. 86*, 012104.1 - 012104.3.

[63] H. K. Kim, D. G. Kim, K. S. Lee, M. S. Huh, S. H. Jeong, K. I. Kim, H. Kim, D. W. Han, and J. H. Kwon, (2004) Plasma damage-free deposition of Al cathode on organic light-emitting devices by using mirror shape target sputtering, *Appl. Phys. Lett. 85*, 4295 - 4297.

[64] S. Chen, Y. Zhao, G. Cheng, J. Li, C. Liu, Z. Zhao, Z. Jie, and S. Liu, (2006) Improved light outcoupling for phosphorescent top-emitting organic light-emitting devices, *Appl. Phys. Lett. 88*, 153517.1 - 153517.3.

[65] M. Agrawal, Photonic design for efficient solid-state energy conversion, *PhD Thesis – Dec. 2008,* Stanford University, (CA) USA, *Chapter – 4, Non-periodic planar photonic structures for enhanced photon outcoupling from an organic light emitting diode*, 138 - 162.

[66] M. Agrawal, Y. Sun, S. R. Forrest, P. Peumans, (2007) Enhanced outcoupling from organic light-emitting diodes using aperiodic dielectric mirrors, *Appl. Phys. Lett. 90*, 241112.1 - 241112.3.

[67] G. W. Ford, W. H. Weber, (1984) "Electromagnetic interactions of molecules with metal surfaces, *Phys. Rep. 113*, 195 - 287.

[68] R. R. Chance, A. Prock, R. Silbey, (1978) Molecular fluorescence and energy transfer near interfaces, *Adv. Chem. Phys. 37*, 1 - 65.

[69] H. J. Peng, Y. L. Ho, X. J. Yu, H. S. Kwok, (2004) Enhanced coupling diodes using nano porous films, *J. Appl. Phys. 96*, 1649 - 1654.

[70] M. K. Wei, I. L. Su, (2004) Method to evaluate the enhancement of luminance efficiency in planar OLED light emitting devices for microlens array, *Opt. Exp., 12*, 5777 - 5782.

[71] H. Peng, Y. L. Ho, X. J. Yu, M. Wong, H. S, Kwok, (2005) Coupling efficiency enhancement in organic light-emitting devices using microlens array-theory and experiment, *J. Display Technol. 1*, 278 - 282.

[72] M. K. Wei, I. L. Su, Y. J. Chen, M. Chang, H. Y. Lin, T. C. Wu (2006) The influence of a microlens array on planar organic light-emitting devices, *J. Micromech. Microeng. 16*, 368 - 374.

[73] Y. Sun, S. R. Forrest (2006) Organic light emitting devices with enhanced outcoupling via microlenses fabricated by imprint lithography, *J. Appl. Phys. 100*, 073106.1 - 073106.6.

[74] J. H. Lee, Y. H. Ho, K.Y. Chen, H. Y. Lin, J. H. Fang, S. C. Hsu, J. R. Lin, M. K. Wei, (2008) Efficiency improvement and image quality of organic light-emitting display by attaching cylindrical microlens arrays, *Opt. Exp., 16*, 21184 - 21190.

[75] M. K. Wei, J. H. Lee, H. Y. Lin, Y. H. Ho, K.Y. Chen, C. C. Lin, C. F. Wu, H.Y. Lin, J. H. Tsai, T. C. Wu, (2008) Efficiency improvement and spectral shift of an organic light-emitting device by attaching a hexagon-based microlens array, *J. Opt. A: Pure Appl. Opt., 10*, 055302.1 - 055302.9.

[76] V. Bulovic, G. Gu, P. E. Burrows, S. R. Forrest, M. E. Thompson, (1996) Transparent light-emitting devices, *Nature, 380,* 29.

[77] J. Y. Kim, K. C. Choi, (2011) Improvement in outcoupling efficiency and image blur of organic light-emitting diodes by using imprinted microlens arrays, *J. Display Technol., 7,* 377 - 381.

[78] W. K. Huang, W. S. Wang, H. C. Kan and F. C. Chen, (2006) Enhanced light out-coupling efficiency of organic light-emitting diodes with self-organized microlens arrays, *Jpn. J. Appl. Phys., 45,* L1100 - L1102.

[79] J. P. Yang, Q. Y. Bao, Z. Q. Xu, Y. Q. Li, J. X. Tang, S. Shen, (2010) Light out-coupling enhancement of organic light-emitting devices with microlens array, *Appl. Phys. Lett., 97,* 223303.1 - 223303.3.

[80] H. J. Peng, Y. L. Ho, C. F. Qiu, M. Wong, H. S. Kwok, (2004) Coupling efficiency enhancement of organic light emitting devices with refractive microlens array on high index glass substrate, *SID 04 Digest,* 158 - 161.

[81] Y. Sun, S. R. Forrest, (2008) Enhanced light out-coupling of organic light-emitting devices using embedded low-index grids, *Nat. Photonics, 2,* 483 - 487.

[82] M. Slootsky, S. R. Forrest, (2010) Enhancing waveguided light extraction in organic LEDs using an ultra-low-index grid, *Opt. Lett., 35,* 1052 - 1054.

[83] M. Slootsky, S. R. Forrest, (2009) Full-wave simulation of enhanced outcoupling of organic light-emitting devices with an embedded low-index grid, *Appl. Phys. Lett., 94,* 163302.1 - 163302.3.

[84] C. H. Lin, T. Y. Cho, C. H. Chang, and C. C.Wu, (2006) "Enhancing light outcoupling of organic light-emitting devices by locating emitters around the second antinode of the reflective metal electrode, *Appl. Phys. Lett. 88,* 081114.1 - 081114.3.

[85] J. Lee, N. Chopra, D. Bera, S. Maslov, S.-H. Eom, Y. Zheng, P. Holloway, J. Xue, F. So, (2011) Down-conversion white organic light-emitting diodes using microcavity structure, *Adv. Energy Mater., 1,* 174 - 178.

[86] V. Bulovic, V.B. Khalfin, G. Gu, P.E. Burrows, D.Z. Garbuzov, S.R. Forrest, (1998) Weak microcavity effects in organic light-emitting devices, *Phys. Rev. B, 58,* 3730 - 3740.

[87] J. Lee, N. Chopra, F. So, (2008) Cavity effects on light extraction in organic light emitting devices, *Appl. Phys. Lett., 92,* 033303.1 - 033303.3.

[88] E. F. Schubert, N. E. J. Hunt, M. Micovic, R. J. Malik, D. L. Sivco, A. Y. Cho, G. J. Zydzik, (1994) Highly Efficient Light-Emitting Diodes with Microcavities, *Science, 265,* 943.

[89] N. Takada, T. Tsutsui, S. Saito, (1993) Control of emission characteristics in organic thin-film electroluminescent diodes using an optical-microcavity structure, *Appl. Phys. Lett., 63,* 2032 - 2034.

[90] A. Dodabalapur, L.J. Rothberg, T. Miller, E.W. Kwock, (1994) Microcavity effects in organic semiconductors, *Appl. Phys. Lett., 64,* 2486 - 2488.

[91] A. Dodabalapur, L.J. Rothberg, R.H. Jordan, T.M. Miller, R.E. Slusher, J.M. Phillips, (1996) Physics and applications of organic microcavity light emitting diodes, *J. Appl. Phys., 80,* 6954 - 6964.

[92] T. Tsutsui, N. Takada, S. Saito, E. Ogino, (1994) Sharply directed emission in organic electroluminescent diodes with an optical-microcavity structure, *Appl. Phys. Lett., 65,* 1868 - 1870.

[93] R. H. Jordan, L. J. Rothberg, A. Dodabalapur, R. E. Slusher, (1996) Efficiency enhancement of microcavity organic light emitting diodes, *Appl. Phys. Lett., 69,* 1997 - 1999.

[94] S. Tokito, K. Noda, Y. Taga, (1996) Strongly directed single mode emission from organic electroluminescent diode with a microcavity, *Appl. Phys. Lett.* 68, 2633 - 2635.

[95] T. A. Fisher, D. G. Lidzey, M. A. Pate, M. S. Weaver, D. M. Whittaker, M. S. Skolnick, D. C. Bradley, (1995) Electroluminescence from a conjugated polymer microcavity structure, *Appl. Phys. Lett. 67,* 1355 - 1357.

[96] J. Grüner, F. Cacialli, R.H. Friend, (1996*)* Emission Enhancement in single-layer polymer microcavities, *J. Appl. Phys. 80,* 207 - 215.

[97] T. Shiga, H. Fujikawa, Y. Taga, (2003) Design of multi wavelength resonant cavities for white organic light-emitting diodes, *J. Appl. Phys. 93,* 19 - 22.

[98] *X. Liu, D. Poitras, Y. Tao, C. Py, (2004) Microcavity organic light emitting diodes with double sided light emission of different colors,* J. Vac. Sci. Technol. 22, *764 - 767.*

[99] T. Nakayama, Y. Itoh, A. Kakuta, (1993) Organic photo- and electroluminescent devices with double mirrors, *Appl. Phys. Lett. 63,* 594 - 595.

[100] C. L. Lin, H. W. Lin, C. C. Wu, (2005) "Examining microcavity organic light-emitting devices having two metal mirrors, *Appl. Phys. Lett. 87,* 021101.1 - 021101.3.

[101] J. R. Tischler, M. S. Bradley, V. Bulovic, J. H. Song, A. Numikko, (2005) Strong coupling in a microcavity LED, *Phys. Rev. Lett. 95,* 036401 - 036404.

[102] H. Peng, J. Sun, X. Zhu, X. Yu, W. Wong, H.-S. Kwok, (2006) High-efficiency microcavity top-emitting organic light-emitting diodes using silver anode, *Appl. Phys. Lett. 88,* 073517.1 - 073517.3.

[103] T. Y. Cho, C. L. Lin, C. C. Wu, (2006) Microcavity two-unit tandem organic light-emitting devices having a high efficiency, *Appl. Phys. Lett. 88,* 111106.1 - 111106.3.

[104] J. Lim, S. S. Oh, D. Y. Kim, S. H. Cho, I. T. Kim, S. H. Han, H. Takezoe, E. H. Choi, G. S. Cho, Y. H. Seo, S. O. Kang, B. Park, (2006) Enhanced out-coupling factor of microcavity organic light-emitting devices with irregular microlens array, *Opt. Exp. 14,* 6564 - 6571.

[105] S.Y. Nien, N. F. Chiu, Y. H. Ho, J. H. Lee, C. W. Lin, K. C. Wu, C. K. Lee, J. R. Lin, M. K. Wei, T. L. Chiu, (2009) Directional photoluminescence enhancement of organic emitters via surface plasmon coupling, *Appl. Phys. Lett., 94,* 103304.1 - 103304.3.

[106] Y. Tanaka, J. Upham, T. Nagashima, T. Sugiya, T. Asano, S. Noda, (2007) Dynamic control of the Q factor in a photonic crystal nanocavity, *Nat. Mater. 6,* 862 - 865.

[107] E. Ozbay, (2006) Plasmonics: Merging photonics and electronics at nanoscale dimensions, *Science, 311,* 189 - 193.

[108] J. Vuckovic, M. Loncar, A. Scherer, (2000) Surface plasmon enhanced organic light emitting diodes, *IEEE J. Quant. Electron. 36,* 1131 - 1144.

[109] P. A. Hobson, J. A. E. Wasey, I. Sage, W. L. Barnes, (2002) The role of surface plasmons in organic light-emitting diodes, *IEEE J. Sel. Top. Quant. Electron. 8,* 378 - 386.

[110] K. Okamoto, I. Niki, A. Scherer, Y. Narukawa, T. Mukai, Y. Kawakami, (2005) Surface plasmon enhanced spontaneous emission rate of InGaN/GaN quantum wells probed by time-resolved photoluminescence spectroscopy, *Appl. Phys. Lett. 87,* 071102.1 - 071102.3.

[111] D. K. Gifford, D. G. Hall, (2002) Extraordinary transmission of organic photoluminescence through an otherwise opaque metal layer via surface plasmon cross coupling, *Appl. Phys. Lett. 80*, 3679 - 3681.

[112] S. Wedge, J. A. E. Wasey, W. L. Barnes, I. Sage, (2004) Coupled surface plasmon-polariton mediated photoluminescence from a top-emitting organic light-emitting structure, *Appl. Phys. Lett. 85*, 182 - 184.

[113] S. Wedge, W. L. Barnes (2004) Surface plasmon-polariton mediated light emission through thin metal films, *Opt. Exp., 12*, 3673 - 3685.

[114] A. Giannattasio, S. Wedge, W. L. Barnes, (2006) Role of surface profiles in surface polariton mediated emission of light through a thin metal films, *J. Modern Opt., 53*, 429 - 436.

[115] C. J. Yates, I. D. W. Samuel, P. L. Burn, S. Wedge, W. L. Barnes (2006) Surface plasmon polariton mediated emission from phosphorescent dendrimer light emitting diodes, *Appl. Phys. Lett. 88*, 161105.1 - 161105.3.

[116] K. Ishihara, M. Fusita, I. Matsubara, T. Asano, S. Noda, H. Ohata, A. Hirasawa, H. Nakada, N. Shimoji, (2007) Organic light-emitting diodes with photonic crystals on glass substrate fabricated by nanoimprint lithography, *Appl. Phys. Lett. 90*,111 - 114.

[117] A. O. Altun, S. Jeon, J. Shim, J. H. Jeong, D. G. Choi, K. D. Kim, J. H. Choi, S. W. Lee, E. S. Lee, H. D. Park, J. R. Youn, J. J. Kim, Y. H. Lee, J. W. Kang, (2010) Corrugated organic light emitting diodes for enhanced light extraction, *Org. Electron. 11*, 711 - 716.

[118] Y. R. Do, Y. C. Kim, Y. W. Song, C. O. Cho, H. Jeon, Y. J. Lee, S. H. Kim, Y. H. Lee, (2003) Enhanced light extraction from organic light emitting diodes with 2D $SiO_2=SiN_x$ photonic crystals, *Adv. Mater. 15*, 1214 - 1218.

[119] Y. J. Lee, S. H. Kim, J. Huh, G. H. Kim, Y. H. Lee, S. H. Cho, Y. C. Kim, and Y. R. Do, (2003) A high-extraction-efficiency nanopatterned organic light-emitting diode, *Appl. Phys. Lett. 82*, 3779 - 3781.

[120] S. Jeon, J. W. Kang, H. D. Park, J. J. Kim, J. R. Youn, J. Shim, J. H. Jeong, D. G. Choi, K. D. Kim, A. O. Altun, S. H. Kim, and Y. H. Lee (2008) Ultraviolet nanoimprinted polymer nanostructure for organic light emitting diode application, *Appl. Phys. Lett. 92*, 223307.1 - 223307.3.

[121] S. Y. Chou, P. R. Krauss, P. J. Renstrom, (1996) Nanoimprint lithography, *J. Vac. Sci. Technol. B, 14*, 4129 - 4133.

[122] H. H. Cho, B. Park, H. J. Kim, S. Jeon, J. H. Jeong, J. J. Kim (2010) Solution-processed photonic crystals to enhance the light outcoupling efficiency of organic light-emitting diodes, *Appl Opt., 49*, 4024 - 4028.

[123] J. G. C. Veinot, H. Yan, S. M. Smith, J. Cui, Q. Huang, T. J. Marks, (2002) Fabrication and properties of organic light-emitting "nanodiode" Arrays, *Nano Lett. 2*, 333 - 335.

[124] Y. C. Kim, Y. R. Do, (2005) Nanohole-templated organic light-emitting diodes fabricated using laser-interfering lithography: moth-eye lighting, *Opt. Exp. 13*, 1598 - 1603.

[125] K. Saxena, D. S. Mehta, V. K. Rai, R. Srivastava, G. Chauhan, M.N. Kamalasanan, (2008) Implementation of anti-reflection coating to enhance light out-coupling in organic light-emitting devices, *J. Lumin., 128*, 525 - 530

[126] H. J. Peng, Y. L. Ho, X. J. Yu, H. S. Kwok, (2004) Enhanced coupling diodes using nano porous films, *J. Appl. Phys. 96*, 1649 - 1654.

[127] S. Y. Hsu, M. C. Lee, K. L. Lee, P. K. Wei, (2008) Extraction enhancement in organic light emitting devices by using metallic nanowire arrays, *Appl. Phys. Lett.* 92, 013303.1 - 013303.3.

[128] S. P. Price, J. Henzie, T. W. Odom, (2007) Addressable, Large-area Nanoscale Organic Light Emitting Diodes, *Small, 3*, 372 - 374.

[129] B. Wang, L. Ke, S. J. Chua, (2006) A nano-patterned organic light-emitting diode with high extraction efficiency, *J. Cryst. Growth, 288*, 119 - 122.

[130] S. A. Choulis, M. K. Mathai, V. E. Choong, (2006) Influence of metallic nanoparticles on the performance of organic electrophosphorescence devices, *Appl. Phys. Lett. 88,* 213503.1 - 213503.3.

[131] C. J. Sun, Y. Wu, Z. Xu, B. Hu, J. Bai, J. P. Wang, (2007) Enhancement of quantum efficiency of organic light emitting devices by doping magnetic nanoparticles, *Appl. Phys. Lett. 90,* 232110.1 - 232110.3.

[132] Y. H. Ho, C. C. Liu, S. W. Liu, H. Liang, C. W. Chu, P. K. Wei, (2011) Efficiency enhancement of flexible organic light-emitting devices by using antireflection nanopillars, *Opt. Exp., 19*, A295 - A302.

In: Light-Emitting Diodes and Optoelectronics: New Research   ISBN: 978-1-62100-448-6
Editors: Joshua T. Hall and Anton O. Koskinen   © 2012 Nova Science Publishers, Inc.

Chapter 2

RELIABILITY ESTIMATION FROM THE JUNCTION TO THE PACKAGING OF LED

Yannick Deshayes[*]
IMS Laboratory, University Bordeaux I
Talence Cedex, France

I. CONTEXT AND OBJECTIVES

Selection tests and life testing can be used to satisfy the requirements of early and long term failures. But besides the fact they are costly and time consuming. Generally they do not give efficient information about failure mechanisms and defect nature in particular for optical modules. In this context, much effort has been conducted to clarify degradation mechanisms and eliminate the causes [1]. Many studies have demonstrated the sensitivity of the LED to ageing tests leading to packaging or semiconductor degradations.

Activation of packaging degradations are traditionally demonstrated using thermal cycles. This category of test induces thermomechanical stresses, in particular, due to the mismatched thermal expansion coefficients of the different materials or process defects (threading or misfit dislocations in semiconductor, cracks or delaminations in solder, …) resulting in an optical beam deviation and the drop of the LED optical coupling performances [2, 3, 4]. Actual qualification standards requirement tends to be 500 thermal cycles (233 K/358 K or 218 K/398 K with different slopes and well parameters) to characterize assembling process quality and packaging robustness. For LED in metal package, like our components, it has been demonstrated that the sensitivity to thermal cycles is relatively weak [2, 5].

One of the other main causes of optical power decrease in LEDs comes from semiconductor degradations, and is especially correlated with defects diffusion in the active zone. For example, crystal defects are introduced during crystal growth and manufacturing process for LED [6, 7]. Accelerated ageing tests based on active storage (current or/and temperature) were yet conducted on chips in order to characterize degradation kinetics related to failure mechanisms. Failures are traditionally activated by an increase of bias current

---

[*] Ph.: ++33 556842857, Fax: ++33 556371545.

and/or junction temperature and degradation kinetics are actually well known for this kind of semiconductor [4, 7, 8]. Ageing tests are generally carried out at different temperatures until 15000 h with high current density leading to an increase of internal junction temperature ranging from 300 K to 620 K in order to identify suitable failure mechanisms and calculate lifetime distribution using high acceleration factors upper than 300-400 [5, 7].

A recent cause of degradation comes from the interface between semiconductors and polymer coating the chip. This manufacturing process was added for: increase the packaged LED output and reduce the vibration impact on the bonding for board system. In this case, the leakage current is observed and the efficient current through the active zone decrease. The optical power losses can reach 40 % in addition with the previous mechanisms. The temperature and the current are the aggravation factor for this failure mechanism. The rupture of the hydrogen link in the polymer matrix explains failure mechanism. This modification of the polymer/semiconductor interface is characterized by the presence of electronic level responsible of the leakage current simulated by Pool-Frenkel theory.

In the case of commercial LEDs, it is important to master the thermal management of LEDs for better estimate failure mechanisms and degradation kinetics. Temperature gradient between the junction of the chip and the body package is generally equal to 150 K according to thermal resistance R_{th} close to 700K/W and for thermal power equal to 150 mW [7]. The thermal resistance is given for LED packages without heat tink. The new applications of LED, centered on the public light, use a bias current around 1 A for each LED and the thermal management becomes critical. The computed aided designs are necessary to study different solution and chose the better one to guaranty the lowest junction temperature and its stability during lifetime of the device. In this case the thermal resistance is near 10 K/W.

So considering junction temperature value during ageing tests previously given, package temperature can reach more than 450 K exceeding the temperature upper range for most of materials, especially interconnection materials such as adhesive conductors and solder joints. For adhesive conductors, a temperature of 450 K represents a condition close to maximal operating temperature (460 K) [9, 10]. For solder joint $Au_{80}Sn_{20}$, generally used for optoelectronic applications, maximal operating temperature is generally close to 500 K and reflow temperature is close to 570 K [11, 12, 13]. With a temperature of 450 K, degradation phenomenon located in solder joints can not related to those observed during operating conditions and reliability investigations could thus be difficult.

All these considerations lead to a drastic decrease of the acceleration factor value and a solution consists in largely increase the number of samples. Indeed, even accepting only two failures over 1000 hours for classical life test with an acceleration factor near 300, more than one hundred of components are necessary. Moreover, the possibility to reach the chosen failure criterion is not guarantied.

On the other hand, the increase of junction temperature is not the only stress condition for accelerated ageing test of LEDs, the increase of bias current is also considered. A high number of papers have already shown that an increase of bias current also represents a useful solution for reliability investigations [1, 4, 14].

Accurate knowledge of time to failure and long-term reliability is one of the main goals of optoelectronic manufacturers and end-users. To assess this challenge even for complex assemblies, process control, design and technological solutions must be optimized to guarantee a minimum of defects during the operating life. Reliability evaluation must also evolve from classical accelerated test for optoelectronic emissive systems, towards the use of more sophisticated laws especially based on physics of failure and process parameters dispersion [15]. This approach has the main advantage to make an easy comparison between different manufacturers [16].

In this chapter, different kinds of ageing tests (storage and thermal cycles) have been considered on different technologies assembly in TO42 package. This paper develops the complete methodology to estimate the reliability of the packages LED (P-LED). An accurate failure diagnostic is performed on commercial LEDs after ageing tests to identify and localize the degradation in the LED. An appropriate model has been developed to determine the correlation between optical power losses and failure mechanisms. In order to simplify this approach, both electrical and optical models are proposed [17, 18]. Each electrical parameter, composing the equivalent electrical circuit of the LED, is related with a part of the component structure and their gradual changes versus ageing time have been monitored.

Finally, degradation laws extracted from a weak drift of monitored electrical and optical parameters have been used in order to calculate times to failure and failure rates only after 1200 hours of accelerated ageing time with less than 20 LEDs on three different technologies [19].

II. State of Art of the LED Module

Figure 1a shows the structure of InGaAs/GaAs 935 nm commercial LED. SEM and EDX analyses have been performed to identify chemical material composition. InGaAs/GaAs chip (300 μm * 300 μm * 150 μm) is reported by Ag-based conductor adhesive (1μm) in a truncated conic cavity realized by chemical etching in Invar (Fe/Ni) submount. Light is emitted in 4π sr side of the chip and collected by two mirrors fabricated by the side of the truncated conic cavity (figure 1b). Mirrors system drives the light in the SiO_2 lens. Their surface is covered by a thin film (1 μm) of Ni material using chemical vapor deposition (CVD) process. Lens and submount are aligned and set by cylinder (Fe/ Ni) for circular light output with a beam angle of 20°.

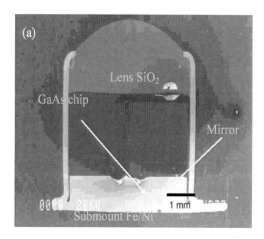 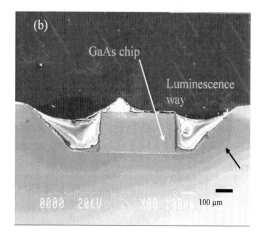

Figure 1. Structure of the LED (a) : package view (× 14) and (b) : chip view (× 20).

Figure 1b shows the inferior and superior contacts of the LED device. Superior contact is realized by a thin film of Au (500 nm)/Ge (500 nm)/Ni(600 nm)/Au (2 μm) deposed by metal organic CVD (MOCVD) on GaAs N$^+$-doped layer (Si - 10^{19}cm^{-3}) [6]. Inferior contact consists in a thin film of Au$_{95}$Zn$_5$ (1 μm) deposed by MOCVD on GaAs P$^+$-doped layer (Zn / 10^{18} cm^{-3}), Ag doped conductor adhesive (1 μm) and Ni (2 μm)/Au (1 μm) thin films deposed by CVD on Invar substrate material. A molded thermoset resin coats the final device in order to improve optical efficiency η_{tot} [7].

Figure 2 shows the structure of an 872 nm AlGaAs/GaAs commercial DH-LED elaborated by molecular beam epitaxy. Infrared light is emitted in 4π sr around the active zone.

The top contact is performed using a thin multi-layer of Au (500 nm)/Ge (500 nm)/Ni (600 nm)/Au (2 μm) deposed by metal organic chemical vapor deposition (MOCVD) on GaAs P$^+$-doped layer (Be/$1.0.10^{18}$cm^{-3}).

The bottom contact consists in a thin film of Au$_{95}$Zn$_5$ (1 μm) deposed by MOCVD on GaAs substrate N$^+$-doped layer (Si/10^{18} cm^{-3}).

Figure 3 shows the typical structure of a 472 nm In$_{0.2}$Ga$_{0.8}$N/GaN Multi quantum wells (MQW) commercial LED [20]. The active zone is elaborated with p-Al$_{0.15}$Ga$_{0.85}$N (50 nm) and n-GaN cladding layers [21, InGaN/GaN (10/40 Å) MQW with indium substitution equal to 0.2 processed by molecular beam epitaxy. Blue light is emitted in 4π steradians around the active zone.

The n-contact is performed using a thin multi-layer of Pt (40 nm)/ Ti (40 nm)/Au (120 nm) deposed by metal organic chemical vapour deposition (MOCVD) on GaN N$^+$-doped layer (Si/$2.5.10^{18}$cm^{-3}) [22]. The p-contact consists in a thin film of Ni (10nm)/Au (10 nm) deposed by MOCVD on GaN P$^+$-doped layer (Mg/2.10^{20} cm^{-3}) [21, 23, 24].

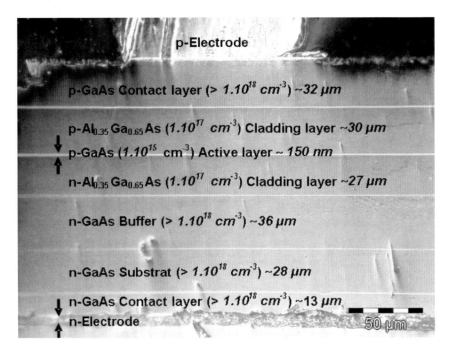

Figure 2. Internal structure of AlGaAs/GaAs LED.

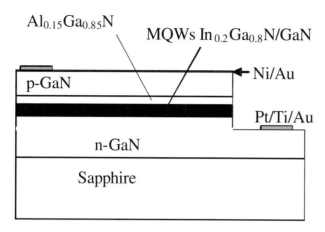

Figure 3. Internal structure of MQW InGaN/GaN LED.

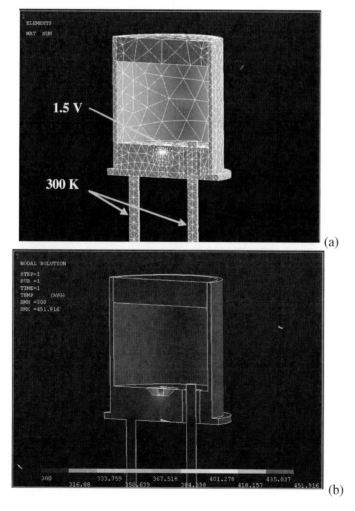

Figure 4. Meshing structure of the TO42 device (a), thermal cartography of the TO42 device for boundary condition @ 300 K (b).

III. MODEL OF THE LED MODULE

1. Thermal Simulation

Current graphic performances of the simulator allow us, without difficulty, to build a complex geometric model generally coming from the design plans of the considered system (figure4a). The most important difficulty resides in the dividing of the geometric model into finite elements. It is necessary to follow a very strict procedure of model building allowing us to insure that we obtain an optimum pattern in the end. In our case, the mixing of the parallelepiped and cylindrical volume drove us to mesh the structure with tetrahedral elements. The number of elements, electrical and thermal is 3000 and the number of nodes is 4500. Elements of this type require the material features properties listed below: thermal conductivity, material density, heat capacity and electrical conductivity.

The boundary conditions are thermal with 300 K fixed on the pin and electrical with 1.5 V bias voltages applied on chip. The electrical model is based on the electrical characteristics developed in the next section. The thermal cartography is shown in figure 4b. The simulation result established that junction temperature is close to 450 K for this condition of thermal management.

The thermal power P_{th} dissipated by a LED depends on the current injected into the diode and also on its voltage drop. A basic model is illustrated in figure. The electrical power, which is dissipated at the junction, produces an increase of temperature that causes a drop in the light output power.

In the case of optoelectronic devices, the junction temperature T_j depends on the electrical working point. This last one is defined by current and voltage bias (I, V), on the package temperature T_p and on the junction-ambient thermal resistance. Junction temperature can be expressed by equation (1).

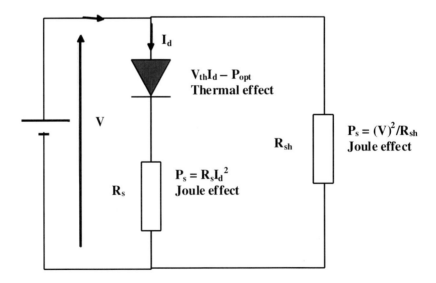

Figure 5. Basic LED model – power assessment.

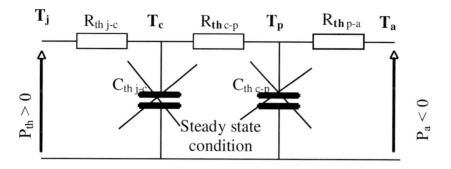

Figure 6.Thermal circuit indicating the thermal resistance from junction to ambient.

$$T_j = T_a + R_{thj-a}\left[V_{th}I - P_{op} + R_sI_d^2 + \frac{V^2}{R_{sh}}\right]$$

$$T_j = T_a + R_{thj-a}P_{th}$$

(1)

The dissipated power of the junction in packaged LEDs corresponds to the difference between the electrical input power (VI) and the light output power (P_{opt}) [25]. The difference of power assessment is due to both the Joule and thermal effects. The Joule effect is located in both series resistance (R_s) and shunt resistance (R_{sh}) responsible to the leakage current. The Joule effect is outside the junction. A great part of power is loss in junction zone by Shockley Read Hall effect and absorption of photons in the cavity and called thermal effect. The electrical parameters are estimate using the electrical characteristics (I-V curves) used in previous study [26].

The thermal circuit from T_j to T_p is presented in figure 6.

The C_{th} parameter represents the thermal capacitance and are neglected because of the steady state condition manage by the LNC system. $R_{th\,j-p}$ is the thermal resistance of junction-package and composed by semiconductor thermal resistance explained by R_{thj-c} and package thermal resistor R_{thc-p}.

2. Optical Models

a. Optical Power Characterizations

Experimental optical power characterizations have been also made. Typical P(I) curve is shown in and is characterized by two different working zones limited by threshold condition (I_{th}) such as exposed in the electrical modeling part. The non-linear part can be considered as quadratic, depending on the forward current.

This behavior is well known and can be understood when considering the continuity equation for the injected electrons n described by equation (2).

$$\frac{dn}{dt} = \frac{J}{qd} - A_{nr}n - Bn^2 - C_{Aug}n^3$$

(2)

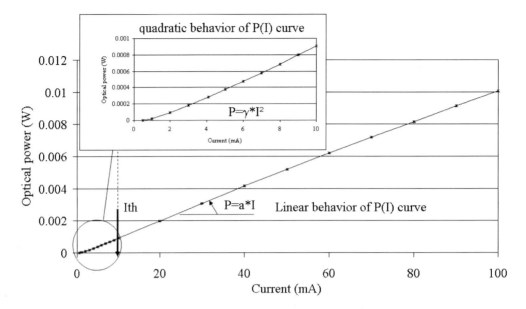

Figure 7. Optical power P(I) characteristic measured at 100 mA and 300 K.

J: the total current density $J = I/S$, S: the active zone area, d: the thickness of active layer, A(B): the non radiative (radiative) recombination coefficient and C_{Aug} : the Auger recombination coefficient.

Non-radiative recombination coefficient A is given by Shockley-Hall-Read recombination rate (3).

$$A_{nr} = 1/2(N_{def}v_{th}\sigma) \tag{3}$$

N_{def} : the defects density located in active zone, v_{th}: the thermal velocity, and σ : the capture cross section [27]. The radiative recombination coefficient B is given by equation (4).

$$B = \frac{1}{\tau_R} \frac{N_j}{N_c N_v} \tag{4}$$

τ_R : radiative lifetime, N_j, N_c and N_v : the effective density of states.

At this level of current ($< I_{th}$) the Auger recombination is negligible due to the large band gap energy of $In_{0.53}Ga_{0.47}As$ (1,26 eV at 400 K). It also assumed that the background carriers' density is negligible, i.e, electrons concentration is equal to holes concentration as explained by B. Winter. Assuming steady-state conditions $dn/dt = 0$, we obtain the light emission L at low forward currents, as long as $An >> Bn^2$.

$$L = Bn^2 \approx \frac{B}{A^2}\left(\frac{J}{qd}\right)^2 \tag{5}$$

Equation (5) reveals that luminescence has a quadratic behaviour versus J when J is lower than a threshold value $J_{th} = I_{th}/S$. Below 10 mA, it has been shown that experimental P(I) curve well fits this quadratic behavior ($\gamma = 3.10^{-6}\ W/A^2$) [27]. The optical power is then expressed by equation (6).

$$P = \eta_{mod} Bn^2 E_c = \frac{1.3 \cdot \eta_{tot} S^2}{dq^2} \frac{B}{A^2} I^2 = \gamma\ I^2 \qquad (6)$$

η_{tot} : the total efficiency of the LED, given between device interface and lens output, and E_c : the central energy given by optical spectrum analysis.

Total efficiency η_{tot} could be considered as a constant for initial measurements and for all current values. γ is proportional to B/A^2 and defined as a luminescence factor (W/A² for $I < I_{th}$) [7].

As the current density J increases the Bn^2 term starts to dominate, light emission becomes linear with J (7).

$$L = Bn^2 \approx \frac{J}{qd} \qquad (7)$$

For $I \square I_{th}$ upper than 10 mA, P(I) curve is described by equation (8) with a linear behavior as shown in equation (8).

$$P = \eta_{tot} Bn^2 E_c = \frac{1.3 \eta_{tot} S^2}{qd} I = \alpha \cdot I \qquad (8)$$

α : a luminescence factor (W/A for $I > I_{th}$).

b. Spectral Characterizations

Spectral emission of solid semiconductor structure can give some information about active zone temperature, material composition and physical phenomenon inducing luminescence [23, 24, 25, 26, 27]. The main objective of this chapter is centered on the demonstration of the material composition of the active zone (GaAs) and extraction of fundamental parameters necessary to build physical law about luminescence in LED considered.

Different hypotheses have been considered to simplify the calculus and evaluate, by analytic equations, both static and exciting state of the structure. The bias current, for spectral analysis, is adjusted to 100 mA with Keithley 6430 analyser to ensure the stability of DC current, about 10 pA, in LED. The number of carrier injected is more than 10^{17} cm^{-3} and the inversion of the population of the carriers can be easily supposed in DH active zone (see figure 7).

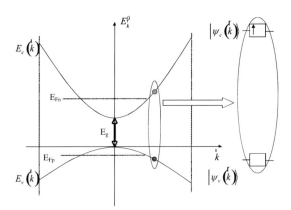

Figure 8. $\overset{\shortmid}{k}$ representation of crystal band scheme.

The specific characteristics of this energy band are that Fermi level of electrons (E_{Fn}) and hole (E_{Fp}) is located in the conduction and valence bands respectively. This last schematic design shows both the importance of DH structure on the inversion of population responsible to the optical gain.

The key of the quantum theory using to explain luminescence in semiconductor is the consideration that semiconductor energy level is the superposition of quantum dots energy. In the case of the semiconductor have a great dimension (200 nm) before single cellular (5, 86 Å), the level separation is very weak (< 0.1 meV). The semiconductor is considered as a superposition of two level elementary structures. Using time dependant perturbation theory, the probability of electronic transition between $|\psi_v(\overset{\shortmid}{k})\rangle$ level to $|\psi_c(\overset{\shortmid}{k})\rangle$ is given by equation (9).

$$P_{12} = \frac{\pi}{2h}\left|\langle\psi_{c,k}|W|\psi_{v,k}\rangle\right|^2 \delta\left(h\omega_{vc} = E_c(\overset{r}{k}) - E_v(\overset{r}{k})\right) \qquad (9)$$

With W: the electromagnetic Hamiltonian of the perturbation (photon), ω_{vc} : the characteristic pulsation of the elementary structure. In the case of luminescence phenomenon, the Hamiltonian of the perturbation is dipole electrical (D.E) detailed by equation (10).

$$W(t) = -qr\overset{uu}{E}(\overset{uu}{r_0},t) = -D\overset{uu}{E}(\overset{uu}{r_0},t) \qquad (10)$$

With D: the electrical dipole, r: the dimension in spherical representation.

In semiconductor and for luminescence amplitude considered, the optical susceptibility is considered as linear and can be evaluated for elementary cellule by equation (11).

$$\chi(\omega) = \frac{q(\chi_{vc}(k))^2 T_2}{\varepsilon_0 h} \frac{(\omega - \omega_{vc}(k))T_2}{1 + (\omega - \omega_{vc}(k))^2 T_2^2}(N_v - N_c) \qquad (11)$$

With $\chi_{vc}(k)$: the element of dipolar matter, T_2 : the phase relaxation time, N_c : the number of carrier on level $|\psi_c(\vec{r})\rangle$, N_v : the number of carrier on level $|\psi_v(\vec{r})\rangle$. The equation (9), (10) and (11) are used on the single system with two levels of energy $|\psi_v(\vec{r})\rangle$ and $|\psi_c(\vec{r})\rangle$ as it shown in the inset figure 7.

Using Einstein balance equations, the elementary spontaneous emission is given by relation (12) with the definition of radiation lifetime.

$$r_{spon} = \frac{1}{\tau_R} f_c(E_c)\left[1 - f_v(E_v)\right]$$
$$\frac{I}{\tau_R} = \frac{q^2 \chi_{vc}^2 n_{op}^3 \omega_{vc}^3}{\pi c^3 h \varepsilon_0}$$

(12)

With $f_c(E_c)$ is the Fermi-Dirac function for conduction band, $f_c(E_c)$ is the Fermi-Dirac function for valence band and τ_R is the radiation lifetime of material. The Fermi-Dirac theory can be used because the number of quantum dots and carriers is high ($> 10^{20}$ cm^{-3}) for 200 nm thickness structures.

The spectral representation about radiation recombination is proposed using both probability of spontaneous emission in semiconductor (13) and coupled state density [26, 28].

$$R_{spon}(h\nu) = \frac{(2m_r)^{3/2}}{\pi h^2 \tau_R} \exp\left(\frac{\Delta E_F - E_g}{kT}\right)(h\nu - E_g)^{1/2}\exp\left[-\left(\frac{h\nu - E_g}{kT}\right)\right]$$

(13)

With m_r: the total effective mass and ΔE_F: the separation of Fermi levels. Considering the model of spectral emission, the smallest value of $h\nu$ energy of the spectrum corresponds to the band gap E_g of active zone material. The model is valuable only for the energy of photon upper than band gap energy.

Experimental measurements have been performed on AlGaAs/GaAs LED at 300 K for bias current equal to 100 mA. Typical optical spectrum is given on figure 8-a where experimental and analytic spectrum, using equation (13), are reported.

A specific experimental procedure has been explored to find band gap energy of the active zone material for 300 K. In our case, the spectral response observed in figure 8-a is only caused by active zone excitation of the LED device. This active zone is size limited by cladding layers (Al$_{0.3}$Ga$_{0.7}$As) as it shown in figure 3. A first experimental optical spectrum is performed at 300 K for packaging temperature. The temperature of active zone is evaluated by analysis of optical spectrum in semi logarithmic slope at high energy (>1.5eV). In this case, the slope of the curve is directly related to active zone temperature. At high level of energy, only spontaneous emission has been observed and the model presented by equation (13) traduces perfectly this physical phenomenon. So, temperature of packaging is adapted to have the best fitting at high energy level as it shown in figure 8-a. The temperature of the component packaging is adjusted by cryostat system to have the active zone temperature equal to 300 K. The beginning of analytic curve corresponds to band gap energy of active zone material and is equal to 1.424 eV assuming that material is GaAs [3, 28]. The physical parameters for GaAs material useful for spectral calculus have been reported in table 1.

An accurate analysis of the superposition of analytic and experimental optical spectrum shows deference in high energy levels in term of amplitude.

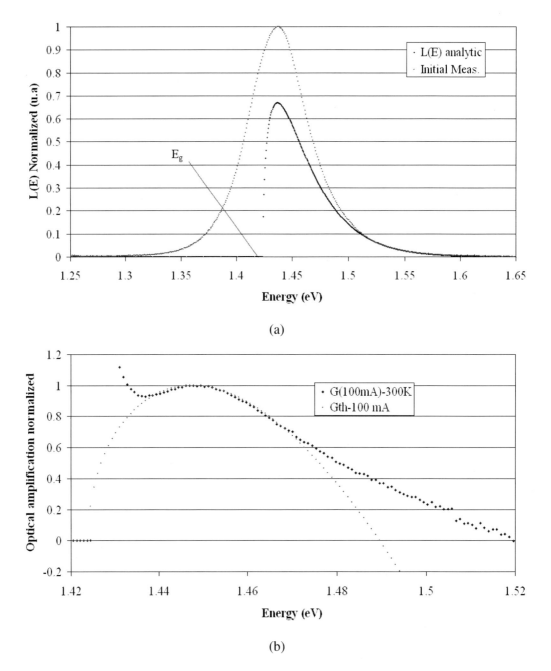

(a)

(b)

Figure 9. Superposition of experimental and modeling curves of Optical spectrum (a) and optical gain (b).

Table 1. Physical parameters of GaAs material for emission spectra calculus

Physical parameters	Numerical Value
B : Radiative recombination coefficient	7.10^{-10} cm^{3}s^{-1}
Eg : Band gap energy	1.424 eV
τ_R : Spontaneous lifetime	132 ps
N^1_c : Effective state density of conduction band	$3.95.10^{17}$ cm^{-3}
N^1_v : Effective state density of valence band	$9.11.10^{18}$ cm^{-3}
N^1_j : Relative effective state density band	$3.28.10^{17}$ cm^{-3}
m$_r$: relative masses	$\dfrac{1}{m_r} = \dfrac{1}{m_c^*} + \dfrac{1}{m_v^*}$
m_c^* : effective electron masses	0.063 m$_e$
m_v^* : effective hole masses	0.51 m$_e$

The first analysis of this difference is that only the spontaneous phenomenon is traduced by equation (13) and the DH structure of the LED induces optical gain in the active zone. The second one considers that the model is based on the hypothesis that atoms of the crystal network are fixed in energy and kinetic moment point of view. In experimental condition, at room temperature for example, the phonons in the crystal induce some perturbations of the band energy responsible to the whitening of the spectral response. The last point is highlight by Stark effect induced by potential difference located in active zone. This last physical effect induces electronic transitions for energy lower than E_g. The impact of the Stark effect is the widening of the optical spectrum emitted by active zone in the lowest energy and will be observed in the experimental measurements.

Considering the emission in the DH LED, it is well established that this kind of structure imposes the carrier inversion of population. The material has a luminescence gain related to the density of carrier located in active zone. In this case, the stimulated emission phenomenon is also observed and the amplification condition described in equation (14) can be established [1, 26, 28].

$$\gamma(h\nu) = \frac{(2m_r)^{3/2}}{\pi h^2 \tau_R} \exp\left(\frac{\Delta E_F - E_g}{kT}\right)(h\nu - E_g)^{1/2}\left[f_c(h\nu) - f_v(h\nu)\right] \qquad (14)$$

The whole functional parameters are directly related with fundamental physics and material properties have been previously defined.

Moreover, the equation (14) is used to access to the Fermi energy level difference ΔE_F. After some adjustment of optical spectrum curves, Fermi energy level difference and optical amplification can be determined and the optimized solution is shown in figure 8b.

The difference of Fermi level ΔE_F, density of carriers and maximal optical amplification of the cavity is given by table 2.

$$N_c = \frac{1}{4}\left(\frac{2m_c^* kT}{\pi h?}\right)^{3/2} \quad ; \quad N_v = \frac{1}{4}\left(\frac{2m_v^* kT}{\pi h?}\right)^{3/2} \quad ; \quad N_j = \frac{1}{4}\left(\frac{2m_r^* kT}{\pi h?}\right)^{3/2}$$

Table 2. Functional parameters evaluated for bias current about 100 mA and at 300 K

Physical parameters	Numerical Value
ΔE_F : separation of Fermi energy level	1,47095 eV
$n = p$: Carriers density	$8.05.10^{17}$ cm^{-3}
γ_{max} : Maximal optical amplification	67.4 cm^{-1}

The γ_{max} parameter is calculated using equation (14) with physical parameters values used bibliographic results of GaAs material. At this step, it is difficult to evaluate the exact amplitude of current density related to active zone thickness and non recombination radiation coefficient A_{nr}. These last parameters have been evaluated using electrical measurement.

To evaluate the density of carriers injected in the structure, the current density is used and defined by relation (15).

$$J = qdA_{nr}n_{nr} + qdBn_{ph}^2 + qdC_{aug}n_{aug}^3 \tag{15}$$

With $A_{nr:}$ the non-radiation recombination coefficient, B: radiation recombination coefficient, C_{aug} is the Auger recombination coefficient and d is the thickness of the active zone. To simplify equation (15), we supposed that Auger phenomenon is weak before the other and the population inversion imposes that density of electrons is equal to holes one. The calculus for evaluate radiation carriers density is given by relation (16).

$$n = \int_{E_c}^{E_{FC}} \rho_c(E)dE = \frac{1}{2\pi^2}\left(\frac{2m_c}{h^2}\right)^{3/2} \int_{E_c}^{E_{Fc}} (E - E_c)^{1/2} dE = \frac{1}{3\pi^2}\left(\frac{2m_c}{h^2}\right)^{3/2} (E_{Fc} - E_C)^{3/2} \tag{16}$$

The analysis of the optical spectrum gave the value of carrier density, the band gap energy and finely the bimolecular (B) coefficient. With equation (17), we have only one equation for two unknown parameters: A_{nr} and d.

$$J = qdA_{nr}n + qdBn^2 \tag{17}$$

To find these two parameters, another experimental measurement must be performed at the same temperature but for different injection of level of current. So, the electrical measurement for injected current less than 10^{-4} A is the complementary analysis to find second equations describing a second boundary condition.

3. Electrical Model

Electrical measurements are complementary measurements for the identification of the physical phenomenon, especially current path in the semiconductor heterostructure. Only low injection level of current is considered. This level of current is chosen because it is possible to express the simple analytic equation to traduce the electrical behavior of the system. Moreover it presents a complete different boundary condition comparing with optical spectrum measurement that is valuable at high injection level of current. The electro-optical characterizations are complementary to establish link between functional parameters and material physical parameters at 300 K.

The main hypothesis is centered on the low injection level of current assumes that the total current of electrons J_n and holes J_p are negligible before the conduction and diffusion current at room temperature (300 K). The weak injection level of current is correlated by the fact that carrier density injected $\Delta n = \Delta p$ is negligible before major carrier density described by the equilibrium thermodynamic condition at 300K. The electric field $\overset{u}{E}$ induced by the combination of potential barrier and external potential applied to the structure, is quasi only located on the active zone. So, external zones are considered as neutral zone with electric field equal to zero and carriers in these zones are essentially moved by diffusion phenomenon. The equation (18) introduces x dependence related to the structure of the LED considered in figure 9. The structure of LED is performed by epitaxy technique to made succession of thin parallel layers. So, it is clearly admitted that electrical phenomenon is only dependant of direction perpendicular to the layers. The reference Ox axe, perpendicular to each layer of LED, is chosen in this study to simplify the expression of current density.

$$J_n = \mu_n \left(qnE + kT \frac{dn}{dx} \right) \approx 0$$

$$J_p = \mu_n \left(qpE - kT \frac{dp}{dx} \right) \approx 0$$

(18)

Considering this hypothesis, the calculus of the carrier density n and p is possible by analytic resolution and the result is proposed in relation (19).

$$np = n_i^2 \exp\left(\frac{qV}{kT} \right)$$

(19)

The structure of LED implies that only recombination phenomenon have been observed in the active zone structure caused by high potential barrier (200 to 300 meV and 200 nm thickness) stopping all diffusion phenomenon and tunnel effects of carrier through the component (figure 9).

To resume the physical effects in the LED, the carriers have been injected in the active zone DH structure by diffusion phenomenon. But the barriers resulting to DH structure stop the carriers and only recombination phenomenon explains the flux of carrier related to the current in the LED. These considerations simplify the total density of current because recombination phenomenon impose that in active zone electrons density is equal to holes one. So, it is possible to write simplify equations of current versus external potential V in equation (20).

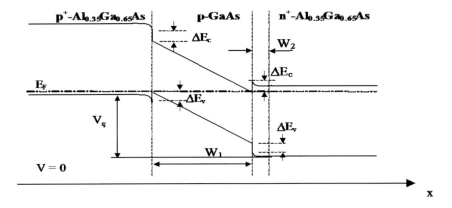

Figure 10. Band Scheme of GaAs LED.

$$n = p = n_i \exp\left(\frac{qV}{2kT}\right)$$

$$\begin{cases} J = qdA_{nr}n_i \exp\left(\dfrac{qV}{2kT}\right) + qdBn_i^2 \exp\left(\dfrac{qV}{kT}\right) \\ J = \dfrac{qdn_i}{\tau_{nr}} \exp\left(\dfrac{qV}{2kT}\right) + qdBn_i^2 \exp\left(\dfrac{qV}{kT}\right) \end{cases} \Rightarrow A_{nr} = \frac{1}{\tau_{nr}}$$

(20)

The first term of the relation (20) is related to non-radiation recombination A_{nr} and the second one traduces the photons emission by electron-hole interaction B. Using Shockley-Read Hall relation, the non irradiative coefficient is related with average carrier lifetime τ_{nr}. Typical measurement of electric characteristics is shown in figure 10.

This study is focused on the central zone using hypothesis of weak injection level of current allowing to extract an electrical model using equation (20) for build analytic I(V) curve. Figure 9 shows the perfect correspondence between experimental and analytic I(V) curves. So, using the equation (20) and the extrapolation of the I(V) curves, it is easy to define the unknown parameters with equations (21).

$$J_{nr} = qdA_{nr}n_i = 1.78.10^{-11} \, Acm^{-2}$$
$$J_{ph} = qdBn_i^2 = 7.56.10^{-25} \, Acm^{-2}$$

(21)

The first estimation of the solution of equations (21) is based on the bibliographic value of B and n_i parameters for GaAs material. Numerical value of parameter d, the thickness of the active zone, is close to 150 nm and A_{nr} value is equal to $3.77.106 \, s^{-1}$. Concerning the thickness of the active zone, lots of references confirm the typical active zone thickness is around 200 nm [1, 19, 29, 30]. The non radiative recombination coefficient A_{nr} is in relation with lifetime of carriers located in depleted zone ($\tau_{nr} = 265$ ns) and non intentional doping concentration ($N_{def} = 1.11.10^{17} \, cm^{-3}$). For this level of injection of current, the number of carrier is weak and so the depleted zone is relatively large. Considering the doping level in active zone P$^-$ ($10^{17} \, cm^{-3}$) and cladding zone N+ ($5.10^{18} \, cm^{-3}$), the depleted zone is totally effective in the active zone. The non radiation recombination rate r_{nr} can be traduced by equation (22) using Shockley-Read- Hall relation.

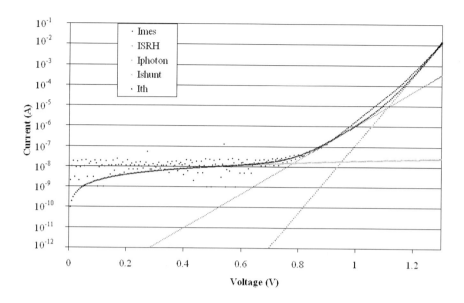

Figure 11. Comparison between experimental measurement and electrical models.

$$r_{nr} = \frac{1}{\tau_{nr}} \frac{pn - n_i^2}{2n_i + p + n}$$

(22)

IV. FAILURE ANALYSES

The conditions of ageing tests depend to the mission profile. The study considers three different applications: the spatial applications, the IR command and the public light. The most active stress is the active storage for all technologies and this one is only considered in this chapter. We consider a drop of 20 % from the initial optical power as failure criterion. The sensitivity of the chip on active storage is established and the complete study is now detailed. The failure criterions are, in one hand, the leakage current, and in other hand, the increase of SRH current. The two failure signatures are observed on GaAs technologies principally used in IR applications. The GaN technologies are actually intensively studied and will be presented on the next work. The methodology exposed in this section is the same for all technologies and present advantage to build a physical degradation laws based on failure signatures analyses.

1. Ageing Test Conditions – Active Storage

In these experiments, a whole batch composed of 26 LEDs of the same technology were used with :

1 LED for electro-optical measurement control (reproducibility and repeatability)

5 LEDs dedicated to SEM and EDX analyses,
20 LEDs submitted to ageing tests (5 LEDs for each condition of test).

A failure criterion has been set to a drop of 20% of the initial output optical power. In this context, we have only used Telcordia specifications to establish stress levels for our ageing tests and then output optical power has been monitored versus time for reliability calculations regarding previous failure criterion [16, 28].

Figure 11 shows that optical power drift, for active storage (110 mA-398 K), is close to 40% for LED A3 and 21% for LEDs A2 and A4.

From figure 11, expressing $P_{opt}(t)/P_0$ an analytical degradation law can be fitted from optical power variations versus ageing time as proposed in equation (23).

$$\frac{P_{opt}(t)}{P_0} = \exp\left(-\sqrt{t/\tau_a}\right)$$

$$\frac{\Delta P_{opt}}{P_0} = \frac{P_{opt}(t) - P_0}{P_0} = \exp\left(-\sqrt{t/\tau_a}\right) - 1$$

(23)

τ_a: characteristic time constant of degradation, calculated for ageing conditions, P_0: initial optical power.

τ_a is calculated for LED A1, A2, A3 and A4 for $I_{acc} = 110$ mA and $T_p = 398$ K as ageing conditions. Lifetime $t_{EOL}(a)$ is directly evaluated in ageing conditions, for each sample, using an extrapolation of the mathematic degradation law given by equation (23), when optical power drop is equal to 20% of reference optical power equal to 4.7 mW, representing the minimum of optical power guaranty by manufacturer. This condition represents the failure criterion for LED considered. All time to failure presented in this section are calculated for this failure criterion. A part of study demonstrates that equation (23) is not only mathematical low but a physical one. Then we could be related degradation law to physics of failure.

2. Analysis of Recombination Current Drift

a)Failure Mechanism Diagnostic Using Electrical Characterizations

The increase of recombination current is characterized by an increase of the slope $p(T,t)$ of LogI(V) curve in zone I and observed in figure 8a at 300 K.

G. Beister has already reported this kind of signature on InGaAs/AlGaAs laser diodes emitting at 980 nm [29]. The LogI(V) slope decreases and its variation Δp, detailed in equation (24), after 600 h and 1200 h of ageing time are both presented in figure 8b.

$$\Delta p = \frac{p(t) - p(t = 0)}{p(t = 0)}\bigg|_T$$

(24)

with $p(T,t)$ corresponds to the slope of LogI(V) curve measured after ageing time, $p(t=0)$ represents the slope of LogI(V) curve measured at initial condition and T is the semiconductor temperature during measurement.

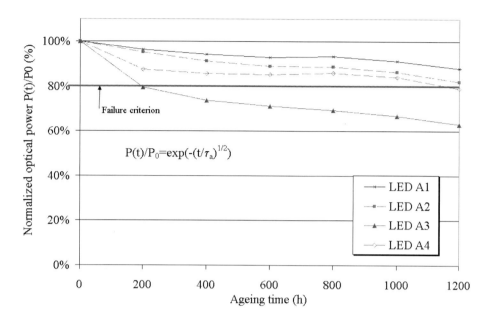

Figure 12. Variations of optical power versus ageing time for active storage (110 mA / 398 K) measured at 100 mA and 300 K.

In the first time, the fit of p(T) between mathematical model based on equation (20) and experimental results is verified by comparison between experimental p(T)-0h and p(T)-simulated curves (figure 8b). Equation (25) expresses the current versus voltage relation in zone I of the I(V) curve. The coefficient 1/2 corresponds to the ideality factor (n_{id}) value in relation with SRH recombination phenomenon ($n_{id} \approx 2$) well-established for electrical transport in DH structure as shown in equation (25).

I_{SRH}: the non radiative coefficient current. After ageing, the p(T) value has changed and the gradual change of Δp demonstrates a temperature dependence. The degradation mechanism is more important at high temperature than at low temperature. The physical phenomenon associated with the I(V) behavior curve is thermally activated with measurement conditions and could be compared to an increase of doping concentration in semiconductor [30]. However, active zone temperature during ageing test, calculated with equation (1), is close to 500 K. This temperature value cannot explained an increase of doping concentration in active zone that is basically due to migration effects starting at annealing temperatures above 1100 K but probably a diffusion inducing an increase of defects concentration [31, 32, 33, 34].

$$I = I_{SRH} \exp\big(p(T).V\big) \Rightarrow \begin{cases} p(T) = \dfrac{q}{n_{id}kT} \approx \dfrac{q}{2kT} \\[2mm] LogI \approx LogI_{hj} + \dfrac{q}{2kT} V \end{cases} \tag{25}$$

In order to understand failure mechanisms, a theoretical study deals with defects diffusion in active zone has been developed according to results of T. Lipinski [35].

Physical equations have been extracted from diffusion phenomenon using Fick's law (26) traditionally applied on semiconductor and we give, in equations (27) and (28), the current distribution versus temperature in the LED taking into account of expression (26) and radiative and non radiative recombination current contributions.

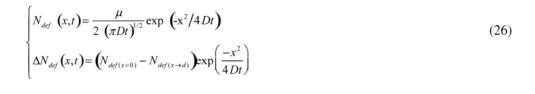

$$
\begin{cases}
N_{def}(x,t) = \dfrac{\mu}{2\,(\pi Dt)^{1/2}} \exp\left(-x^2/4Dt\right) \\[2mm]
\Delta N_{def}(x,t) = \left(N_{def(x=0)} - N_{def(x\to d)}\right)\exp\left(\dfrac{-x^2}{4Dt}\right)
\end{cases}
\tag{26}
$$

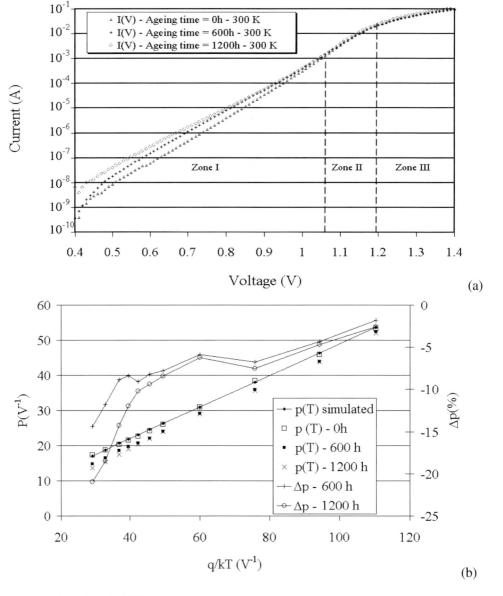

(a)

(b)

Figure 13. Typical electrical failure signature in zone I versus ageing time.

Table 3. Experimental results giving I_{hj} and b gradual change versus ageing time and measurement temperature

Measurement temperature (K)	300	300	398	398
t (time/h)	600	1200	600	1200
a (A)	$1.74 \ 10^{-11}$	$1.74 \ 10^{-11}$	$1.74 \ 10^{-8}$	$1.74 \ 10^{-8}$
b (V^{-1})	6	5.5	6	5.5

With $\Delta N_{def}(x)$: defects concentration variation in active zone: $N_{def(x=0)}$ for P$^-$-InGaAs/N$^+$-GaAs interface and $N_{def(x \to d)}$ for P$^+$-GaAs/P$^-$-InGaAs interface, μ : the surface mass of defects, D : diffusion coefficient and x : direction of diffusion. In InGaAs/GaAs heterostructure, the diffusion of defects is essentially activated along the normal of the epitaxial plane [2, 6].

$$\begin{cases} I_{hj} = qn_i d \dfrac{1}{\tau_r + \tau_{nr}} \\ \dfrac{1}{\tau_r + \tau_{nr}} = k(N_{def(x \to d)} + \Delta N_{def}(x)) \end{cases} \Rightarrow \begin{cases} I_{hj} = a \exp\left(-\dfrac{d^2}{4Dt}\right) + I_{hj0} \\ a(T) = \dfrac{qn_i kd \left(N_{def(x=0)} - N_{def(x \to d)}\right)}{2} \end{cases} \tag{27}$$

With n_i : intrinsic doping level, τ_r : radiative lifetime, τ_{nr} : non-radiative lifetime, d : active layer thickness, a(T) : the temperature dependant saturation current, I_{hj0} : the sum of radiative recombination current I_{rr} and non-radiative recombination current I_{nr} evaluated at the initial conditions.

$$\begin{cases} I = \left(I_{hjcr} + I_{hj0}\right) \exp\left(\dfrac{qV}{2kT}\right) \\ I_{hjcr} = a(T)\exp(-bV) \end{cases} \Rightarrow \begin{cases} p(t) \approx -b(t) + \dfrac{q}{kT} \\ I_{hj} = I_{hj0} + I_{hjcr} \end{cases} \tag{28}$$

With b(t) corresponds to the drift of p(t) parameter that is ageing time dependent.

Experimental gradual changes of I(V) curves before and after ageing test, respectively 600 h and 1200 h, have been fitted using equations (26), (27) and (28) ; and the values of parameter "a" and "b" have been calculated and given in table 3.

Different values of parameters "a" and "b" is calculated for two different temperatures (300 K and 398 K) and ageing times (600 h and 1200 h). The monitoring of these two parameters allows the validation of the temperature dependence for "a" and the ageing time dependence for "b".

Defects diffusion in active layer is generally effective because molecular beam epitaxy can produce a lot of defects located at the interface GaAs/InGaAs [7, 6, 34, 35]. Figure 13 illustrates the gradual change of defects concentration versus the x-direction, corresponding to the normal of epitaxial planes (P$^+$-GaAs/P$^-$-InGaAs/N$^+$GaAs).

Our assumption is based on the fact that defects are located in P$^-$-InGaAs/N$^+$-GaAs interface considering an initial concentration of defects called $N_{def \ x \to 0}$. After 600 h and 1200 h, defects concentration has increased in the active zone as shown in figure 14 according equation (26). With equation With equation (28), it is possible to estimate the increase of defects density in the middle of active zone. Considering a theoretical study developed by H. Mathieu, the most active defects in the junction are located in the middle of depleted zone for recombination current injection level [36]. In our case, the LED voltage corresponding to recombination current injection level and the length of depleted zone is equal to active zone thickness because of the low doping level (10^{15} cm^{-3}) [26, 27]. Finally, the average increase of defects density N_{def} localized in the middle of active zone is close to 30 % using equations (26), (27) and (28).

b. Failure Mechanism Diagnostic Using Optical Characterizations

In order to confirm previous electrical results, optical power P(I) curves gradual changes have been observed after 600 and 1200 h (figure 15).

Experimental value of parameters γ and α have been extracted after 600 and 1200 hours of ageing time showing that:

γ decreases of 40 % after 1200 hours,
α is constant.

Considering equations (5) and (6), it is possible to conclude that only the term (B/A^2) decreases while γ and η_{tot} has not evolved insuring that only active zone of LED device is degraded. B/A^2 is expressed by equation (29).

$$\frac{B}{A^2} = \frac{4N_j}{\tau_R N_c N_v N_{def}^2 v_{th}^2 \sigma^2} \tag{29}$$

All parameters have previously defined in equations (4) and (12). N_j, N_c and N_v parameters are assumed to be constant as proposed by S.M. Sze [27]. Radiative lifetime τ_R is detailed in equation (30).

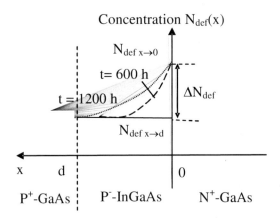

Figure 14. Representation of defects concentration profile gradual change before and after ageing time.

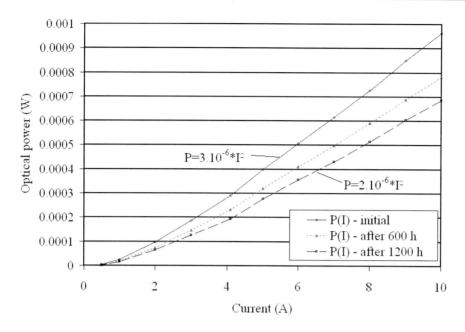

Figure 15. Optical power P(I) before and after ageing time measured at 100 mA and 300 K.

$$\tau_R = \frac{2\pi c^3 h^3 \varepsilon_0 m_e}{q^3 n_{op} E_g E_p} \tag{30}$$

E_g is energy gap level, E_p is Kane Energy and n_{op} is optical index. E_g and n_{op} are only temperature dependent as explained S.M. Sze and their value is related to InGaAs material properties [27]. It is well-known that emission spectra characteristics are strongly dependent on band gap energy and it has been demonstrated by V. Rakovics that centre wavelength or emission spectra broadening can be associated to dopants diffusion or non-uniform active layer composition observed on InGaAsP/InP infrared LED grown by Liquid phase epitaxy (LPE) [36]. In our case, we consider that E_g is constant. Experimental optical spectrums have been measured versus ageing time on our LEDs. It is clearly shows that only intensity has changed but spectral characteristics, in particular central wavelength E_c shift and emission spectra broadening ΔE, have not changed. Considering equations (12), (13) and (14), and optical spectrum results, the thermal resistance R_{th} and temperature of active zone is constant concluding that E_g and n_{op} are constant before and after ageing tests.

For a bias fixed condition, the variation of the Kane Energy (E_p) can only be explained by the energy bands variation. This variation is directly associated to a doping change in the active layer. But internal active zone temperature during ageing condition is close to 500 K according to equation (12) and cannot stimulate dopants diffusion. Indeed the necessary temperature to activate dopants diffusion must be upper than 1000 K [27, 31, 33]. Finally the last parameter allows explaining the drop of γ parameter that is due to an increase of defects density N_{def} located in the active layer. After 1200 hours of ageing time, the decrease of γ parameter is close to 33% and the calculated increase of average defects density is close to

30%. This result shows the good accordance with electrical results previously exposed and highlighted the complementarities of electrical and optical measurements for:

location of degradations,
failure mechanisms identification.

In conclusion, we have demonstrated that the main failure mechanism of our LEDs is driven by defects diffusion in the active zone. The presence of such defects at initial condition and after ageing time has been already reported by O. Pursiainen, N Gulluoglu and B. Darek and experimentally highlights using I(V) measurements and Transmission Electron Microscopy (TEM) by means of complex selective chemical preparations around active zone [35, 65]. The native defects located in P⁻-InGaAs/N⁺GaAs interface have been induced by a large difference of lattice parameters between the two layers (0.21Å ≈ 4%) [6, 38]. The tensile strength located at the P⁻-InGaAs/N⁺GaAs interface is commonly at the origin of defects propagation in the active zone.

3. Analyses of Leakage Current

Other failure signatures observed on AlGaAs/GaAs LED are the leakage current (increasing about 3 decades at 300 K), appears at low bias in the current-voltage curves. The electrical failure signature obtained versus temperature is shown in figure 16. The temperature dependence of the I(V) curve is well adapted to extract physical phenomenon explaining leakage current drift: the Pool-Frenkel phenomenon. This kind of conduction appears between defect and conduction band of the semiconductor as it shown in the figure 17.

The conduction theory based on the diffusion of carrier on the defect drive us to consider equation (31) to simulate the I(V) behavior.

$$J_{PF} = q\mu_p \left(\frac{N_c N_d}{2} \right)^{1/2} F \exp\left(-\frac{\phi_{PF}}{2kT} \right) \exp\left(\frac{\beta_{PF} F^{1/2}}{2kT} \right) \tag{31}$$

μ_p: the carriers mobility in active zone, N_c: the effective density of state in conduction band, N_d: the defect density, Φ_{PF}: the Pool-Frenkel barrier and β_{PF}: the Pool-Frenkel coefficient. This model is based on the quantum theory explain the local disturbing in the structure [39].

In this model, the density of defects is weak (less than 10^{10}cm^{-3}) with no recovering wave function between two defects [40]. The conduction mechanism is the thermoelectronic emission of carrier between defect centre and conduction band. The potential barrier is defined by $\Phi_{PF} = E_c - E_T = 130$ meV in our case.

The first parameter estimated is the Pool-Frenkel coefficient using $\frac{I}{V} = f\left(\sqrt{V} \right)$ curves

versus temperature presented in figure 18. β_{PF} is extracted from the slope of the curves and defect densities is extrapolated to the amplitude of I/V for V=0V. The number of defect is very weak and close to 10^8 cm^{-3}. The hypothesis concerning the density of defect is clearly respected and the model of current transport is validated.

The second parameter Φ_{PF} is extract using the slope of the curves exposed in figure 19. The slope β is linear dependent with $\sqrt{F}$ and we determine the value of Φ_{PF} = 130 meV.

The spectral failure signature is plotted in figure 20 and shows the decrease of the optical spectrum intensity about 20%. The extraction of physical parameters gives us the same value excepted for ΔE_F and F.

After neutrons irradiation ΔE_F is close to 1.4835 eV for I_{bias} = 100 mA. With equation (17), we find the equivalent current in the structure is close to 75 mA. This information conducts us to consider that 25 mA of current is loss outside the active zone. The defect responsible to degradation of DH-LED is located in the side of the chip outside the active zone.

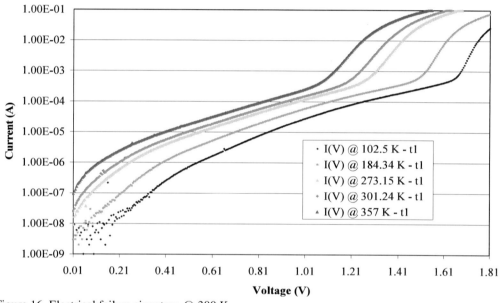

Figure 16. Electrical failure signature @ 300 K.

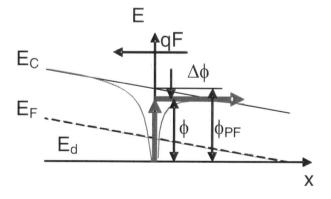

Figure 17. Schematic design of the electron transport with trap centre.

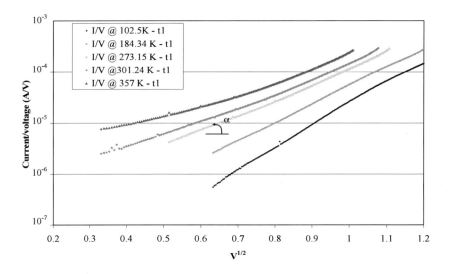

Figure 18. I/V = f(√V) curves for temperature ranging from 102 K to 337 K.

The first analysis of electrical and optical signatures indicates that active zone is not affected because principally A_{nr} and B coefficient not changed before and after neutron irradiation. The semiconductor/glue interface might be affected. The transport of carriers is driven by the diffusion phenomenon related to defects located in this interface semiconductor/package [6]. A Pool-Frenkel model has been used to explain the leakage current induced by electron trapping by an energy level of 130 meV. The strong interest of temperature measurements ranging from 79 to 360K is used to fit a Pool-Frenkel effect.

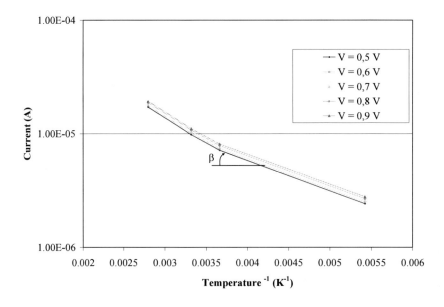

Figure 19. I = f(1/T) curves for voltage ranging from 0.4V to 1V.

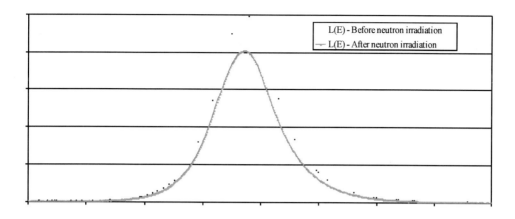

Figure 20. Spectral failure signature @ 300 K.

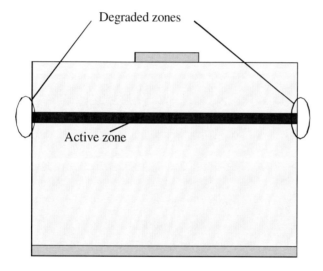

Figure 21. Localization of degraded zones.

This last is well justified by the increase of defects density in the side of the chip caused by chemical link transformation between glue and semiconductor (figure 21).

The next experience, in progress, consists to generate defects in the semiconductor active zone only. In this case, first simulation of neutron irradiation gives the energy is close to 10 MeV and the direction of neutrons is perpendicular to epitaxy planes.

V. RELIABILITY ESTIMATION

In this paper, different conditions of ageing tests (storage and thermal cycles) have been considered and only active storage using various values of bias current are the most predominant ageing test for the studied technology considering a drop of 20 % from the initial optical power as failure criterion (figure 22). This result shows the robustness of the package related to thermal cycle tests showing a decrease of optical power lower than 5% after 700 cycles. The sensitivity of the chip on active storage is established and the complete study is now detailed.

Our previous results have shown that ageing tests have triggered more degradation at chip level than at package level. The main objective of this last section is to establish the fundamental relation between the degradation law extrapolated from optical power drift versus ageing time, assuming defects propagation in the active zone, and lifetime distribution for this technology working in operating conditions.

According to manufacturer specifications, the failure criterion used for lifetime estimation is a decrease of 20% of 4.7 mW (see table 4) [41, 42, 43].

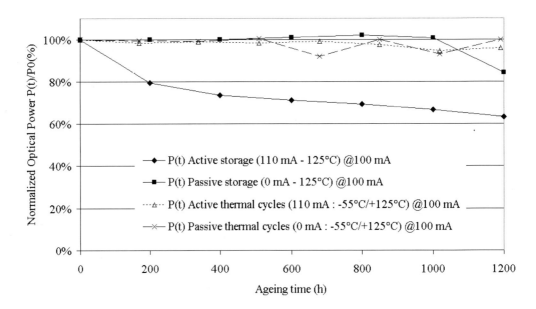

Figure 22. Variations of optical power versus ageing time measured at 100 mA and 300 K for different kinds of ageing test (5 LEDs used per test).

Table 4. Operating conditions

Manufacturer parameters	Typical value @ 300 K
P_0 : initial optical power	P_{min}=4,7 mW, V = 1,47 V (I = 100 mA)
T_{wk} : operating temperature range	218 K to 398 K
I_f : nominal forward current	100 mA
E_c : central energy	1.32 eV
ΔE : spectral energy bandwidth	62 meV
θ_e : beam angle	20 °
τ : rising or fall time	0,7 μs

This last data corresponds to the lower value of optical power specified by the manufacturer: bias current I = 100 mA, voltage V = 1.47 V (at 300 K).

For ageing tests, a higher junction temperature or higher bias current is generally applied on these LEDs. Two different acceleration factors related to failure rates are essentially considered: a thermal one (AF_T) ruled by an Arrhenius behaviour, and a current one (AF_I). In cases of multi-stress environments, these factors are multiplicatively applied (equation 32).

$$AF_T = \exp\left[\frac{E_a}{k}\left(\frac{1}{T_{op}} - \frac{1}{T_{acc}}\right)\right]$$

$$AF_I = \left(\frac{I_{acc}}{I_{op}}\right)^{\beta} \tag{32}$$

with E_a: thermal activation energy, T_{op}: operating junction temperature, T_{acc}: junction temperature for accelerated test, I_{acc}: current for accelerated test, I_{op} : operating current and β: exponent of current dependency [1].

In literature, experimental lifetesting already conducted on LED chips during more than 10000 hours are usually defined using a bias current (I_{bias}) close to 100 mA and a junction temperature of InGaAs/GaAs LED chip (T_j) close to 300 K; in these conditions, the extrapolated mean time to failure (MTTF) is estimated to 2.10^5 hours (23 years) [1, 7]. In order to reduce the duration of tests, an increase of the acceleration factor AF_T is basically proposed by an increase of the active layer temperature. For instance, considering $I_{bias} = 100$ mA and $T_j = 500$ K conditions, the calculated MTTF is consequently reduced to 8.10^4 hours.

For GaAs compounds used for light emission, activation energy (E_a) of gradual degradation is close to 0.5 eV, not depending on the operating current density. In this case, the corresponding acceleration factor AF_T is about 25 [7].

Present levels of extremely low failure rates cannot be demonstrated by usual techniques of selection of finished products, nor shown by the classical way of statistical exploitation using life tests aiming to evaluate the mean life span to 50 % of failures. Indeed, even if two failures on 1000 hours are only accepted, a standard test with an acceleration factor largely over 400 and several hundreds or thousands of components are necessary to the composition of the test sample. But now, industrial qualification processes are more and more mandated and require shorter tests (up to 1000 hours). So the active layer temperature of InGaAs/GaAs LED chip must increase up to 620 K leading to a drastic increase of the acceleration factor AFT until 1050 always considering the same origin of failure mechanism. Indeed, during ageing test at elevated temperature (T_j upper than 450 K), it was sometimes observed that Dark Line Defects and/or Dark Spot Defects were introduced in the light-emitting region generally conducting to a homogeneous decrease in light output with time and is usually relative with defects diffusion (dislocation loops of irregular sizes) uniformly distributed in the active zone accelerated by an increase of ageing current density. In our study, LEDs are packaged and it has been previously demonstrated that only an increase of bias current is preferable to diminish the acceleration factor value, signifying that a current acceleration factor (AF_I) will be also considered in the next part to respect the main objective of this paper. So, AF factor considered in the next part is given by equation (33):

$$AF = AF_I \times AF_T = \left(\frac{I_{acc}}{I_{op}}\right)^{\beta} \exp\left[\frac{E_a}{k}\left(\frac{1}{T_{op}} - \frac{1}{T_{acc}}\right)\right] \tag{33}$$

1. Relation between Degradation Law and Lifetime Estimation

From figure 23, expressing $P_{opt}(t)/P_0$ an analytical degradation law can be fitted from optical power variations versus ageing time as proposed in equation (34) :

$$\frac{P_{opt}(t)}{P_0} = \exp\left(-\sqrt{t/\tau_a}\right) \tag{34}$$

$$\frac{\Delta P_{opt}}{P_0} = \frac{P_{opt}(t) - P_0}{P_0} = \exp\left(-\sqrt{t/\tau_a}\right) - 1$$

with τa: characteristic time constant of degradation, calculated for ageing conditions, P_0: initial optical power.

τa is calculated for LED A1, A2, A3 and A4 for $I_{acc} = 110$ mA and $T_p = 398$ K as ageing conditions. Lifetime t_{EOL} (a) is directly evaluated in ageing conditions, for each sample, using an extrapolation of the degradation law given by equation (34), when optical power drop is equal to 20% of reference optical power equal to 4.7 mW, representing the minimum of optical power guaranty by manufacturer. This condition represents the failure criterion for LED considered. All time to failure presented in this section are calculated for this failure criterion.

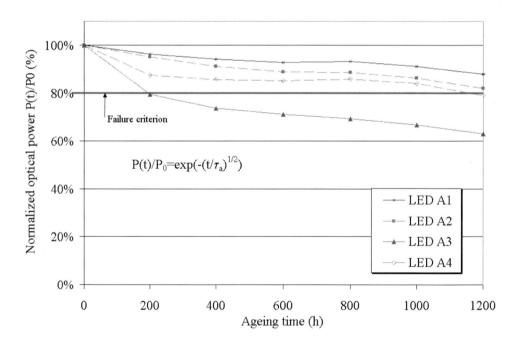

Figure 23. Variations of optical power versus ageing time for active storage (110 mA / 398 K) measured at 100 mA and 300 K.

A value of mean lifetime (MTTF), representing the mean lifetimes of the LED A1, A2, A3 and A4, is calculated and equal to 4.10^4 hours. The goal of this part is to evaluate lifetime distribution (t_{EOL}) and MTTF of the LEDs in operating conditions (100 mA – 398 K). This extrapolation requires now the evaluation of the relationship between t_{EOL} and the degradation factor, in our case, bias current (I_{acc}). The relation between $t_{EOL}(a)$ and bias current I_{acc}, applied during ageing tests, has been determined performing ageing tests with bias current respectively fixed at 125 mA and 150 mA. This complementary study uses 10 new LEDs of the same technology and same batch with 5 LEDs per bias current. The $t_{EOL}(a)$ has been assessed with the same methodology than previously exposed.

Electrical and optical failure analyses performed for I_{acc} at 125 mA and 150 mA demonstrate that the failure mechanism is the same for the different bias currents with a constant β value. Thus, it is possible to extract the relation between t_{EOL} and the bias current I_{acc}, expressed in mA, represented in figure 24.

An extrapolation of the law expressed in equations (35) leads to deduce a MTTF value of $1.8.10^5$ hours in operating conditions (100 mA-398 K).

$$t_{EOL\,min} = 2.155.10^9 \exp(-0.1.I_{acc})$$
$$MTTF = 4.790.10^9 \exp(-0.1.I_{acc}) \tag{35}$$
$$t_{EOL\,max} = 9.975.10^9 \exp(-0.1.I_{acc})$$

From these results, it is possible to calculate acceleration factors (AF), defined in equations (32), for MTTF extrapolation:

AF = 2 for I_{acc} = 110 mA
AF = 15 for I_{acc} = 125 mA
AF = 240 for I_{acc} = 150 mA.

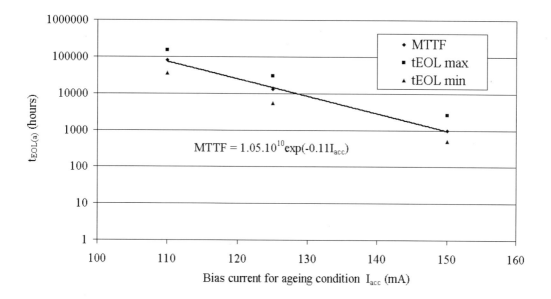

Figure 24. Variations of time to failure versus bias current.

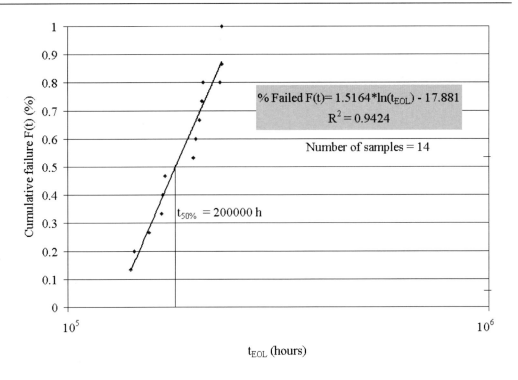

Figure 25. Log-normal distribution of cumulative failures F(t) *[X-axis in log. scale]*.

Figure 4 shows bias current dependence for minimal lifetime t_{EOLmin}, mean lifetime (MTTF) and maximal lifetime t_{EOLmax} according to equations (4). This extrapolation methodology shows that factor β, defined in equation (32), can be considered as constant versus current using activation energy E_a close to 0.5 eV [2, 7]. The calculation of β for maximal (t_{EOLmax}) and minimal (t_{EOLmin}) lifetimes leads to a value of 7.3-7.4 in good agreement with values, generally ranging from 7 to 8 as reported in literature [2, 44]. This result confirms that the assumed failure mechanism is the same for bias current ranging from 100 mA to 150 mA. For operating conditions, extrapolations of maximal and minimal lifetimes are respectively equal to $2.3.10^5$ h (26 years) and $1.4.10^5$ h (15 years).

2. Reliability Investigations Using Technological Dispersion

Cumulative failure distribution F(t) has been calculated and plotted versus time in figure 25 showing a logarithmic dependence (R^2 # 1)[45, 46, 47]. The $t_{50\%}$ value, calculated for 50% of failures from F(t), is equal to 2.10^5 hours and the failure rate corresponds to 550 FITs for 15 years in accordance with results reported by numerous papers on GaAs unpackaged LED technology [1, 2, 7].

Classical MTTF demonstration for InGaAs/GaAs LEDs have granted a value of 2.10^5 hours for 100 mA bias current and a junction temperature of about 500 K. In our case, MTTF is calculated for 100 mA and 400 K but 400 K corresponds to the body package temperature. Equation (36) gives the thermal resistance used to calculate the junction temperature, close to 500 K, when package temperature is close to 400 K.

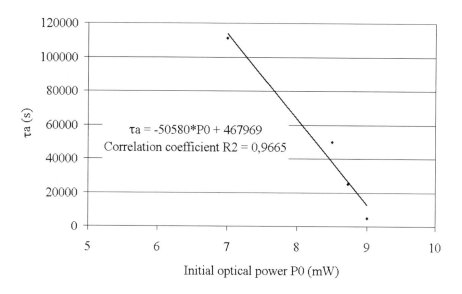

Figure 26. Characteristic time constant of degradation.

$$\Delta E = 1.8 \times kT_j \tag{36}$$

In order to satisfy reliability needs, there is a strong motivation to know more about the whole lifetimes distribution of products rather than classical prediction of MTTF or median life. Demonstration of this exigency through qualification procedures requires many inputs among which main conditions must be considered [48].

An original approach focuses on the introduction of material and/or process dispersions in order to evaluate their impact on product lifetime distribution. This last point has to be combined with mission profile distribution.

In this last section, our goal is now to link the characteristic time constant of degradation (τ_a) with optical power distribution of a LED (P_0 : the main parameter for a LED) represented in tables 1 and determined at initial time on LEDs of the same batch. A relation between τ_a (in seconds) and P_0 (in mW) has been found using τ_a parameter for all LEDs aged in the same conditions than LED A3 (110 mA-398 K). This parameter is calculated from the fitting of curves plotted in Fig. 3 and taking into account of equation (4). The linear behaviour of τ_a versus P_0 is shown in figure. 26 and equation (37) traduces the relation between P_0 and τ_a:

$$\tau_a = -50580.P_0 + 467969 \tag{37}$$

The two last equations (4) and (5) allow to determine lifetime distribution ($t_{EOL}(a)$) of LEDs for ageing conditions (110 mA-398 K) versus initial optical power P_0. Experimental $t_{EOL}(a) = f(P_0)$ curve (P_0 in mW) has been Gaussian-like fitted by equation (38) for ageing conditions (110 mA-398 K):

$$f(P_0) = -t_0 + 2t_{max} \exp\left(-\frac{P_0 - 7.10^{-3}}{P_{ref}}\right)^2 \tag{38}$$

t_0: $t_{EOL}(a)$ at σ of maximal $t_{EOL}(a)$ (26280 hours), t_{1max}: maximal $t_{EOL}(a)$ (42486 hours) and P_{ref}: reference optical power (mW) for 935 nm InGaAs/GaAs LED (4.7 mW) given by manufacturer as the minimum of optical power @ I = 100 mA and Tcase = 298 K.

The same methodology has been also applied for the other ageing tests considered: 125 mA/398 K and 150 mA/398 K demonstrating the same Gaussian-like trend that of (110 mA-398 K). Using an extrapolation from equations (38), the relationship between lifetime of LEDs is provided for operating conditions (100 mA/398 K) and initial optical power P_0 allowing to select LEDs considering initial optical power P_0 distribution called $f(P_0)$. This method could lead to implement a go-no go test to qualify production lots.

Figure 27 plots lifetime distribution (t_{EOL}) for operating conditions (100 mA-398 K) versus initial optical power P0 which has been determined for all the samples (14 LEDs) taking into account of equation (39).

$$f(P_0) = -1,8.10^5 + 2 \times 2,3.10^5 \exp\left(-\frac{P_0 - 7.10^{-3}}{P_{ref}}\right)^2 \qquad (39)$$

with constant t_0: t_{EOL} at σ of maximal t_{EOL} (180000 hours) and t_{1max}= maximal t_{EOL} (230000 hours).

The relation between t_{EOL} and P_0 can thus be evaluated and it is possible to determine the lifetime distribution versus P_0.

Finally, table 5 reports all reliability results using statistic computations on 14 LEDs aged using active storage.

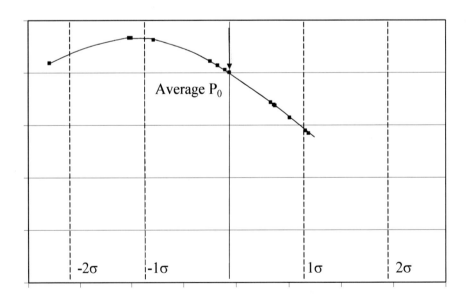

Figure 27. Lifetime distribution versus initial optical power calculated for operating conditions (100 mA-398 K).

Table 5. Reliability results coupled to statistic analyses of 14 LEDs samples

Optical power distribution	F(t)	Mean P_0	Maximal P_0	P_0 value at σ	P_0 value at 2σ
t_{EOLmax} (h)	2300000	200000	-	230000	230000
MTTF (h)	200000	-	-	186000	180000
t_{EOLmin} (h)	140000	-	140000	145000	140000
FITs at 15 Years	550	-	-	550	550

CONCLUSION

In this study, electro-optical analyses are made using current-voltage LogI(V) and optical spectrum L(E). The strong interest of L(E) measurements, complementary to I(V) characteristics, was demonstrated in this study to improve material and structure characterization of component in final environment. In particular, electrical and optical parameters, relative to the active zone of the LED, can be determined from comparisons between I(V) and L(E) experimental curves. After this step, physical models were build based on methodologies reported in few papers [25, 31, 32, 33]. The main results about the material properties is the thickness of the active zone close to 150 nm and the non radiation coefficient $A_{nr} = 3.77.10^6$ s^{-1} for GaAs material DH structure.

The study is one of the first one giving the optical gain of the commercial LED close to 150 cm^{-1} with $5.46.10^{17}$ cm^{-3} of carrier density. The key result shows the interest of DH structure for inversion populations of carriers before homojunction structure using 2.10^{18} cm^{-3} density of carriers for same optical gain. All analytic models have been validated using both experimental measurements and analytic calculus checking. The main interest of this model is the simplification of the different approaches to access directly to the physical parameters of active zone using quantum theory. One of the parallel works is to develop finite element model that is driven by both by analytic model and experimental results. This last methodology will be developed for comparison with analytic method and ensure the close working of optoelectronic source. Basically, electrical and optical measurements have been successfully used to characterize optoelectronic source and can be now used to highlight with a strong accuracy degradation mechanisms located in the chip.

After modeling process, a complete methodology was described to propose accurate reliability models based on experimental determination of failure mechanisms on GaAs material.

The first step considers the scheme of a simplify model in relation with the structure of $Al_{0.3}Ga_{0.7}As$ and $In_{0.53}Ga_{0.47}As$ on GaAs substrate for packaged LED of 872 and 935 nm centre wavelength. Intrinsic parameters of the model have contributed to improve the determination and the location of degradations into the structure after short-time ageing tests using active storage (current activation). It has been demonstrated that optical power losses are directly caused by the diffusion of defects in the active layer and leakage current. In particular, native defects located in P$^-$-InGaAs/N$^+$GaAs interface may be induced by a large difference of lattice parameters between the two layers (0.21Å $\approx$ 4%) as already reported in previous papers. The tensile strength located at the P$^-$-InGaAs/N$^+$GaAs interface may explain defects propagation in the active zone. It has been clearly shown that ageing tests have activated more degradation at the chip level than at the package level for this technology.

A relation between the degradation law extrapolated from optical power drift versus ageing time and the impact on reliability predictions is proposed. In this configuration, lifetime has been estimated considering degradation law $\Delta P_{opt}/P_0$ extrapolation at different bias current values ranging from 100 mA to 150 mA. Cumulative failure distribution has been calculated and plotted versus time presenting a logarithmic lifetime dependence. MTTF and failure rates of these LEDs are respectively estimated to $1.8.10^5$ hours and near 550 FITs for 15 years in operating conditions (100 mA – 398 K).

Reliability aspects are also investigated considering the relation between initial optical power distribution and lifetime. The both methodologies used to extrapolate lifetime are in accordance considering MTTF, minimal lifetime, maximal lifetime and failure rates for GaAs LED technology. The strong interest of this study is that MTTF and failure rates can be evaluated with a small number of components from the measurement of optical power at initial condition. So, for end-users it is possible to select GaAs LED from initial optical power distribution.

Outgoing activities are now focused on the use of such methodology for reliability estimation of another technologies in particular GaN compounds for different applications.

REFERENCES

[1]   M.R. Matthews, B.M. MacDonald and K.R. Preston, Optical components-the new challenge in packaging, *IEEE Transactions on Components, Hybrids, and Manufacturing Technology* 1990 13 (4); pp. 798-806.

[2]   O. Wada, *Optoelectronic integration : physics, technology and application,* Kluwer Academic Publisher, Boston 1994.

[3]   W.M. Sherry, C. Gaebe, T.J. *Miller and R.C. Schweizer, High performance optoelectronic packaging for 2.5 and 10 Gb/s Laser modules,* ECTC 1996; pp. 620-627.

[4]   D. Sauvage, D. Laffitte, J. Perinet, Ph. Bertier and JL Goudard, *Reliability of optoelectronic components for telecommunications, Microelectronics Reliability* 2000 40; pp. 1701-1708.

[5]   Honeywell technical datasheet, *Summary of SE450 GaAs:Si IRED chip long-term life study,* 1986; pp. 368-376.

[6]   O. Ueda, *Reliability and degradation of III-V optical devices,* Artech house, Boston London 1996.

[7]   M. Fukuda, *Reliability and degradation of semiconductor Lasers and LEDs,* Norwood MA Artech House, London 1991.

[8]   M. Fukuda, Optical semiconductor device reliability, *Microelectronics Reliability 2002* 42 (4-5); pp. 679-683.

[9]   G. Sarkar, S. Mridha, T.T. Chong, W.Y. Tuck and S.C. Kwan, Flip chip interconnect using anisotropic conductive adhesive, *Journal of Materials Processing Technology* 1999 89-90 (19); pp. 484-490.

[10] M. P. Rodriguez and N. Y. A. Shammas, Finite element simulation of thermal fatigue in multilayer structures: thermal and mechanical approach, *Microelectronics Reliability* 2001 41 (4); pp. 517-523.

[11] Q. Tan and Y.C. Lee, Soldering technology for optoelectronic packaging, *46th Electronic Components and Technology Conference* 1996; pp. 26-36.

[12] M. Itoh, J. Sasaki, A. Uda, I. Yoneda, H. Honmou, K. Fukushima, Use of AuSn solder bumps in three-dimensional passive aligned packaging of LD/PD arrays on Si optical benches, *46th Electronic Components and Technology Conference* 1996 pp. 1-7.

[13] ASM international, *ASM engineered materials reference book,* 2ème Edition - ASM international 1994.

[14] Osamu Ueda, Reliability issues in III–V compound semiconductor devices: optical devices and GaAs-based HBTs, *Microelectronics Reliability* 1999 39 (12); pp. 1839-1855.

[15] *Generic Reliability Assurance Requirements for Optoelectronic Devices Used in Telecommunications Equipments.* Telcordia GR-XXX-CORE 1998.

[16] J. -L. Goudard, P. Berthier, X. Boddaert, D. Laffitte and J. Périnet, New qualification approach for optoelectronic components, *Microelectronics Reliability* 2002 42(9); pp. 1307-1310.

[17] F. Magistrali, D. Sala, G. Salmini, F. Fantini, M. Giansante and M. Vanzi, *ESD induced degradation mechanisms of InGaAsP/InP Lasers,* ESREF, Bordeaux 1991; pp. 261-268.

[18] T.Takeshita, M.Sugo, T.Nishiya, R.Iga, M.Fukuda and Y.Itaya; *Microelectronics Reliability 1998* 38 (6-8); pp. 1211-1214.

[19] S. Konishi and G. Kitagawa, Asymptotic theory for information criteria in model selection—functional approach, *Journal of Statistical Planning and Inference* 2003 114(1-2); pp. 45-61.

[20] Jun Liu, W.S. Tam, H. Wong, V. Filip, Temperature-dependent light-emitting characteristics of InGaN/GaN diodes, *Microelectronics Reliability* 49, (1), 2009, pp. 38-41.

[21] S.J. Chang, Y.C. Lin, Y.K. Su, C.S. Chang, T.C. Wen, S.C. Shei, J.C. Ke, C.W. Kuo, S.C. Chen, C.H. Liu, Nitride-based LEDs fabricated on patterned sapphire substrates, *Solid-State Electronics* 47 (2003) 1539–1542.

[22] L. Dobos,, B. Pecz, L. Toth, Zs.J. Horvath, Z.E. Horvath, B. Beaumont, Z. Bougrioua, *Structural and electrical properties of Au and Ti/Au contacts to, Vacuum* 82 (2008) 794–798. n-type GaN

[23] J-M. Hwang, K.-F. Lee, H-L. Hwang, Optical and electrical properties of GaN micron-scale light-emitting diode, *Journal of Physics and Chemistry of Solids* 69 (2008) 752–758.

[24] J. Lee, D-H. Kim, J. Kim, H. Jeon, GaN-based light-emitting diodes directly grown on sapphire substrate with holographically generated two-dimensional photonic crystal patterns, *Current Applied Physics* 9 (2009) 633–635.

[25] J. H. Han and S.W.Park, "*Theorical and experimental Study on Junction Temperature of Packaged Fabry-Perot Laser Diode*", IEEE-TDMR (2004), Vol. 4 N°2, pp. 292-294.

[26] Y. Deshayes, I. Bord, G. Barreau, M. Aiche, P.H. Moretto, L. Béchou, A.C. Roehrig, Y. Ousten, Selective activation of failure mechanisms in packaged double-heterostructure light emitting diodes using controlled neutron energy irradiation, *Microelectronics Reliability* 48 (2008) pp. 1354–1360.

[27] S.M. Sze, *Physics of semiconductor Devices,* John Wiley & sons, New York 1981.

[28] L. Mendizabal, J. L. Verneuil, L. Bechou, C. Aupetit-Berthelemot, Y. Deshayes, F. Verdier, J. M. Dumas, Y. Danto, D. Laffitte, J. L. Goudard and Y. Hernandez, Impact of 1.55 μm laser diode degradation laws on fibre optic system performances using a system simulator, *Microelectronics Reliability* 2003 43(9) .pp 1743-1749.

[29] G. Beister, J. Maege, G. Erbert and G. Tränkle, Non-radiative current in InGaAs/AlGaAs laser diodes as a measure of facet stability, *Solid-State Electronics* 1998 42 (11); pp. 1939-1945.

[30] Dieter Bimberg, Richard Bauer, Detlef Oertel, Jerzy Mycielski, Karl-Heinz Goetz and Manijeh Razeghi, Recombination lifetime of In0.53Ga0.47As as a function of doping density, *Physica B+C,* 1998 134 (163); pp. *399-402*

[31] T. Mozume, N. Georgiev and H. Yoshida, Dopant-induced interface disorder in InGaAs/AlAsSb heterostructures lattice matched to InP grown by molecular beam epitaxy, *Journal of Crystal Growth* 2001 227-228; pp. 577-581.

[32] A. Stadler, T. Sulima, J. Schulze, C. Fink, A. Kottantharayil, W. Hansch, H. Baumgärtner, I. Eisele and W. Lerch, Dopant diffusion during rapid thermal oxidation, *Solid-State Electronics* 2000 44 (5); pp. 831-835.

[33] M. Ihaddadene, S. Koumetz, O. Latry, K. Ketata, M. Ketata and C. Dubois, A model for diffusion of beryllium in InGaAs/InP heterostructures, *Materials Science and Engineering B.* 2001 80 (1-3); pp. 73-76.

[34] A. N. Gulluoglu and C. T. Tsai, Dislocation generation in GaAs crystals grown by the vertical gradient freeze method, *Journal of Materials Processing Technology* 2000 102 (1-3); pp. 179-187.

[35] B. Darek and T. Lipinski, Effect of structural defects on degradation of diffused GaAs LED's, *Microelectronics and Reliability* 1980 20 (3); pp. 384-385.

[36] H. Mathieu, *Physique des semiconducteurs et des composants électroniques*, trd Edition, Paris : Masson, 1996.

[37] V. Rakovics, S. Püspöki, J. Balázs, I. Réti and C. Frigeri, Spectral characteristics of InGaAsP/InP infrared emitting diodes grown by LPE, *Materials Science and Engineering B.* 2002 91-92; pp. 491-494.

[38] R.J. Malik, *III-V semiconductor materials and devices – Material processing – Theory and practice – Volume 7,* Edition - Amsterdam : F.F.Y.Wang, 1989.

[39] N.W. Ashcroft et D. Mermin, Physique des solides, , Ed. EPS Sciences, 1998.

[40] C. Cohen-Tannoudji, B. Diu, F. Laloë, *Mécanique quantique,* Col. Enseignement des sciences, 2000.

[41] O. Ueda, Degradation of III-V optoelectronic devices, *J.Electrochem.Soc.* 1988; pp. 11-22.

[42] M. Fukuda, Historical overview and future of optoelectronics reliability for optical fiber communication systems, *Microelectronics Reliability* 2000 40; pp. 25-35.

[43] J.D.G. Lacey, D.V. Morgan, Y.H. Aliyu and H. Thomas, The reliability of (AlxGa1-x)0.5In0.5P visible light-emitting diodes, *Quality and Reliability Engineering International* 2000 16 (1) 45-49.

[44] J. Van de Casteele, D. Laffitte, G. Gelly, C. Starck and M. Bettiati, High reliability level demonstrated on 980nm laser diode, *Microelectronics Reliability* 2003 43 (9-11); pp. 1751-1754.

[45] P.P. Vollertsen, Thin dielectric reliability assessment for DRAM technology with deep trench storage node, *Microelectronics Reliability* 2003 43 (6); pp. 865-878.

[46] T. Tomasi, I. De Munari, V. Lista, L. Gherardi, A. Righetti and M. Villa, Passive optical components: from degradation data to reliability assessment – preliminary results, *Microelectronics Reliability* 2002 42 (9-11); pp. 1333-1338.

[47] V. Lista, P. Garbossa, T. Tomasi, M. Borgarino, F. Fantini, L. Gherardi, A. Righetti and M. Villa, *Degradation Based Long-Term Reliability Assessment for Electronic Components in Submarine Applications, Microelectronics Reliability* 2002 42 (9-11).pp. 1389-1392.

[48] J.P. Landesman, Micro-photoluminescence for the visualization of defects, stress and temperature profiles in high-power III–V's devices, *Materials Science and Engineering* B. 2002 91-92; pp. 555-611

In: Light-Emitting Diodes and Optoelectronics: New Research   ISBN: 978-1-62100-448-6
Editors: Joshua T. Hall and Anton O. Koskinen   © 2012 Nova Science Publishers, Inc.

Chapter 3

THE NEXT-GENERATION INTELLIGENT AND GREEN ENERGY LED BACKLIGHTING 3D DISPLAY TECHNOLOGY FOR THE NAKED EYE

Jian-Chiun Liou

Industrial Technology Research Institute,
Electronics and Optoelectronics Research Lab, Taiwan

ABSTRACT

Autostereoscopic displays provide 3D perception without the need for special glasses or other headgear. Drawing upon three basic technologies, developers can make two different types of autostereoscopic displays: a 2D/3D switchable display, head-tracked display for single-viewer systems and a time- multiplexed multiview display that supports multiple viewers. This study investigated the method of using an autostereoscopic display with a synchro-signal LED scanning backlight module to reduce the crosstalk of right eye and left eye images, enhancing data transfer bandwidth while maintaining image resolution. In the following we introduce how LED backlight and optical film elements are used in autostereoscopic 3D display designs including two-view and multi-view designs.

Glasses-free 3D Display Technology, two ways of manufacturing a two-view spatially multiplexed autostereoscopic display. (a) Lenticular: An array of cylindrical lenslets is placed in front of the pixel raster, directing the light from adjacent pixel columns to different viewing slots at the ideal viewing distance so that each of the viewer's eyes sees light from only every second pixel column. (b) Parallax barrier: A barrier mask is placed in front of the pixel raster so that each eye sees light from only every second pixel column.

Lenticular lens technology chiefly involves adding periodic arrays of lenticular lens to the outside of a general display. The design must be optimized in accordance with viewpoints and pixel size in order to produce images that can then be transmitted in varying directions. Since each sub-pixel occupies a different position, its off-axis distance to the lenticular lens will vary. With proper design, parts of the sub-pixels can appear in distinct viewpoints. 3D display can then be realized by employing pictures or images taken at these different viewpoints. Since this technology creates parallax through lens optics, which provides a very high optical efficiency, the only brightness loss comes from interface reflection, lens material transmittance, and light scattering.

A parallax barrier is a device placed in front of an image source, such as a liquid crystal display, to allow it to show a stereoscopic image without the need for the viewer to wear 3D glasses. Placed in front of the normal LCD, it consists of a layer of material with a series of precision slits, allowing each eye to see a different set of pixels, so creating a sense of depth through parallax in an effect similar to what lenticular printing produces for printed products. A disadvantage of the technology is that the viewer must be positioned in a well-defined spot to experience the 3D effect. Another disadvantage is that the effective horizontal pixel count viewable for each eye is reduced by one half; however, there is research attempting to improve these limitations.

A novel time-multiplexed autostereoscopic multi-view full resolution 3D display is proposed. This capability is important in applications such as cockpit displays or mobile, portable, or laptop systems where brightness must be maximized but power conserved as much as possible. The effects are achieved through the creation of light line illumination, by means of which autostereoscopic images are produced, and by simultaneously concentrating the light emitted by the display toward the area the viewer's head is. By turning different illumination sources on and off, it is possible to aim both the concentration area and the 3D viewing area at the observer's head as the observer moves. A variation on the system allows two or more persons to be tracked independently. Cross talk (ghosting) can be reduced to the point of imperceptibility can be achieved.

1. INTRODUCTION

Autostereoscopic displays are those that do no require the observer to wear any device to separate the left and right views and instead send them directly to the correct eye. This removes a key barrier to acceptance of 3D displays for everyday use but requires a significant change in approach to 3D display design. Autostereoscopic displays using micro-optics in combination with an LCD element have become attractive to display designers and several new 3D display types are now available commercially.

Recently, as the luminance efficiency of light emitting diode (LED) has been improved and the cost of LED is going down, the LED is the substitutive solution for the backlight source. Moreover, since LED has many advantages such as long lifetime, wide color gamut, fast response, and so on, LEDs are expected to replace the conventional fluorescent lamps for backlight source of LCD in near future. Although, LED backlight driving systems have been developed and introduced to the market, further reduction of power consumption and cost reduction are still demanded to be widely used as backlight source. Many types of LED backlights are applied to 2D or 3D displays [1–8]. To date, research on 3D display systems has generally focused on providing uniform, collimated illumination of the LCD, rather than addressing low crosstalk issues.

Intelligent and Green Energy LED Backlighting Techniques

(1) Scanning backlight method. The setup is as following:

The backlight unit is separated into several regions. Let's take 4 segmented regions as the example as in Figure 1. the pixel response time is less than three fourths of the frame time when the illumination period is one quarter of the frame time. Arranging for the required illumination period to end just before the new image is written into the panel provides the most relaxed requirement for panel response time; i.e., the response time can be longer (Figure 1). A novel controlled circuit architecture of scanning regions for 120Hz high frequency and high resolution stereoscopic display is shown in Figure 2. Setup all the parameters of scanning backlight method by counting the amount to decide turning time between 4-regions、and 2-regions LED backlight type. If counted times equal to 100 then jump to next backlight segmented region.

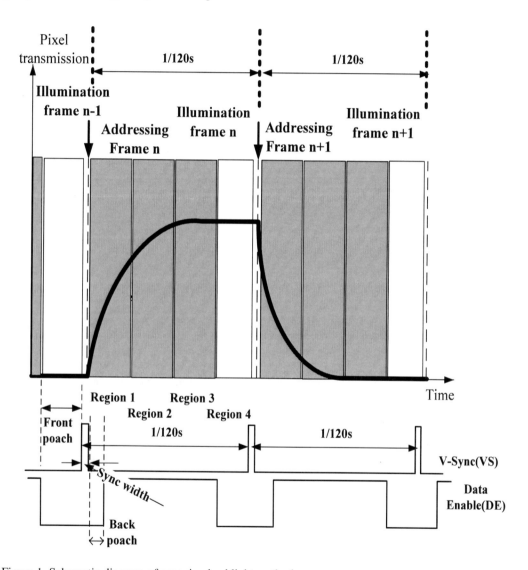

Figure 1. Schematic diagram of scanning backlight method.

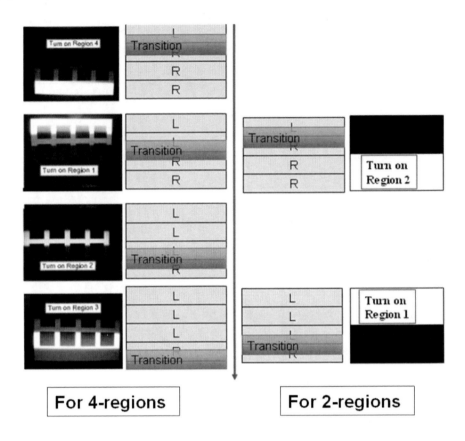

Figure 2. Synchronization Signal LED backlight architecture.

For 4-region scanning backlight method, when the panel is filled in segmented regions 1, 2, 3 and 4 by the new image, the backlight lights up are in the corresponding regions 4, 1, 2 and 3. In anticipation of an image for a left eye and right eye is shown in the region 1 of the panel, we turned on region 3 of the backlight unit. Analogize the image shown in region 2 and turned on region 4 of the backlight unit. For 2-region scanning backlight method, when the panel is filled in regions 1 and 2 by the new image, the backlight lights up in the corresponding regions 2 and 1. For avoiding seeing both L-image and R-image at the same time, the backlight regions R1 have to be off until R1 filled up the image. Analogize the backlight regions R2 have to be off until R2 filled up the image.

(2)In backlight strobe method as shown in Figure 3, Setup the parameters of backlight strobe method by counting the amount to decide turning on time of full screen (full screen of one frame 1/120sec counted amount equal to 400). Setup the parameters of backlight strobe duty time by counting the amount to decide turning time on full screen backlight regions. If counted times equal to (400 - 400*9/10) then jump to next full screen backlight region. The backlight is turned off when the image data refreshes. The backlight only turns on at the system time, or at most a little bit longer than the system time. But the system time is short compared with the time between two adjacent vertical synchronization signals (less than 10%), the display brightness operated under this method is probably quite small.

In this research, we have successfully designed and demonstrated a decent performance with 120Hz optimized synchronization signal between LED brightness/darkness flash and adjusted shutter glasses signal. It has been demonstrated that the 120Hz scanning characteristic from upper row to lower row of the horizontally arranged of stereoscopic image. A quadrate image for a left eye is projected by the light from the left eye image file and a circle image for a right eye is projected by the light from the right eye file through a liquid crystal panel.

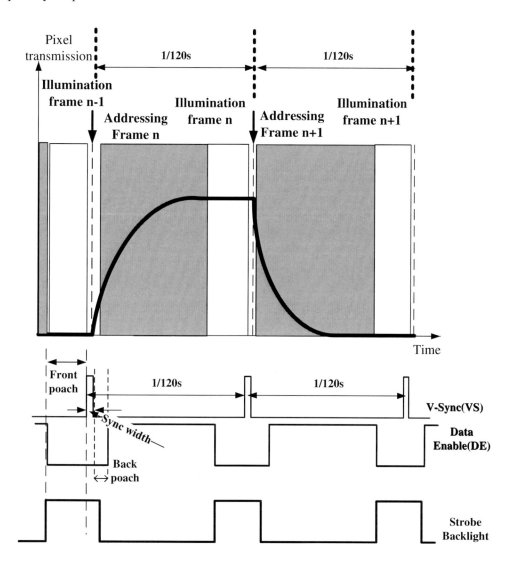

Figure 3. Time scheme of backlight strobe method.

LED scanning backlight stereoscopic display is provided to realize stereoscopic image viewing even in a liquid crystal display. In a frame time, some kinds of brightness/darkness characteristic from upper row to lower row of the horizontally arranged rows of LED in the backlight module, cooperating with the scanning of the LCD, to thereby realize an effect similar to scanning. The general strategy that we employ is to integrate all relatively small-signal electronic functions into one ASIC to minimize the total number of the components. This strategy demonstrates that both the cost is lowered and the amount of the printed circuit board area is reduced. Based on this concept, a smart three dimensional multiplexed driver for LED switching chip with more than 640 LEDs are proposed and the circuit architecture is shown in Figure 4. It is difference from the traditional two dimensional arrays driven by scanning scheme. Three lines are employed to control one LED, including voltage, shift register, and data line. Each LED requires a voltage line for the driving current and shares the same ground with the other resistors. The resistors are individually addressable to provide unconstraint signal permutations by a serial data stream fed from the controller. The shift register is employed to shift a token bit from one group to another through AND gates to power the switch of a LED group. The selection of a LED set is thus a combined selection of the shift register for the group and the data for the specific LED. Such an arrangement allows encoding one data line from the controller to provide data to all of the LEDs, permitting high-speed scanning by shortening the LED selection path and low IC fabrication cost from the greater reduction of circuit component numbers.

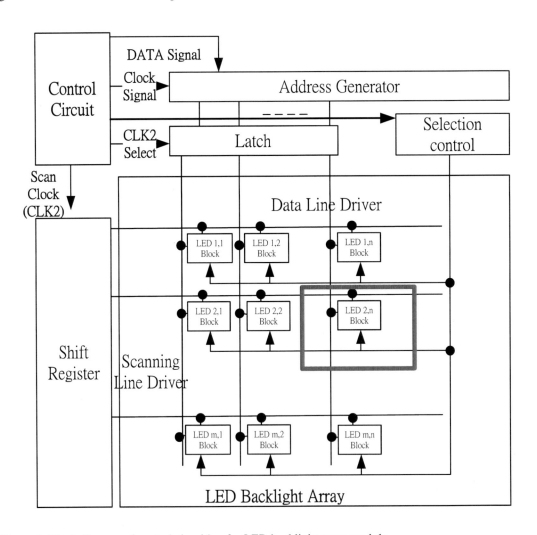

Figure 4. Block diagram of control algorithm for LED backlight array module.

2. 3D DISPLAY BACKLIGHT AND APPLICATION

Throughout the proposed system, lower power consumption are successfully obtained as well as high contrast ratio even with less number of drivers than that of conventional local dimming method. This chapter also contains a new adaptive dimming algorithm and image processing technique for the proposed stereo LCD backlight system.

Recent progress in stereo display research has led to an increasing awareness of market requirements for commercial systems. In particular areas of display cost and software input to the displays are now of great importance to the programme. Possible areas of application include games displays for PC and arcade units; education and edutainment; Internet browsing for remote 3D models; scientific visualisation and medical imaging.

Intelligent and green power LED backlighting techniques of two-dimensional (2D) to three-dimensional (3D) convertible type, mult-view time multiplexed naked eye type, and multi-viewer tracking type for stereo liquid crystal displays are shown as follows.

2.1. Localized 2D/3D Switchable Maked-eye 3D Display

Convertible two-dimensional-three-dimensional display using an LED array based on modified integral imaging as shown in Figure 5. This type propose a two-dimensional (2D) to three-dimensional (3D) convertible display technique using a light-emitting diode (LED) array based on the principle of modified integral imaging. This system can be electrically converted between 3D and 2D modes by using different combinations of LEDs in the LED array without any mechanical movement. The LED array, which is controlled electrically, is used for backlight, and a lens array is used for making a point light source array with higher density[9-14]. We explain the principle of operation and present experimental results.

2.2. Mult-view Time Multiplexed Autostereoscopic Displays

Three-dimensional displays which create 3D effect without requiring the observer to wear special glasses are called autostereoscopic displays. A number of techniques exist – parallax barriers, spherical and lenticular lenses, the latter being the most common one. Depending on the design parameters, various tradeoffs between screen resolution, number of views and optimal observation distance exist. The most popular ones, so called multiview 3D displays, work by simultaneously showing a set of images ("views"), each one seen from a particular viewing angle along the horizontal direction. Such effect is achieved by adding an optical filter, which alters the propagation direction for the information displayed on the screen.

Currently, several 2D/3D switched displays had been proposed such as switched barrier and LC-lens. However, both of the parallax barrier and the cylindrical lens arrays still has the issues of narrow viewing angle and low resolution when displaying the 3D images. Besides, in order to balance the horizontal versus vertical resolution of an autostereoscopic a display, a slanted lens array is used. This causes the high crosstalk of stereo display and the subpixels of a view to appear on nonrectangular grid. This type describes the work aimed at developing optical system; active barrier dynamic backlight slit multi-view full resolution and lower crosstalk 3D panel as shown in Figure 6. The panel of 240Hz displays the corresponding images of the four viewing zones by the same time sequence according to temporal multiplexed mechanism.

All modern multiview displays use TFT screens for image formation. The light generated by the TFT is separated into multiple directions by the means of special layer additionally mounted on the screen surface. Such layer is called "optical layer", "lens plate" and "optical filter". A characteristic of all 3D displays is the tradeoff between pixel resolution (or brightness or temporal frequency) and depth. In a scene viewed in 3D, pixels that in 2D would have contributed to high resolution are used instead to show depth. If the slanted lenticular sheet were placed vertically atop the LCD, then vertical and horizontal resolution would drop by a factor equal to the number of views.

This type addresses the specific technological challenges of autostereoscopic 3D displays and presents a novel optical system that integrates a real-time active barrier dynamic backlight slit system with a naked eyes multi-view stereo display. With 240Hz display and tunable frequency LED backlight slits, only a pair of page-flipped left and right eye images was necessary to produce a multi-view effect. Furthermore, full resolution was maintained for the images of each eye. The loading of the transmission bandwidth was controllable, and the binocular parallax and motion parallax is as good as the usually full resolution multi-view autostereo display.

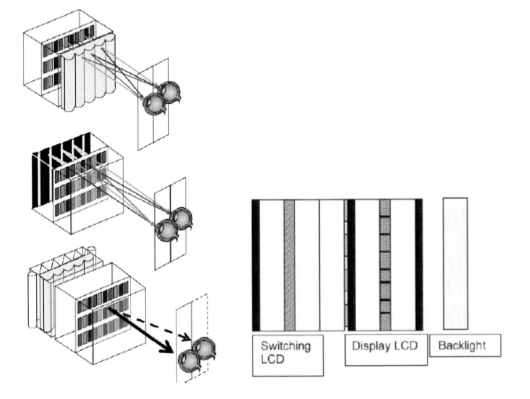

Figure 5. Display configuration.

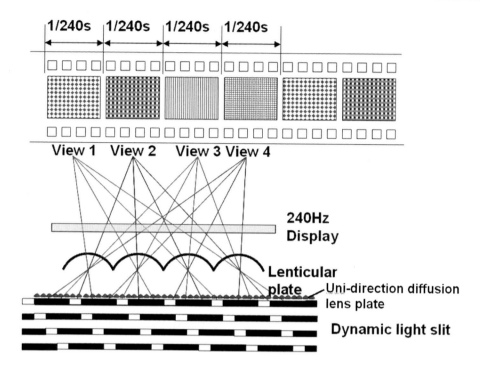

Figure 6. The structure of the proposed multi-view 3D display.

A lenticular-based 3D display directs the light of neighboring sub-pixels into different directions by means of small lenses placed immediately in front of the sub-pixels. In this manner different pictures can be transmitted into different directions. Usually a multitude of directions is chosen, e.g. 4 different views. Two of these views can be seen by the left and right eye respectively, and as such create a stereoscopic (3D) image. Figure 6. shows the structure of the proposed multi-view 3D display. Only one eye individually receives the image at one corresponding viewing zone at its displaying time period, such as 1/240 second. As a result, the 3D image can be created for the viewer by naked eyes. Each view is 60Hz.

2.3. Multi-Viewer Tracking Stereoscopic Display

Many people believe that in the future, autostereoscopic 3D displays will become a mainstream display type. Achievement of higher quality 3D images requires both higher panel resolution and more viewing zones. Consequently, the transmission bandwidth of the 3D display systems involves enormous amounts of data transfer. This type integrated a viewer-tracking system and a synchro-signal LED scanning backlight module with an autostereoscopic 3D display to reduce the crosstalk of right/left eye images and data transfer bandwidth, while maintaining 3D image resolution. Light-emitting diodes (LED) are a dot light source of the dynamic backlight module as shown in Figure 7. When modulating the dynamic backlight module to control the display mode of the stereoscopic display, the updating speed of the dynamic light-emitting regions and the updating speed of pixels were synchronal. For each frame period, the viewer can accurately view three-dimensional images, and the three-dimensional images displayed by the stereoscopic display have full resolution.

The stereoscopic display tracks the viewer's position or can be watched by multiple viewers. This type demonstrated that the three-dimensional image displayed by the stereoscopic display is of high quality, and analyzed this phenomenon. The multi-viewer tracking stereoscopic display with intelligent multiplexing control of LED backlight scanning had low crosstalk, below 1%, when phase shift was 1/160s.

Due to the LCD with physical delay characteristic (low response speed), both images are alternately switched from one to the other by switching the light emitting full panel. Thus, stereoscopic images are shown to a viewer. As a result, all kinds of low cross-talk stereoscopic display with an intelligent multiplexing control LED scanning backlight have low crosstalk below 1% through a liquid crystal display.

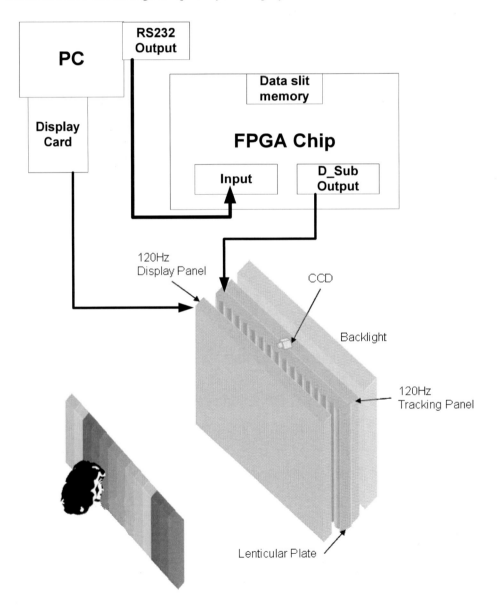

Figure 7. Viewer-tracking 3D display system.

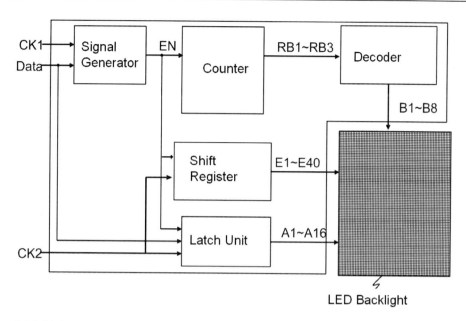

Figure 8. Multiplexer circuit system.

Intelligent and green energy LED backlighting techniques of stereo liquid crystal displays have been successfully designed and demonstrated a decent performance of all kinds of stereo types display system with optimized synchronization signal between LED brightness/darkness flash and adjusted driving signal enabled by a 3-Dimensional controlling IC. At high scanning rate from upper row to lower row, the system demonstrated horizontally arranged clear stereoscopic images. This method of backlighting also allows dimming to occur in locally specific areas of darkness on the screen. This can show truer blacks and whites at much higher dynamic contrast ratios, at the cost of less detail in small bright objects on a dark background, such as star fields.

3. EXPERIMENTAL AND RESULTS

3.1. Localized 2D/3D Switchable Maked-eye 3D Display

As LED backlight of flat panel display become large in format, the data and gate lines turn into longer, parasitic capacitance and resistance increase, and the display signal is delayed. Three dimensional architecture of multiplexing data registration integrated circuit method is used that divides the data line into several blocks and provides the advantages of high accuracy, rapid selection, and reasonable switching speed.

The design concept can be easily scaled up for large LED backlight array format TFT-LCD elements system without much change in the terminal numbers thanks to the three dimensional hierarchy of control circuit design, which effectively reduces the terminal numbers into the cubic root of the total control unit numbers and prevent a block defect of the flat panel. The TFT-LCD unit lights, line(s) in the vertical or horizontal axis appear dim, but not completely on or off. These defects are generally the result of a failure in the row (horizontal) or column (vertical) drivers or their connections. The TFT-LCD includes an extension part defect such as an extension piece overlapping with a pixel electrode of boundary pixels at a boundary data line applying a data signal to the boundary pixels.

3.1.1. Three Dimensional Data Registration LED Backlight

A LED backlight of flat panel display with three dimensional architecture of multiplexing data registration integrated circuit having a plurality of scanning electrodes, a plurality of data electrodes extending perpendicularly to the scanning electrodes, and liquid crystal filling a space between the scanning electrodes and data electrodes, pixels being formed at each intersection of the scanning and data electrodes together with the liquid crystal, the display panel being divided into an even row part and a odd row part; a scanning control circuit for scanning the scanning electrodes by sequentially supplying scanning voltages to each scanning electrode and by maintaining the same for a predetermined period, the scanning electrodes located in the even row part of the panel and the scanning electrodes located in the odd row part being scanned separately but simultaneously in the same directions from upper to lower of the panel; an image data control circuit for sequentially supplying image data voltages to the data electrodes in synchronism with scanning of the scanning electrodes, the scanning electrodes are scanned in such a manner that the image data is written on the pixels in a selecting period, the written image data is held on the pixels in a holding period and the image data is eliminated in an eliminating period; In traditional control circuit design for TFT-LCD elements array system, each TFT-LCD element requires one driver switch. As a result, when the TFT-LCD backlight of LEDs' pixels scale up into a large array, the numbers of input/output ports will increase enormously. To handle large array of driving circuits for such large pixels array, 2D circuit architecture was employed for the traditional driving circuit to reduce the IO number from n*n into 2n+1. However, firstly, this reduction still can not meet the requirement for high speed signal scanning with low data accessing points when switch numbers greater than 640×480 pixels. It would be necessary to increase the display frequency to 240 Hz or higher to eliminate flicker. If the display frequency is 240 Hz, a period of time for writing one frame is 4.17 ms. Assuming the number of scanning electrodes is 480, a period of time available for writing one line is only 8.7 microseconds. The number of scanning electrodes has to be larger than 480 to display a high resolution image, making the writing period further shorter. Secondly, no technology is ever completely perfect, of course, and the LCD can still suffer from some defects in the displayed image. In this technology, though, most defects in the basic electronics, such as failure of the backlight or the row or column drivers, result in a completely unusable display, and so when such occur in production they are easily detected and corrected. It is extremely rare for a product to ship with any such problems.

To achieve this, In this study, a three dimensional data registration LED backlight of flat panel display scheme(Figure 8, Figure 9) to reduce the number of data accessing points as well as scanning lines for large array TFT-LCD element with switch number more than 640×480 is proposed(Figure 10). The total numbers of data accessing points will be $N=3\times\sqrt[3]{Y}$ +1, which is 68 for 640×480 switches by the 3D novel design, the scanning time is reduced up to 30% (The scanning speed is also increased by 3 times) thanks to the great reduction of lines for 3D scanning, instead of 2D scanning. Figure 11 is shown pad connections from 1D, 2D, and 3D control circuits. Figure 12. is localized dimming LED backlight.

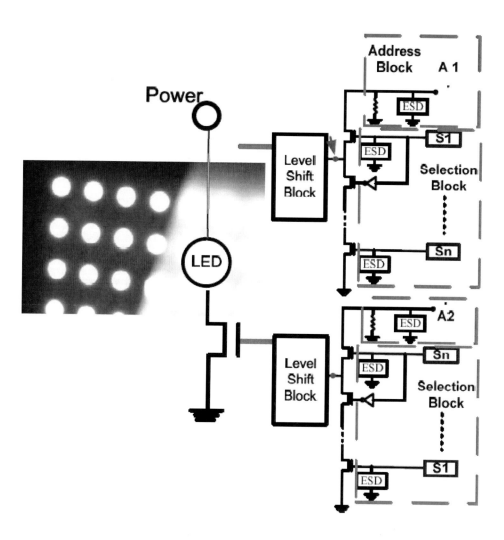

Figure 9. Photograph of LEDs backlight sets.

3.1.2. Localized 2D/3D Switchable Naked-eye 3D Display

The patented 2D/3D control module is made of a low-resolution panel with micro-retarder film, the LCD panel and backlight dynamically controlled form a patented architecture that can switch 2D/3D mode display area, provides three-dimensional image web application. Area by demand, 2D/3D switchable, 2D coexists with 3D on the one screen, 2D area shows small character clearly and 3D area shows multi-view naked-eye stereo image.

For simultaneous display of 2D and 3D information, we have develop an integrates naked-eye 2D/3D display window technology, called integrated 2D/3D windows (i2/3DW), that can display 2D and 3D images with flexibility and best quality on the same screen. As for stereo 3D gaming, there are dialog windows or pop-up windows in the stereo 3D game frames. Without an integral 2D/3D display, 2D texts in the 3D mode windows appear in broken and blurred characters. This situation is very much annoying especially for small fonts. But, with i2/3DW's localized 2D/3D switchable display technology, 3D gaming becomes true joy because 2D texts will be as clear as they are on a 2D screen while the 3D game scenes would still be the same fascinating as on a 3D device. Figure 13 shows the construction of an i2/3DW display made according to technology from ITRI. It comprises three "primary "component layers." The first at left is a conventional liquid crystal display panel (LCD panel).

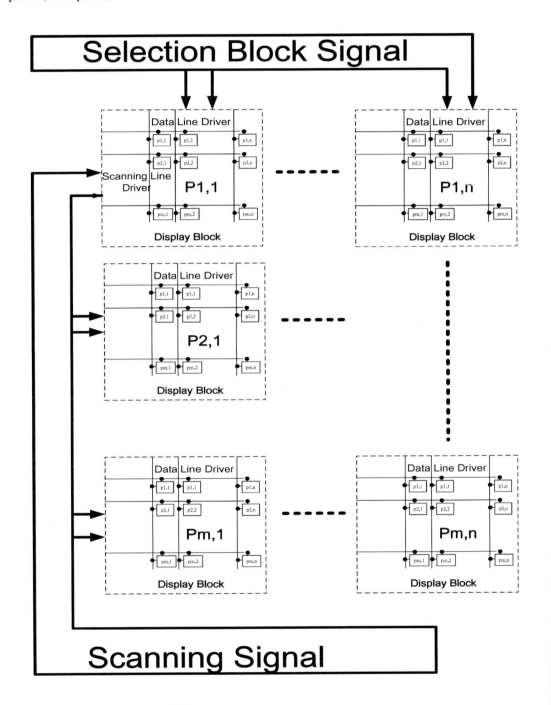

Figure 10. 3D scanning display block array.

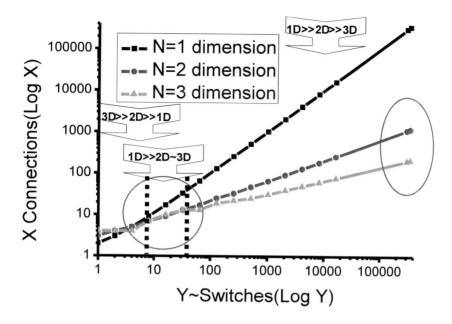

Figure 11. Pad connections from 1D, 2D, and 3D control circuits.

Figure 12. Localized dimming LED backlight.

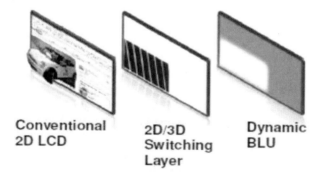

Figure 13. The structure of 2D/3D switching design for integrated 2D/3D display.

The third at right is a dynamic back-light unit (DBLU), one that is similar, for example, to an LED matrix-based BLU now seen in many laptop PC display panel but with brightness of each (or at least groups) of the LED elements in the matrix separately controllable. There is a second component layer, the 2D/3D switching component, inserted between the first and third that is responsible for the automatic switching of individual pixels in the first between its 2D and 3D display mode. This technology uses a microretarder-based switching device to partially switch various parts of the display screen between 2D mode and 3D mode. Structurally, a microretarder plate has an interleaved pattern of half-wavelength-retardation and zero-retardation stripes. Working together with the microretarder is a liquid crystal (LC) switching panel inserted between the microretarder and the polarized backlight of the device. Each "pixel" of the LC switching panel functions as the switching unit cells between 2D and 3D mode regions. As a general rule of thumb, the smaller these "pixels" are, the smaller the 2D/3D switching cells can be. Though, for any practical application, the number of units required for the 2D/3D switching LC panel lands in where barely enough but with the lowest cost case, for example, 16 by 10.

3.3. Mult-View Time Multiplexed Autostereoscopic Displays

Multi-view displays use TFT screens for image formation . The light generated by the TFT is split into multiple directions by the means of special optical layer (called also lens plate or optical filter) mounted in front of the TFT. The intensity of the light rays passing through the filter changes as a function of the angle, as if the light is directionally projected. There are two important points to note when considering multi-view screens: (1)They are not directly compatible with standard stereo-scopic footage or software. This is because, rather than display two distinct views to the viewer (as with most other stereoscopic displays), they provide up to 5, allowing the viewer to walk around the screen whilst maintaining the 3D effect. (2)Due to the display showing 4 or 5 views simultaneously, each view contains only 1/4 or 1/5 of the standard resolution of the display panel. An optimal lens, designed to handle a single situation, would be shaped to contain only as many distinct views as local participants; maximizing each views resolution would require a lens width sufficient to cover the same number of subpixels as views[15-22].

Our current prototype system uses a lenticular lens multiview technique with a time-multiplex autostereoscopic display based on active directional backlight (active dynamic backlight). Throughout the proposed system, lower power consumption are successfully obtained as well as high contrast ratio even with less number of drivers than that of conventional local dimming method. This architecture also contains a new adaptive dimming algorithm and image processing technique for the proposed stereo LCD backlight system. Recent progress in stereo display research has led to an increasing awareness of market requirements for commercial systems. In particular areas of display cost and software input to the displays are now of great importance to the program. Possible areas of application include games displays for PC and arcade units; education and edutainment; Internet browsing for remote 3D models; scientific visualisation and medical imaging. Intelligent and green power LED backlighting techniques of two-dimensional (2D) to three-dimensional (3D) convertible type, shutter glasses type, multi-view time multiplexed naked eye type, and multi-viewer tracking type for stereo liquid crystal displays are applied for 3D display system.

According to the time sequence for turning the groups of the light source, multiple viewing zones at multiple directions are created. To meet the requirements of different one-eye images, we propose that the real-time active barrier dynamic backlight slit system on stereo-display. To confirm our design workable, we did the optical simulation using ASAP software. The detector is set at the convergent point, and the intensity profiles of the four views are shown in Figure 14. The intensity profiles are evolving every 1/240 sec, and the separation of peaks is about 60 mm, quite close to the design value, 65 mm. The small inaccuracy resulted from the absorption of black matrix, and it makes the dead zone explicit. Setting the pixel size larger, or the black matrix smaller would improve the results. On the other hand, the center-viewing group has the least crosstalk, and side lobe groups have larger crosstalk, especially when viewing groups departs from center very much. The increase of crosstalk arises from the abbreviation when light is not close to the optical axis. Thus, the profiles of the four views confirm our design workable.

3.3.1. Experimental Results

The photographs of displayed images used the luminance meter (Konica Minolta CS-200) as shown in Figure 15. The distance from the optical sensor to the center of LED backlight panel is in 60 cm, which is the normal range of distance for watching a 3D computer monitor. Measurements of angle are done from observation point -50 degree to observation point +50 degree; view 2 is the central view. The illuminance meter has an analog output to the oscilloscope and the illuminance signal can be recorded and processed by a computer.

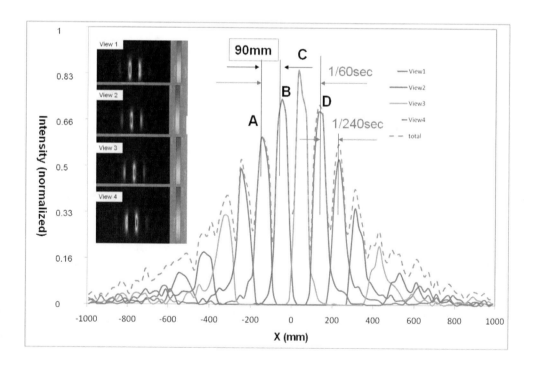

Figure 14. The optical simulation using ASAP software.

In this research, we observed backlight light stain structure for 3D image display based on lenticular lens array. In Figure 16, the photo is illustrated for four viewing zones 1-4 located at the viewing location. Each viewing zone uses 1/240 second to display one image. The light source at specific location is grouped corresponding to each lenticular lens of the lenticular lens array. For the four viewing zones, each lenticular lens has four groups 1-4 of light sources corresponding to four viewing zones 1-4. The four groups of light are sequentially turned on for 1/240 second. The group 1 of light source is turned on, and then the group 2 of light source is turned on next for 1/240 second. Likewise, the groups 3 and 4 of light source are sequentially turned on for 1/240 second. Generally, the multiple viewing zones equally shares 1/60 second for one image frame. The viewable zone area from first viewing zone to be contiguous to second viewing zone is 90 mm, and for 4 viewing zone of viewable area is 360 mm(The separation of viewing zones is about 90 mm, and overall width of viewing group is 360 mm) as shown in Figure 15.

Yellow stripe pattern is created by phosphor of yellow color. White LEDs are blue LED chips covered with a phosphor that absorbs some of the blue light and fluoresces with a broad spectral output ranging from mid-green to mid-red. So, the backlight modular was taken on yellow stripe.

The configuration of uni-direction diffusion lens plate is shown in Figure 16(b). The panel of 240Hz displays the corresponding images of the four viewing zones by the same time sequence according to temporal multiplexed mechanism. The uni-direction diffusion lens plate can condense the light individually belonging to each the lenticular lens at transverse direction. The lenticular lenses of the lens array receive the light and deflect the light into each viewing zone in a time sequence, respectively.

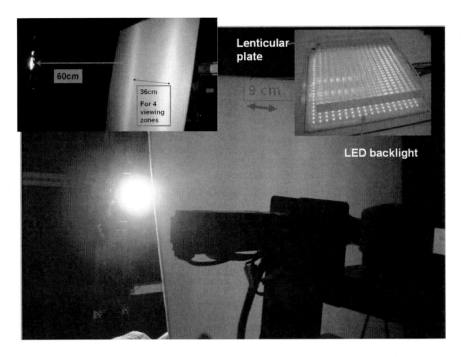

Figure 15. Four views backlight.

Lenticular/LED

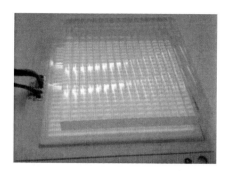
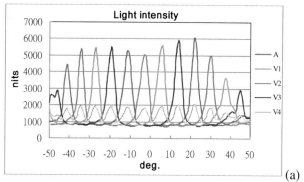

(a)

Lenticular/Optical film/LED

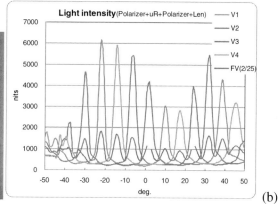

(b)

Figure 16. The crosstalk under different observation scanning angles.

According to the time sequence for turning the groups of the light source, multiple viewing zones at multiple directions are created. To meet the requirements of different one-eye images, we propose that the real-time active barrier dynamic backlight slit system on stereo-display. The center viewing group has the least crosstalk, and side lobe groups have larger crosstalk, especially when viewing groups departs from center very much. The crosstalk under different observation scanning angles is showed from data in Figure 16, including the cases of 4-views field scanning. The crosstalk of view 1 is about 5% respectively, the results are better than slanted lenticuler lens type.

3.4. Multi-viewer Tracking Stereoscopic Display

This study integrated an autostereoscopic display with a viewer-tracking system. Figure 17. illustrates the basic structure of the display. In the proposed structure, a retarder inserted between the image panels rotated the light beam at 90°; simultaneously, a lenticular plate adjusted the light direction to show the light slit from the tracking display. Retarder film is a clear birefringent material that alters the phase of a polarized beam of light. A quarter wave plate can convert linearly polarized light (oriented at 45° from the direction of the fast/slow axis) into circularly polarized light. Conversely, the wave plate can convert a circularly polarized beam into linearly polarized light.

In this study, when the polarization direction of the incident light formed an included 45° angle with the optical axis of the retarder, the polarization of the light passing through the $\lambda/2$ retardation regions rotated by 90° and became orthogonal to the polarization of the light passing though the 0° retardation regions. The molding method fabricated the lenticular plate with polymeric film as the substrate material. One of the light slit pattern pairs adjusted the direction of light from the tracking panel to the viewer's eyes through the lenticular plate.

In this display, the PDLC panel played an important role in the function of the 2D/3D switch. When the PDLC panel was turned to clear state, the microretarder interacted with the polarizers to form a parallax barrier pattern as shown in Figure 18, making the display autostereoscopic. In a case where the PDLC panel is in a diffusive state, the light passing through the PDLC destroys the polarization. The microretarder then loses its function as a parallax barrier, and the display becomes a general 2D display.

This study developed autostereoscopic display apparatus and a display method. The autostereoscopic display apparatus included a display panel, a backlight module, a tracking slit panel and an optical lens array. In a frame time, the display panel and tracking panel share the same synchronization signal for the display panel. The tracking panel controls the light of the backlight module. The tracking panel features tracking slit patterns and switches the slit patterns according to the synchronization signal. Until all screen data is updated, the backlight module is inactive during the frame time. A light provided by the part of the backlight regions passes through the tracking slit set, optical lens array, and the display panel in such a way that each eye separately perceives images. As shown in Figure 19, when the viewer moves to the left, the tracking slit set changes its pattern to display the correct image.

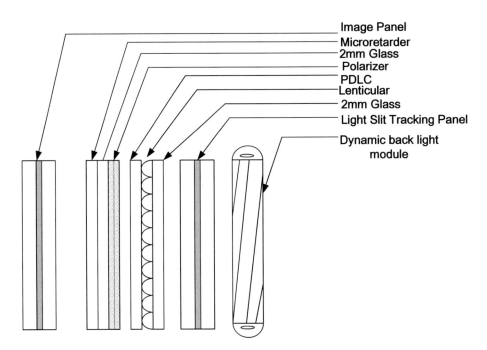

Figure 17. The structure of the proposed viewer-tracking display panel.

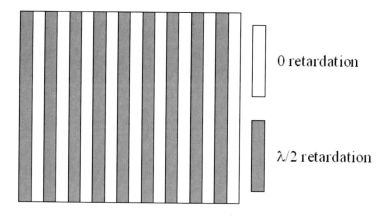

Figure 18. The pattern of a microretarder.

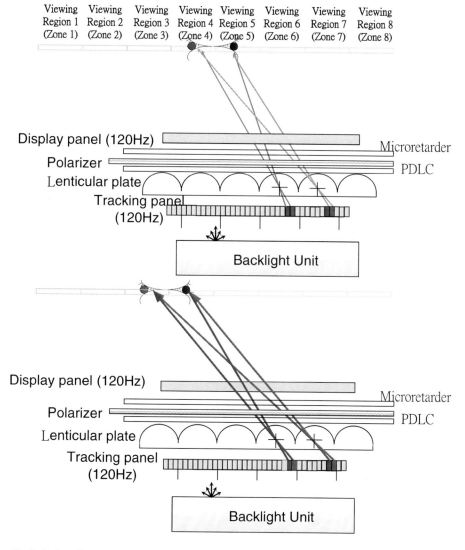

Figure 19. Relations between viewer and tracking panel.

The autostereoscopic display integrated a webcam as the real-time detection device for tracking of the viewer's head/eye positions, so that the display showed left and right eye images correctly[23-31]. The computer vision-based tracking method detects viewer's eyes over a specific range and under conditions of low and fluctuating illumination. By capturing the image of the viewer in front of the display, the viewer's position is calculated and the related position data is transferred to the field programmable gate array (FPGA) controller through RS232. When the viewer recognizes that he/she is standing at the borders of the viewing zones, analyzing the captured viewer images determines the border positions of the viewing zones. The resulting eye reference pattern allows the tracker to locate the viewer's eyes in live video images. If an observer moves away from his original position, the tracking slit will vary its pattern according to the viewer's new position. The viewer still perceives two eye images separately before exceeding the webcam detection range. Figure 20 shows the viewer-tracking system.

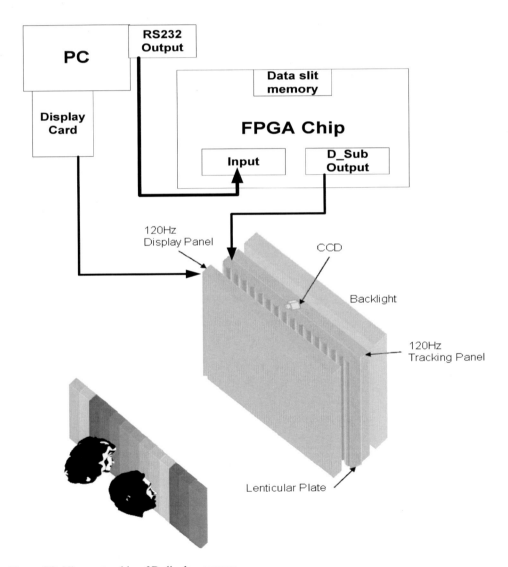

Figure 20. Viewer-tracking 3D display system.

This research addresses the specific technological challenges of autostereoscopic 3D displays and presents a novel system that integrates a real-time viewer-tracking system with an autostereoscopic display. Our successfully designed prototype utilized a FPGA system to synchronize between a display panel and tracking slit panel. With 120Hz display and tracking panels, only a pair of page-flipped left and right eye images was necessary to produce a multi-view effect. Furthermore, full resolution was maintained for the images of each eye. The loading of the transmission bandwidth was controllable, and the binocular parallax and motion parallax is as good as the usually lower resolution multi-view autostereo display.

(B) LED Backlight Architecture

Many types of LED backlights are applied to 2D or 3D displays. To date, research on 3D display systems has generally focused on providing uniform, collimated illumination of the LCD, rather than addressing low crosstalk issues. This study investigated the method of using an autostereoscopic multi-viewer tracking 3D display with a synchro-signal LED scanning backlight module to reduce the crosstalk of right eye and left eye images, enhancing data transfer bandwidth while maintaining image resolution. Figure 21A is a schematic view of a stereoscopic display. Figure 21B is a block diagram illustrating the stereoscopic display; the stereoscopic display can track the viewer's position and be watched by multiple viewers.

The backlight module of the stereoscopic display is a dynamic backlight module featuring many light-emitting regions R(1)~R(4). Figure 26A excludes the control unit and optical lens array. In the stereoscopic display, the graphic card outputs and transmits the vertical synchro-signal to the control unit. After receiving the synchro-signal, the control unit outputs the synchro-signal to control (turn on or off) the light-emitting regions R(1)~R(4).

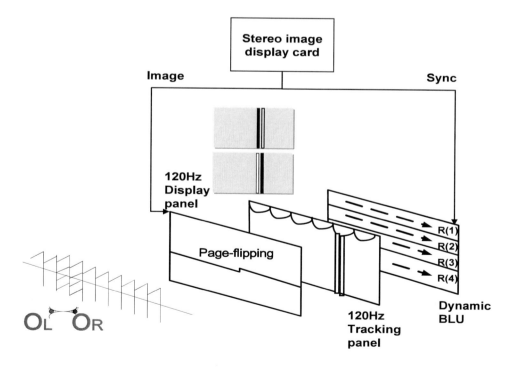

Figure 21A. The schematic view illustrating a stereoscopic display.

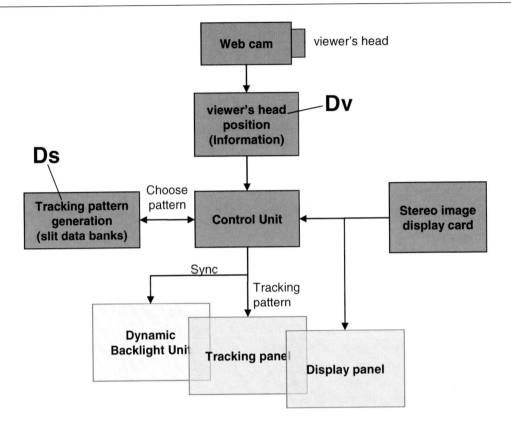

Figure 21B. The block diagram illustrating the stereoscopic display.

To meet the requirements of different one-eye images, we propose that the dynamic LED backlight tracking panel has have many backlight slit sets. According to the position information of viewer O and the vertical synchro-signal, one of the slit sets of the tracking panels is selected and turned-on. Each slit set includes either left or right eye slits. Light emitted from the dynamic backlight module passes through the either left or right eye slit and the display panel, and projects onto one eye of viewer O. Similarly, light emitted from the dynamic backlight module passes through the either left or right eye slit and the display panel, and projects onto the other eye of viewer O. In this way, the pair images are projected to the two eyes of viewer O, who can see accurate three-dimensional images. For example, light emitted from the dynamic backlight module passes through the left eye slit of the slit set and the display panel, and projects onto the left eye OL of viewer O. Similarly, light emitted from the dynamic backlight module passes through the right eye slit of the slit set and the display panel, and projects onto the right eye OR of viewer O. The one-eye slits are stripe-shaped and the lengths of the one-eye slits are approximately equal to the longitudinal length of the display panel.

When the display panel displays an image based on the vertical synchro-signal, the slit set of the tracking panel is enabled. Meanwhile, pixels in the updated region of the display panel display a left-eye image, but pixels in the non-updated region of the display panel still display the previous right-eye image. Light passing through the slit set of the tracking panel and the non-updated region of the display panel can be projected onto left eye OL of viewer O (i.e. a crosstalk phenomenon) if no alternative methodology is applied. This research proposes using a dynamic backlight module to suppress the crosstalk. The light-emitting regions R1~R4 of the dynamic backlight module are separately controlled according to the vertical synchro-signal.

During a frame period, the light-emitting regions R(1) and R(2) corresponding to the updated region are turned on and the light-emitting regions R(3) and (4) corresponding to the non-updated region are turned off. In this way, only the light-emitting regions R(1) and R(2) provide light, so that no light passes through the slit set of the tracking panel and the non-updated region of the display panel. This reduces the crosstalk phenomenon of the stereoscopic display system.

As shown in Figure 21A and Figure 21B, the display method of the stereoscopic display comprises the following steps:

First, slit data banks (Ds) corresponding to the many viewing angles of the stereoscopic display apparatus is established. Next, the control unit receives information (Dv) on the position of the viewer. The control unit compares the position information and the slit data banks stored in advance. Meanwhile, the control unit outputs the vertical synchro-signal from the graphic card to control the output mode of the dynamic backlight module and operation mode of the tracking panel. The display panel is driven to display images (i.e. image updating) according to the vertical synchro-signal output from the graphics card. Many of the light-emitting regions (R(1)~R(4)) of the dynamic backlight module are stripe-shaped and the light-emitting regions R(1)~R(4) extend across the slits of the tracking panel. The extending direction of the light-emitting regions R(1)~R(4) is perpendicular to the extending direction of the slits of the tracking panel. Many of the light-emitting regions (R(1)~R(4)) of the dynamic backlight module are array in an arrayed manner.

3.4.1. Crosstalk Analysis

To avoid ghost images, the backlight modular provides backlight control signals which are dependent on the position of an associated part of the panel. The system is provided for controlling synchronization timing between backlighting and pixel refresh, in dependence of a location of a section within the display panel. The backlight unit is separated into several regions. Let's take 4 regions as the example, the pixel response time is less than three fourths of the frame time when the illumination period is one quarter of the frame time. Optical sensor and CS-100 Spot Chroma Meter of luminance crosstalk measurement of the 4-regions 、 2-regions scanning and strobe backlight method without lenticular. Frame sequential(page flip, temporal multiplexed) process, the process is referred to as alternate frame sequencing.

Crosstalk is a critical factor determining the image quality of stereoscopic displays. Also known as ghosting or leakage, high levels of crosstalk can make stereoscopic images hard to fuse and lack fidelity. Crosstalk is measured by displaying full-black and full-white in light-emitting regions R(1)~R(4) of the display system without lenticular and using an optical sensor to measure the amount of leakage between channels.

The crosstalk is given by equations (1) and (2):

$$CL = \frac{BW - BB}{WB - BB}$$

(1)

and

$$CR = \dfrac{WB - BB}{BW - BB} \qquad (2)$$

Where WB represents a video stream with all-white as left-eye images, all-black as right-eye images),

BW represents a video stream with all-black as left-eye images, all-white as right-eye images),

BB represents a video stream with all-black for both left and right eyes.

CL and CR represent the crosstalk experienced by the left eye and right eye

The CS-100 Spot Chroma Meter is used in this research to measure all the luminance values. The images are displayed in page-flipping mode using the resolution and color-depth set in Stereo/Page-flip Setup.

For example, the optical sensor is placed at the left eye position (either behind the left eye of 3D glasses, or in the left eye viewing zone for an autostereoscopic display) and measurements are taken for the four cross-combinations of full-white and full-black in the left and right eye-channels. An additional reading is also taken with the display in the off state. These readings can then be used in the crosstalk equations described above. This metric can be called black-and-white crosstalk and this metric is often used because maximum crosstalk occurs when the pixels in one eye-channel are full-black and the same pixels in the opposite eye-channel are full-white. According to the results, the brightness of the 4R scanning backlight and the widened backlight strobe is about half of the 2R scanning backlight. But the crosstalk performance of the widened backlight strobe crosstalk is the best of three methods. Only about 1.68% left. Although the 4R scanning backlight display crosstalk is higher than widened backlight strobe, it still performs better than the 2R scanning backlight one.

In the study, the CS-100 Spot Chroma Meter was used to measure the brightness of the backlights, which can be controlled using the duty cycle of backlight signal. Moreover, photodiode s3072 was used to measure the optic characteristics of the display device.

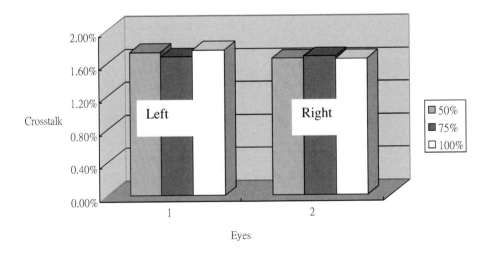

Figure 22. The crosstalk under different brightness conditions.

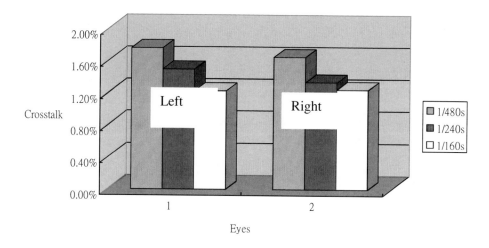

Figure 23. The crosstalk under different phase shift conditions.

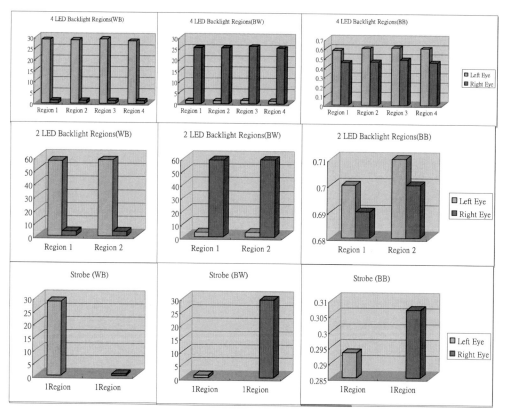

Figure 24. CS-100 Spot Chroma Meter of luminance measurement of the 4-regions 、 2-regions scanning and strobe backlight method.

To view the correct image from the tracking display, the synchronization relationship between image display and backlight requires calibration. Figure 22 is the crosstalk of right eye and left eye under three different brightness conditions. The phase of the V-sync signal exceeds the phase of the backlight signal in 1/480s. It is too dark if the duty cycle is lower than 50%, so the three chosen duty cycles all exceeded 50%. According to Figure 22, the differences in brightness do not significantly affect the crosstalk. The performances of both eyes were approximately in agreement. The experiment selected maximum brightness.

Figure 23 shows the crosstalk of the right eye and left eye with different phase shifts between the V-sync signal and backlight signal, where the duty cycle of backlight signal is 100%. The lowest crosstalk only occurs when phase shifts are 1/160s, not both 1/480s and 1/160s. Here, light leaking to other regions and the response time of the liquid crystal affect the crosstalk. The horizontal axis is time (5 ms per grid) and the vertical axis is voltage (20 mV per grid). One display frame is 1/120 second, approximately equal to 8 milliseconds. The response waveform can be divided into four sections (2ms per section). The waveform of the liquid crystal still rises (section II) when the phase of V-sync signal exceeds the backlight signal in 1/480s ($\doteqdot$2ms); here, the phase does not reach a bright state. But, the phase shifts of 1/240s ($\doteqdot$4ms) and 1/160s ($\doteqdot$6ms), located at region III and region IV, respectively, gradually near the bright state; this explains the difference in crosstalk.

The luminance of the 3D LCD is measured and recorded as WB, BW and BB charts. They are shown in Figure 24. The measurement distance is 1 meter and the recording unit is "nit". The crosstalk is calculated by the equations 1 and 2. As a result, the luminance of a 4R scanning backlight display for one viewing zone is 28.825 nits, that of a 2R scanning backlight display for one viewing zone is 57.25 nits, and that of backlight strobe display is 28.45 nits.

The crosstalk under different scanning conditions is calculated from data in Figure 24, including the cases of 4-Region and 2-Region scanning backlight method, and backlight strobe method (1 Region). The crosstalk of 4R and 2R scanning backlight displays are 3.2% and 5.08% respectively, and that of backlight strobe display is 1.68%.

According to the results of Figure 24 and 25, the brightness of the 4R scanning backlight and the widened backlight strobe is about half of the 2R scanning backlight. But the crosstalk performance of the widened backlight strobe crosstalk is the best of three methods. Only about 1.68%. Although the 4R scanning backlight display crosstalk is higher than widened backlight strobe, it still performs better than the 2R scanning backlight one.

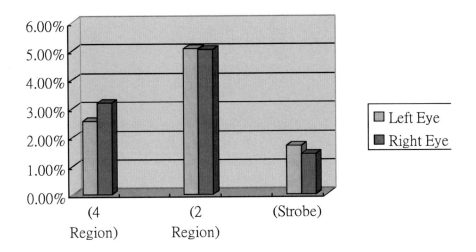

Figure 25. Results of crosstalk calculation of different dynamic backlight methods.

3.4.2. Measured Results

(A) The Optical Properties Measurement

Detailed and quantitative measurements were used in the autostereoscopic display. To measure the borders and the performance of the viewing zones, a luminance meter, Minolta CS-200, was located at the designed viewing distance (630mm from the display), which reduced the display panel's optical interference during measurement. Only the backlight module, including a backlight, tracking panel, and lenticular plate, was used in optical luminance measurement experiments. In this backlight structure, no additional brightness was lost in the optical path.

Figure 26 shows the results of the luminance intensity experiment. The entire measuring process was completed in a darkroom, which provided measurements of fairly high quality. In the luminance intensity experiment, only 40 degrees on both sides of the center of the backlight module was measured; the intensity value was captured every 0.5 degree. When measuring, only one viewing zone of the tracking panel was switched on. The luminance meter was used to scan the viewing zones horizontally. The maximum peak luminance value of the viewing zone was approximately 514 cd/m^2 when CS-200 detected the luminance intensity value near the center of the backlight module. The minimum peak luminance intensity value of the backlight was approximately 364 cd/m^2 at the edge of the viewing group. The luminance intensity range of the viewing group in front of the backlight module ranged between 264 and 514 cd/m^2. The intersection point between two adjacent luminance intensity curves may determine the borders of the viewing zones.

After luminance intensity measurement of viewing zones 1 to 8 was completed, data could be combined to yield a luminance intensity distribution figure to verify the optical design parameters. The peak luminance intensity value of viewing zone 4 is in front of the center of the backlight module. The tracking panel was slightly misaligned with the lenticular plate. Due to the light intensity distribution of the tracking panel and the entrance angle difference of the light path between the backlight and the lenticular lens, a stronger luminance intensity curve was measured in the central part of the viewing group. The viewing group of the 3D display system was approximately 53 cm wide at a viewing distance of 63 cm, indicating that each viewing zone is 6.625 cm wide on average.

In building a viewer-tracking-based autostereoscopic display, the viewer's position and border positions of the viewing zones are the key parameters. To accurately define the viewer's position, the black and white pictures for both eyes were displayed as calibration images. The positions of the borders in the viewing zones could be determined by analyzing the images of the viewer captured while the viewer reported to be at the border positions.

(B) Motion Parallax Function Result

For a 3D display to simulate the natural vision of human beings, both binocular parallax and motion parallax are required. For a multi-view autostereoscopic display system, viewers can see stereoscopic images with binocular parallax and motion parallax within a group of viewing zones. However, a high number of viewing zones is necessary to achieve smooth motion parallax for a sufficiently large view. This normally causes significant reduction of image resolution. While maintaining good image resolution, we implemented smooth motion parallax by adopting viewer tracking function and real-time image updating in a two-view autostereoscopic display system.

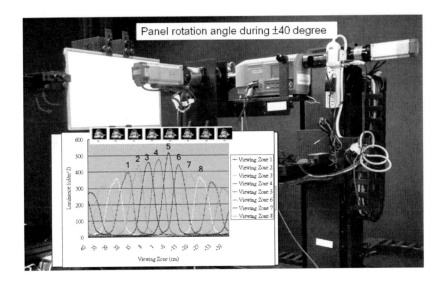

Figure 26. Luminance intensity distribution of lenticular-type BLU.

This study used an ad-boost algorithm capable of evaluating important features to quickly track viewers. If the viewer's eyes were detected in specific viewing zones in front of the display, the viewer's position would determine the images of the corresponding viewing angles shown. When the viewer's eyes move inside the same viewing zones, full stop needs moving, the images for the new viewing angles are fed into the same viewing zones. The viewer experiences the motion parallax because he/she sees different images from different angle.

When the viewer's eyes continue to move and finally cross the border of the viewing zones, the images of the new angle are reversed left-and-right and presented in real-time on the display.

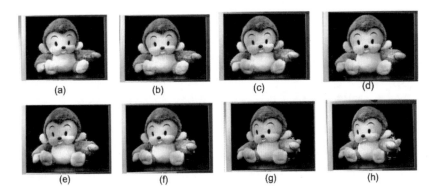

Figure 27. The viewer-tracking-based 2D/3D switchable autostereoscopic display. (a) 35∘image content, (b) 40∘image content, (c) 45∘image content, (d) 50∘image content, (e) 55∘image content, (f) 60∘image content, (g) 65∘image content, (h) 70∘image content.

In this study, tracking stability was good with viewing angles ranging from -15 degrees and 15 degrees. The refresh rate achieved 30 frames per second when the resolution of the capturing image was set to 160×120. To match the resolution of the display, the resolution of image content for each eye was 840×1050. The image content was rendered from the 3D model built from 3D Max or directly captured using cameras. For the webcam coordinates, the accuracy of one pixel was about 6.25 mm according to the viewing angle of the webcam indicated and the designed viewing distance in the autostereoscopic display. Therefore, the rendering or capturing angle was set to 0.5 degree according to the accuracy of one pixel. As shown in Figure 27, the 2D/3D switchable auto- stereoscopic display correspondingly provides about 160 pairs of stereo images to the viewer moving in the viewing angle of the system. The viewer is consistently able to experience the reality of motion parallax.

CONCLUSION

The design of the three dimensional hierarchy with control circuit for large LED backlight array, which effectively reduces the terminal numbers into the cubic root of the total control unit numbers and prevent a block defect of the flat panel. The display panel is divided into many scanning block parts, each part is separately and simultaneously scanned in the same directions to write images on the pixels on the respective scanning electrodes. These defects are generally the result of a failure in the row (horizontal) or column (vertical) drivers or their connections. We have reached the advantages of high accuracy, rapid selection, and reasonable switching speed flat panel.

In full resolution multi-view autostereoscopic display research, we have successfully designed and fabricated the optical system, high density active barrier dynamic LED backlight, the slit pitch is 700um, and the LED chip size is 10×23mil for full resolution multi-view autostereoscopic display. From the measurement results, the dynamic LED backlight optical system can yield ideal parabolic curvature and the crosstalk is lower than 5%. Besides, the lenticular lenses of the lens array optical system was successfully received the light and deflected the light into each viewing zone in a time sequence, which could be one of the candidates for future full resolution time-multiplexed 3D applications.

A viewer-tracking-based auto-stereoscopic display of a synchro-signal LED scanning backlight system that can correspondingly send different pairs of stereo images based on the viewer's position. Additionally, an 8-view autostereoscopic display was implemented with full resolution in the display panel, achieving high 3D image quality in the preliminary configuration. Further modifications, e.g. higher precision for viewer-tracking positioning and design of more viewing zones in the display system, may improve system performance.

A 2R scanning backlight is brighter than a 4R one. Nevertheless, due to better separation of a 4R scanning backlight, the crosstalk of it is less than that of a 2R scanning backlight. However, from the other aspect, the uniformity of a scanning backlight method is usually not as good as than backlight strobe method. For a higher luminance and lower crosstalk, it is suggested to combine the 4R scanning backlight method and backlight strobe method. In this way, a 120Hz LCD can be made a very good performance auto-stereoscopic display with viewer-tracking-based.

REFERENCES

[1]   U. Vogel, L. Kroker, K. Seidl, J. Knobbe, C. Grillberger, J. Amelung, and M. Scholles, "OLED backlight for autostereoscopic displays," *SPIE*, vol. 7237, pp. 72370U-1–9, (2009).

[2]   N. Raman and G. J. Hekstra, "Content Based Contrast Enhancement for Liquid Crystal Displays with Backlight Modulation," *IEEE Trans. Consumer Electron.*, vol. 51, no. 1, pp. 18-21, Feb. (2005).

[3]   Won-Sik Oh; Daeyoun Cho; Kyu-Min Cho; Gun-Woo Moon; Byungchoon Yang; Taeseok Jang;"A Novel Two-Dimensional Adaptive Dimming Technique of X-Y Channel Drivers for LED Backlight System in LCD TVs", *Journal of Display Technology*, Volume: 5 , Issue: 1 , Page(s): 20 – 26(2009).

[4]   M. Doshi and R. Zane, "Digital architecture for driving large LED arrays with dynamic bus voltage regulation and phase shifted PWM," in *Proc. IEEE Appl. Power Electron. Conf. (APEC)*, pp. 287-293 (2007).

[5]   S.-Y. Yseng et al., "LED Backlight Power System with Auto-tuning Regulation Voltage for LCD Panel," in *Proc. IEEE Appl. Power Electron. Conf.*(APEC), pp. 551-557 (2008).

[6]   Y. Hu and M. M. Javanovic, "LED Driver With Self-Adaptive Drive Voltage,"*IEEE Transactions on Power Electronics,* Vol. 23, No 6, pp. 3116-3125,Nov.(2008).

[7]   P.de Greef and H. Groot Hulze, "Adaptive Dimming and Boosting Backlight for LCD-TV System,"*SID Symposium Digest Tech Paper* 38, 1332-1335(2007).

[8]   E.H.A. Langendijk, R. Muijs, and W. van Beek."*Quantifying Contrast Improvements and Power Savings in Displays with a 2D-Dimming Backlight*",IDW(2007).

[9]   Jian-Chiun Liou and Fan-Gang Tseng,"120Hz Display with Intelligent LED Backlight Enabled by Multi-Dimensional Controlling IC , *Displays,* Vol. 30, No. 3, pp. 147-153, 2009.

[10] C. vanBerkel and J.A. Clarke "Characterisation and Optimisation of 3D-LCD Module Design" *Proc SPIE* Vol. 3012, (1997).

[11] S. Pastoor and M. Wopking 3-D displays: a review of current technologies. *Displays* 17, (1997).

[12] L. Lipton Synthagram: autostereoscopic display technology. *Proceedings of the SPIE,* Vol. 4660, (2002).

[13] S.S. Kim, K.H. Cha, J.H. Sung," *3-D Display" SID'02 Digest*, pp.1422,(2002).

[14] Hamagishi, M. Sakata, A. Yamashita, K. Mashitani, E. Nakayama, S. Kishimoto, and K. Kanatani, "New stereoscopic LC displays without special glasses," *Asia Display* '95, pp. 791–794 (1995).

[15] P. Surman, K. Hopf, I. Sexton, W.K. Lee, R. Bates, "Solving the 3D problem - The history and development of viable domestic 3-dimensional video displays", In (Haldun M. Ozaktas, Levent Onural, Eds.), *Three-Dimensional Television: Capture, Transmission, and Display* (ch. 13), Springer Verlag, 2007.

[16] Jian-Chiun Liou and Fo-Hau Chen, "Design and fabrication of optical system for time-multiplex autostereoscopic display", *Optics Express*, Vol. 19, Issue 12, pp. 11007-11017 (2011).

[17]   J. Son and B. Javidi, "Three-Dimensional Imaging Methods Based on Multiview Images," *J. Display Technol.* Vol.:1 Issue:1, pp.125-140 (2005).

[18]   C. van Berkel, D. Parker and A. Franklin, "Multiview 3D LCD," in *Proc. SPIE* Vol. 3012, pp. 32-39, 1996.

[19]   A. Schmidt and A. Grasnick, "Multi-viewpoint autostereoscopic displays from 4D-vision", in Proc. SPIE Photonics West 2002: *Electronic Imaging,* vol. 4660, pp. 212-221, 2002.

[20]   C. van Berkel, "Image preparation for 3D-LCD," in *Proceedings of SPIE,* 1999, vol. 3639, pp.84-91,(1999).

[21]   J. Konrad and P. Angiel, "Subsampling models and anti-alias filters for 3-D automultiscopic displays", *IEEE Trans. Image Processing,* vol.15, no.1, pp. 128-140, Jan. 2006.

[22]   Xiaofang Li, Qiong-hua Wang, Yuhong Tao, Dahai Li, and Aihong Wang,"Cross-talk reduction by correcting the subpixel position in a multiview autostereoscopic three-dimensional display based on a lenticular sheet", *Chinese Optics Letters,* Vol. 9, Issue 2, pp. 021001- (2011).

[23]   Jian-Chiun Liou, Kuen Lee, Chun-Jung Chen, "Low Cross-Talk Multi-Viewer Tracking 3-D Display of Synchro-Signal LED Scanning Backlight System", *IEEE/OSA Journal of Display Technology,* VOL. 7, NO. 8, AUGUST 2011,pp.411-419 (2011).

[24]   J.-Y. Son, V. V. Saveljev, Y.-J. Choi, J.-E. Bahn, and H.-H. Choi, "Parameters for designing autostereoscopic imaging systems based on lenticular, parallax barrier and IP plates*," Opt. Eng.,* vol. 42, no. 11, pp.3326–3333 (2003).

[25]   M.Sakata, G.Hamagashi, A.Yamashita, K.Mashitani, E.Nakayama "3D displays without special glasses by image-splitter method", pp.48-53, *Proc. 3D Image Conference 1995,* Kogakuin University, 6/7 July (1995).

[26]   E. Kurutepe, M. R. Civanlar, and A. M. Tekalp, "Interactive transport of multi-view videos for 3DTV applications,"Journal of Zhejiang University SCIENCE A: *Proc. Packet Video Workshop* 2006, vol. 7, no. 5, pp. 830–836(2006).

[27]   Y. Huang and I. Essa. "Tracking multiple objects through occlusions." In *IEEE Conf. on Computer Vision and Pattern Recognition*, pages II: 1051-1058 (2005).

[28]   F. Jurie and M. Dhome. "Real time tracking of 3D objects with occultations." In I*nt'l Conf. on Image Processing*, pages I: 413-416 (2001).

[29]   Thomas B. Moeslund, Adrian Hilton, and Volker Kruger. "A survey of advances in vision-based human motion capture and analysis." *Computer Vision and Image Understanding,* 104(2):90-126 (2006).

[30]   J. Pan and B. Hu. "Robust occlusion handling in object tracking." In *IEEE Workshop on Object Tracking and Classification Beyond the Visible Spectrum,* pages 1-8 (2007).

[31]   Rajwinder Singh Brar, Phil Surman, Ian Sexton, Richard Bates, Wing Kai Lee, Klaus Hopf, Frank Neumann, Sally E. Day, and Eero Willman," Laser-Based Head-Tracked 3D Display Research", *Journal of Display Technology, Volume:* PP , Issue: 99 , Page(s): 1 − 14(2010).

In: Light-Emitting Diodes and Optoelectronics: New Research   ISBN: 978-1-62100-448-6
Editors: Joshua T. Hall and Anton O. Koskinen   © 2012 Nova Science Publishers, Inc.

Chapter 4

INTERSUBBAND TRANSITION IN CDS/ZNSE QUANTUM WELLS FOR EMISSION AND INFRARED PHOTO DETECTION

S. Abdi-Ben Nasrallah[*], *N. Sfina, S. Mnasri and N. Zeiri*

Unité de Recherche de Physique des Solides, Département de Physique,
Faculté des Sciences de Monastir, Monastir, Tunisia

ABSTRACT

Theoretical study can make a significant contribution to experimental studies and may have profound consequences as regards practical applications of devices.

A great attention has been devoted to the intersubband transition and semiconductor based quantum wells or quantum dots are promising candidates for high-speed intersubband optical devices relying on the quantum confinement of carriers. In fact, the current microelectronics technology does not satisfy the demands of future communication for faster data transfer and a new breakthrough is needed. With the advent of modern epitaxial growth techniques, it has become possible to grow II-VI low dimensional structures and tailor the band to achieve the desired properties for device applications. The band gap engineering allows the design of new optoelectronic devices operating in a specific spectral range. In particular, efforts to improve the optical performance of the CdS/ZnSe SCs has been a major subject of research in the field of II-VI optoelectronic due to the promising applications of this family based light emitters and detectors.

With this motivation, we have proceeded, since a few years, to expand our studies in the areas of compact tunable lasers, infrared photodetectors, and applications in non-linear optics based on CdS/ZnSe system. Various devices are proposed and investigated.

[*] E-mail: samiaabdi@myway.com.

INTRODUCTION

Since the first observation of intersubband transitions (ISBTs) in quantum wells (QWs) by means of infrared spectroscopy [1], they have gained considerable attention. The ISBTs in nanostructures open a new field in fundamental physics and also offer a wide range of potential applications for several optoelectronic devices. Owing to their large values of dipole matrix elements and the possibility of achieving the resonance conditions, the optical processes in these structures are widely investigated [2–5]. The fast relaxation process of ISBT in low-dimensional quantum structures which is one of its technological advantages is involved in several optoelectronic devices such as quantum cascade lasers, infrared detectors and optical modulators [6–8].

However, in order to take advantage of the fast relaxation time for applications in the optical fiber transmission window around λ~1.55μm, ISB transition with high energy is required, so the use of materials with large conduction band offset is needed [9–14].

Many attempts have been made to shorten the ISB transition wavelength. Smet et al. [15] have used InGaAs/AlAs QW while Gmachi et al. [12] have considered a GaN/AlGaN heterostructure. But A great interest has been focused on the intersubband optical transition in wide band gap II-VI semiconductor quantum heterostructures [16–17] which are good candidates for several important optoelectronic devices, such as infrared photodetectors and waveguide switches working in the near-infrared range. For the same reason, Akimoto et al. [13] proposed a ZnSe /BeTe based structure showing an ISBT around 1.6 μm. The ZnSe /BeTe system displays features in structure and banding similar to those of III-V semiconductor compounds. It was developed a few years ago in an attempt to get better lifetime in II-VI optoelectronic devices by means of BeTe and to improve the reliability of ZnSe based devices. The ZnSe/BeTe QW is important for ISBT switches due to its very large conduction band discontinuity of 2.3 eV [18]. Moreover, the higher ionicity of ZnSe leads to an ultra fast carrier relaxation in ZnSe /BeTe QW comparable to that of III-V nitrides. However, the use of ultrafast pulses requires that the optical pulse occupies an extremely short distance in space and this means that the overall the optical device and circuit is very compact. The quantum confinement in thin ZnSe QW make the conduction-band state X of the BeTe barrier lower than the first excited state in the ZnSe well (see Figure 1a). To surmount this difficulty and ensure the ISBT in this structure, we have inserted into the middle of the thin ZnSe well a CdS layer. In fact, CdS II–VI SC has attracted special attention in recent years from both experimental [19–23] and theoretical [24–27] points of view. CdS has an ideal band gap and high absorption coefficient which makes it one of the strong contenders for high efficiency thin-film solar cell devices. The material is also attractive for nonlinear optical devices [28–29], heterogeneous solar cells [30–31] and other optoelectronic devices. At normal temperature and pressure, this material crystallizes in wurtzite crystal structure. However, a cubic CdS structure with a zinc blende lattice can be obtained with the careful control of temperature and pressure [32–35]. This later structure increases the potential applicability of CdS for optoelectronic devices [36–37]. In our work, we have assumed the zinc blende structure of CdS. Although extensive studies on the optical properties of cubic CdS epilayers have been performed to date there have been very few reports on the optical properties of cubic CdS at ambient temperature [20,37–39] and many fundamental physical properties of the CdS-based semiconductor systems are not well understood.

In the present chapter, we pay our attention to ZnSe/CdS/ZnSe QWs surrounded by BeTe barriers, and different CdS/ZnSe based devices are proposed and investigated.

THEORY

The first step of the simulations is to calculate the conduction band profile and to account for the band bending across the active QWs. The electronic states of this system can be obtained using the effective mass theory and the local density approximation [40]. In the one band version of the envelope function approximation, the quantized energy levels E_v for subbands v and their related wave functions Φ_v can be achieved by the self-consistent solution of the one-dimensional Schrödinger and Poisson equations:

In the Schrödinger equation:

$$\left[\frac{-\hbar^2}{2}\frac{d}{dz}\frac{1}{m^*(z)}\frac{d}{dz} + V_H(z) + V_{XC}(z) + V_B(z) - E_v\right]\Phi_v(z) = 0 \tag{1}$$

$m^*(z)$ is the layer-dependent electron effective mass, z is the growth direction, $V_B(z)$ is a stepwise function representing the conduction band offset at the interface, $V_{XC}(z)$ is the local exchange-correlation potential, which accounts for the many body interaction and whose expression is given by the formula [41]:

$$V_{xc}(z) = -0.985\frac{e^2}{4\pi\varepsilon_0\varepsilon}n^{1/3}(z)\left\{1 + \frac{0.034}{a_H^* n^{1/3}(z)}\ln\left[1 + 18.37a_H^* n^{1/3}(z)\right]\right\} \tag{2}$$

The carrier charge density n(z) is deduced for the wave functions of the carriers in the layered quantum system. This is introduced into Poisson's equation as a source term together with the positive charge distribution of ionized donors:

$$\varepsilon_0\frac{d}{dz}\left[\varepsilon_r(z)\frac{d}{dz}V_H(z)\right] = e^2\left[N_D(z) - n(z)\right] \tag{3}$$

where $N_D(z)$ is the total density of ionized doping, ε_0 is the permittivity of vaccum, $\varepsilon_r(z)$ is the relative dielectric constant and e denotes the absolute value of the electron charge. The resulting potential arising from the redistributed charges is obtained by solving Poisson's equation. This potential alters the initial band edge potential with flat bands, and Schrödinger's equation is solved once again for the new total potential energy. This cycle of solving the two differential equations is iterated to convergence.

Note that n(z) is the local density of confined electrons established by the equation:

$$n(z) = \sum_v n_v \left| \Phi_{v,k_z}(z) \right|^2 \tag{4}$$

with n_v the density of electrons for the subband v deduced by the relation:

$$n_v = \frac{m^* k_B T}{\pi \hbar^2} Ln\left[1 + \exp \frac{E_F - E_v}{k_B T} \right] \tag{5}$$

where E_F is the Fermi level verifying $N_D L_d = \sum_v n_v$, L_d being the thickness of the doped zone. Numerically, the problem was treated using the finite differential method.

MODELLING AND SIMULATION OF DEVICES

1. Electronic Properties of ISBT for Fiber Optic Emissions (1.33→1.55 μm)

In view of the fact that the bandwidth required in optical fiber communication systems will exceed 100 Gb s^{-1}, ultrafast optical switching and modulation devices with high efficiency must be developed. Given that intersubband transitions (ISBT) in quantum wells (QWs) are one of the important ultrafast phenomena, we think of ZnSe/BeTe structure in the purpose to achieve ISBT with wavelengths in the range of 1.33-1.55 μm essential for optical fiber communication systems. However, the use of ultrafast pulses requires that the optical pulse occupies an extremely short distance in space and this means that the overall the optical device and circuit is very compact. Nevertheless, the quantum confinement effects in a thin ZnSe QW can raise the Γ conduction-band state of the well (level E_2 in Figure 1.a) above the conduction-band state X of the BeTe barrier (state E_1' in Figure 1.a) leading to an indirect band-gap configuration with a slow relaxation time. This generates a carrier relaxation in two steps: $E_2 \rightarrow E_1'$ then $E_1' \rightarrow E_1$. In this case, the carrier dynamics is controlled by a slow relaxation to the Γ state. This affects negatively the overall relaxation time and the ISBT is not detected [42]. To surmount this difficulty and ensure the ISBT in this structure, we have inserted into the middle of the thin ZnSe well a CdS layer. The structure modeled consists of a few monolayers of CdS embedded into two 0.6 nm ZnSe layers. This region is n-doped and limited by thin BeTe barriers; the whole is surrounded by two 7.5 nm ZnSe layers as spacers. This system combines advantages of both high n-dopability in the CdS/ZnSe and high conduction-band offset at ZnSe/BeTe (ΔE_c=2.3eV). A self-consistent analysis is made to achieve the desired properties and device applications. To achieve our calculation, values of the direct Γ band gaps $E_{g,\Gamma}(ZnSe)$ =2.82 eV and $E_{g,\Gamma}(BeTe)$ = 4.53 eV [43] are used. For band gaps associated with the X valley we have used $E_{g,X}(ZnSe)$ =4.54 eV and $E_{g,X}(BeTe)$ =2.6 eV [43]. For the electron effective masses we have used $m^*_{e,\Gamma}(ZnSe)$=0.19 [44], $m^*_{e,\Gamma}$ (BeTe)=0.19 [44] and $m^*_{e,\Gamma}(CdS)$=0.19 [45], all masses are considered in unit m_0 with m_0 the free electron mass. Note that, because of its direct band gap nature, CdS has the same parameters whether the structure is cubic or hexagonal. The relative dielectric constant $\varepsilon_r(z)$ is equal to 5.4 for ZnSe [46], 4.5 for BeTe [44] and 5.5 for CdS [45]. The total density of donors in ZnSe and CdS layers $N_D(z)$ is taken equal to $1.5 \ 10^{19} \ \text{cm}^{-3}$ and leads to highly n-type doped QWs.

First, we have varied the ZnSe well width for ZnSe/BeTe QW ($\Delta E_c=2.3$ eV) (Figure 1.a) in an attempt to achieve the ISBT with 1.55 µm. However, in the above structure, achieving the ISBT and attaining the range of wavelengths essential in optical fiber communication systems are competitive processes. In fact, by decreasing the ZnSe well thickness sufficiently in order to drop the two Γ subbands E_1 and E_2 below the X one $E_1^{'}$ so that the ISBT can appear, the emission energy (the separation between the E_1 and E_2 levels) $\Delta E= E_2-E_1$ becomes much lower than 0.8 eV and the shortest ISBT wavelength due to an optical transition between the first and the second levels E_1 and E_2 is 2.4 µm. On the other hand, the ISBT can occur in the CdS/ZnSe QW [47] which interface band offset is $\Delta E_c=0.8$ eV, but we can never reach an emission energy ΔE higher than 0.4 eV i.e. a wavelength shorter than 2.9 µm. Taking account these considerations, we thought of using the CdS/BeTe QW. Unfortunately, a trial growth of such a structure has shown a degraded surface. Consequently, we have opted for the (CdS/ZnSe)/BeTe structure (Figure 1.b). ZnSe layer inserted between BeTe and CdS is imperative since it enhances the interface smoothing and improves the structural quality of devices. It is essential to mention the importance of the BeTe barrier in the structure. In fact, for the (CdS/ZnSe)/BeTe QW, the energy of the first subband is situated below the ZnSe conduction band edge while this first level is located near the midgap state unexpectedly incorporated in the BeTe barrier for ZnSe/BeTe QW. R. Akimoto et al. [42] have explained that this midgap state in BeTe compensates the carrier in the well and avoids the ISBT with aimed range wavelengths in ZnSe/BeTe QW.

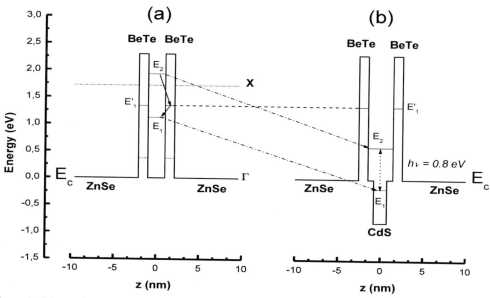

Figure 1. Schematic of band alignment (a) ZnSe/BeTe and b) (CdS/ZnSe)/BeTe quantum wells. The introduction of a CdS layer in ZnSe/BeTe heterostructure can eliminate the slow carrier relaxation

process of Γ(ZnSe)-X(BeTe) transfer which is observed in ZnSe/BeTe QW (a). Strong ISB transition at the wavelength of 1.55 μm (ΔE=0.8 eV) is shown.

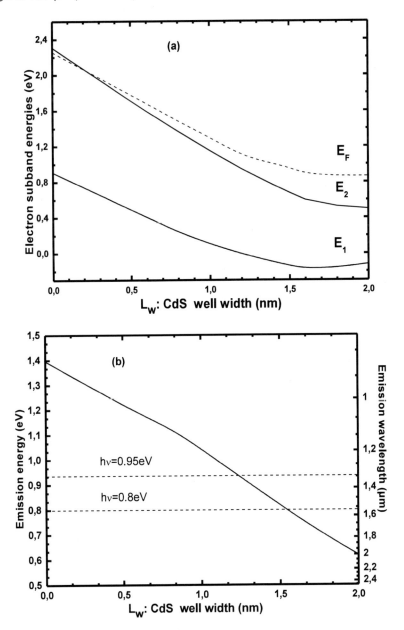

Figure 2. (a) Confinement energies of electron states E_1 and E_2 as well as Fermi level (eV) **(b)** Emission energies and emission wavelengths (eV) versus the CdS well width L_w (nm).

By varying the CdS QW thickness in our structure, we can tailor the band alignment achieving ISBT emission in this range. The results of numerical calculations for the energy electron levels E_1 and E_2 as well as the Fermi level versus the CdS well width are presented in Figure 2a). The confinement energy decreases when L_w increases and as expected, larger confinement can be achieved in thin QWs. However, for very thin CdS wells (L_w lower than

0.2 nm), only the fundamental Γ subband of ZnSe is occupied by electrons. For a larger CdS well, the first excited Γ subband of ZnSe is more and more populated. The decrease of E_1 and E_2 levels makes them lower than the conduction-band state X of the BeTe barrier and the fast direct ISBT Γ-Γ involving E_1 and E_2 wave functions occurs indicating that the Γ(ZnSe)-X(BeTe) electron transfer is suppressed.

To investigate this ISBT, the CdS well width-dependent emission energy and emission wavelengths λ are calculated. As can be seen in Figure 3, the emission energy decays from 1.4 eV without the CdS well to 0.6 eV (0.83 to 2 μm) covering the optical fibre communication wavelengths (1.33→1.55 μm). To make use of this wavelength range, the CdS layer thickness increases from 1.20 to 1.54 nm.

The scheme potential of this design for 1.55 μm (0.8 eV) wavelength emission is shown in Figure.1.

To study and carry out the quantum efficiency of our design, we have calculated the oscillator strength of the transition which is defined by the expression [48]:

$$f_{i\to f} = \frac{2m_0}{\hbar^2}(E_i - E_f)\left|\langle\phi_i|z|\phi_k\rangle\right|^2 \tag{6}$$

where $(E_i\text{-}E_f)$ is the energy difference between the initial and finite states and $\left|\langle\phi_i|z|\phi_f\rangle\right|$ is the dipole matrix element of the transition $\left|M_{fi}\right|$.

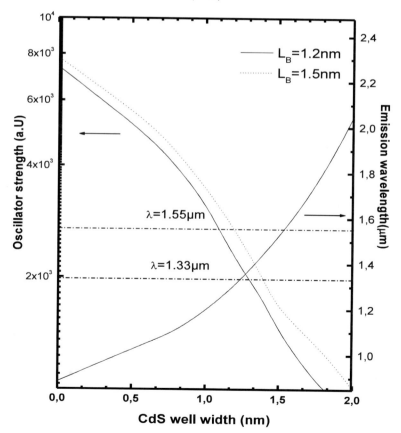

Figure 3. The oscillator strength (for two BeTe barrier widths I_B) and the wavelength λ (μm) versus CdS well width L_w for intersubband transition $E_1 \rightarrow E_2$.

In Figure 3, we have represented the emission wavelengths λ and the oscillator strengths for ISBT $f_{E1 \rightarrow E2}$ versus the CdS well width for two BeTe barrier thicknesses. As can be seen, $f_{E1 \rightarrow E2}$ decreases as L_w increases. But it is shown that this decrease is not significant and the transition oscillator strengths remain relatively high. This means that the efficiency of radiative recombination in active material of (CdS/ZnSe)/BeTe heterostructure is highly preserved (2.707 10^3 and 1.976 10^3 respectively for 1.55 and 1.33μm) and the quantum efficiency of luminescence is good. For the desired emission wavelength, we can obtain from the figure the optimum parameters of the modelled design. For wavelengths solicited in optical communication and covering the range 1.33$\rightarrow$1.55 μm, thicknesses values between (1.5–1.6) nm and (1.2–1.5) nm are needed respectively for CdS well and BeTe barrier.

Figure 4 shows the conduction band, electron energy levels E_1 and E_2 with their related wavefunctions for our modeled device with λ=1.55 μm (ΔE=0.8 eV).

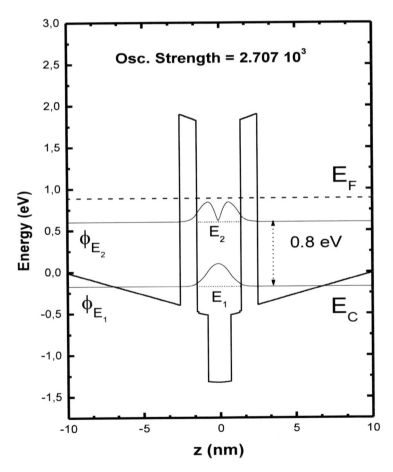

Figure 4. Band structure and wave functions of the two first states for modeled (CdS/ZnSe)/BeTe QW.

2. ABSORPTION COEFFICIENT OF ISBT AT 1.55μM IN (CDS/ZNSE)/BETE QUANTUM WELLS

Several theoretical analyses on the changes of the absorption coefficient associated with intersubband optical transitions in single-quantum wells and multiple-quantum well structures are presented in literature [49–51]. Electroabsorption modulators have been widely used in fiber optic communication systems, because of their small size, low driving voltage and high modulation efficiency. In addition, due to matching of material systems, electroabsorption modulators can be easily integrated with optical components, such as semiconductor lasers, semiconductor optical amplifiers, and attenuators [52].

On the way towards a quantum well based device operating at room temperature with 1.55μm, our investigation focuses on ZnSe/CdS/ZnSe well surrounded by BeTe barriers. The electrical and optical properties are investigated and the absorption coefficient is studied for photon energies around the band gap. The ISBT between the fundamental level and the first excited one in ZnSe/CdS/ZnSe QW can be tuned from near infrared (IR) to middle infrared region of the spectrum covering the optical fibre communication wavelengths (1.33-1.55 μm). This makes this structure flexible to extend application such as IR detector and laser based on II-VI compounds in a wide wavelength range.

Since the accurate subbands, the wave function confined in the channel and the Fermi level are obtained by Schrodinger-Poisson calculations, we can use them to estimate the intersubband optical absorption spectra. We consider the ISBT between the lowest two energy levels only. The intersubband absorption coefficient $\alpha(\omega)$ can be expressed as [53, 54]:

$$\alpha(\omega) = \frac{\omega \mu c e^2}{n_r} \left| M_{fi} \right|^2 \frac{m^* k_B T}{L_{eff} \pi \hbar^2} \times \ln \left\{ \frac{1 + \exp[(E_F - E_i)/k_B T]}{1 + \exp[(E_F - E_f)/k_B T]} \right\}$$

$$\times \frac{\hbar / \tau_{in}}{(E_f - E_i - \hbar\omega)^2 + (\hbar / \tau_{in})^2} \tag{7}$$

where μ denotes the permeability of the well material, n_r is the refractive index, c is the speed of light in free space, L_{eff} the effective width of well, E_i (E_f) is the initial (final) state energy, τ_{in} is the relaxation time, ω is the angular frequency of optical radiation, E_F is the Fermi energy and $\left| M_{fi} \right|$ is the dipole matrix element.

It is assumed that the incident electromagnetic wave propagates in the layer plane with an electric field vector parallel to the growth axis of the structure, which satisfies the intersubband optical transition. By considering an external field to investigate the electro absorption effect, a term -eF is added to the total potential in equation 1 where F the applied external field.

The structure considered here consists on a undoped (0.6nm) ZnSe/CdS/(0.6nm) ZnSe QW surrounded by (1.2nm) BeTe barriers. The thickness L_w of CdS QW is treated as a parameter. We have solved the electron states and calculated the linear intersubband optical absorption coefficient for ZnSe/CdS/ZnSe strained single QW. The relaxation time is $\tau_{in} = 0.1$ ps.

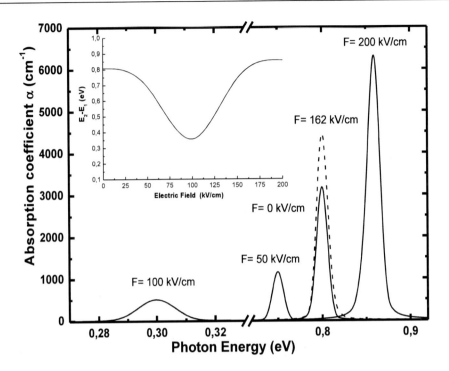

Figure 5. The variation of the absorption coefficient (cm^{-1}) with the photon energy for different values of applied electric field. The inset shows the transition energy E_2-E_1 as a function of applied electric field.

The calculated energies and wavelengths corresponding to the transition between the fundamental and the first excited states are investigated as a function of the CdS QW width L_w. The ISBT in ZnSe/CdS/ZnSe QW can be tuned from near infrared (IR) to middle-IR region of the spectrum covering the optical fibre communication wavelengths (1.33-1.55 µm). The aimed wavelength 1.55 µm (0.8 eV) is achieved with CdS QW width of 1.6 nm at N_D=1.5 10^{19}cm^{-3}. Furthermore, owing to the overlap of the fundamental and the first electron wavefunctions, intra-layer transitions accompanied by optical absorption are possible. Hence, it is essential to examine how these optical transition events contribute to the absorption in such a novel QW system. We apply the equation (7) to calculate the optical absorption coefficient induced by E_2-E_1 intersubband transition and the applied electric field effect on the intersubband optical absorption is studied. Results of absorption coefficient as a function of photon energy are shown in Figure 5 for the system with L_w=1.6 nm.

When an electric field is applied, the conduction band will changes its value and thus leads to variations in the subband states near the interface. It is interesting to note that, for small applied electric field range as much as a critical value of around 100kV/cm, the ITSB optical absorption peak shows red shift with the increasing applied electric field corresponding to the decrease of transition energies between the lowest two levels. We can explain this behaviour as follows: by increasing applied electric field the QW becomes sharper and the electrons moves towards the left side of the well region. The critical electric field is important due to the high conduction-band offsets in this work, ΔE_c=2.3eV at ZnSe/BeTe interface and ΔE_c=0.8eV at CdS/ZnSe interface. However, for high applied electric field range, the absorption peak shifts to the blue region with increasing electric field.

In fact, the high applied electric field moves the electrons to the same side of the well for all states and depletion or saturation will occur for the channel electrons. As a result, the energy difference E_2-E_1 for the structure is larger owing to the much deeper triangular well and leading to a blue shift of the absorption peak. The same tendency is also exhibited in the inset, which gives the transition energy E_2-E_1 dependence of applied electric field. By taking into account this behaviour, it is possible to increase or decrease the energy difference E_2-E_1 by means of applied the electric field. Moreover and as seen in the figure, by increasing the applied electric field the intensity of absorption spectra decreases up to a critical value and then increases when the electric field is further increased. For structures without applied electric field but with different barrier width L_b, energy difference between the lowest two subbands augments when L_b decreases as shown in Figure 6. The absorption coefficient peak increases and shows a blue shift, this behaviour is similar that when the electric field over the critical value increases. It implies that for large value of applied field, control coming from electric field is similar to that which derives from barrier width.

It is interesting to note that the transition energy E_2-E_1 does not change for low values of electric field and high values of barrier width, the variation of coefficient absorption magnitude is mainly due to the dipole matrix element M_{fi}. In Figure 7, the calculated linear absorption coefficient of the ZnSe/CdS/ZnSe QW is plotted as a function of the electron energy for different electron concentrations N_D. We find that the value of the absorption peak is enhanced if the electron density increases. This can be completely understood based on equation (7). The physical reason is because the Fermi level E_F increases with the electron density increasing. Taking into account that the optical absorption coefficient mainly depends on $\left|M_{21}\right|$ and the position of E_F relative to E_1 and E_2 and owing to the much deeper triangular well, electron confinement is enhanced and distance between discrete levels is enlarged. Hence the magnitude of the absorption peak should increase if increasing the electron density and shows a blue shift.

In view of Figures 5, 6 and 7, the ISB absorption covers the 1.55µm wavelength within its spectral half width. One may note that the full widths at half maximum (FWHM) about 40 meV are much narrower than that given by B. S. Li et al. [17] whose found about 90 meV. The spectral width may be influenced by several factors. One of them is the important carrier confinement due to the parabolic approximation used in the calculations. In fact, it is expected that the effects of non-parabolicity are small in a wide gap material; in addition, in vicinity of Γ point with zero wave vector, we can disregard these effects. One can also note that the huge conduction band offset of 3.1 eV providing better confinement make the FWHM in II-VI materials thinner than the results previously reported in III-V material systems [55].

3. ELECTRO-OPTIC PERFORMANCE OF (CDS/ZNSE)/BETE BASED QUANTUM WELL INFRARED PHOTODETECTOR

Photodetectors operating in the infra-red region [56–59] and in particular the spectral region 3-5µm have wide ranging applications including thermal imaging, remote gas sensing, medical imaging etc. This wavelength range is not well covered by existing detectors. These photodetectors have been generally based on the inter-band transitions. Superlattice infrared photodetectors using InAs/GaSb heterostructure have been proposed thanks to several theoretical advantages induced by its type II band alignment [1,60].

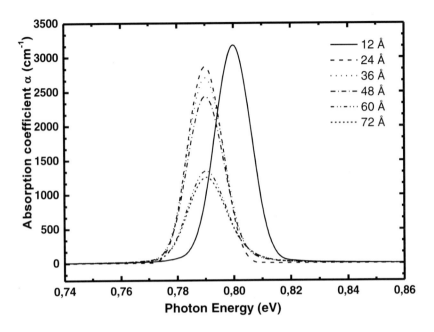

Figure 6. The variation of the absorption coefficient (cm⁻¹) with the photon energy for different barrier widths.

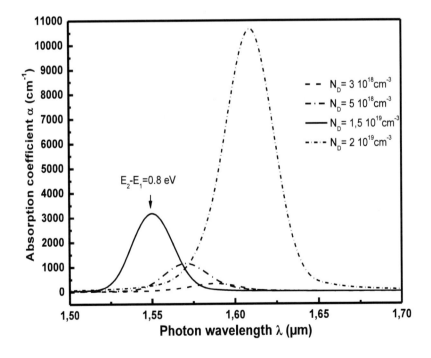

Figure 7. The variation of the absorption coefficient (cm⁻¹) with the photon energy for different values of doping concentrations N_D.

Later, photodetectors based on electron and hole ISBTs in single, multiple quantum wells and superlattices have been studied extensively [61,17]. In single QW, the overlap of the active layer is low due to the limited thickness of the QW layer, but in multiple QWs, strong optical absorption [62] has been observed experimentally. Such structures offer larger flexibility in the design of operating wavelength of the device compared to the photodetector based on the interband transition where the operating wavelength is determined by the energy gap and therefore can be varied only by using alloy semiconductors such as $Cd_xHg_{1-x}Te$. In addition, the functioning wavelength of ISBT in multiple QWs, may be changed by using different quantum well-widths or by using different combinations of well and barrier materials.

One can note that intersubband photodetectors have been designed based on both transitions in the valence as well as conduction band. However, in general, those based on electron transitions show greater detectivity compared to the hole-based photodetectors. One of the reasons for this advantage is their lighter effective mass.

The studies of ISBT in QWs have been mainly concentrated on III-V semiconductor based devices. Nevertheless, the progress in growth technology makes possible the development of low-dimension structures based on II-VI semiconductors. The II-VI-based QWs are promising candidates for ultrafast devices in the mid- or near-infrared regions [41,53]. And specially, (BeTe /ZnSe/CdS/ZnSe /BeTe) is a family technologically little controlled, less than GaAs and even than GaSb/InAs, but have attracted a great interest in last years not only for 1.55 μm photoconduction but also for higher wavelengths.

Our purpose is the study of an infrared photodetector operating at 3.3μm based on a BeTe/ZnSe/CdS/ZnSe/BeTe QW. The QWs and the shapes of the structures are carefully tailored; the absorption coefficient is studied in order to optimize the modeled structure. Then, the performance of the infrared photodetector is tested by the calculation of the dark current.

Once the quantum structure is designed, the electro-optic performances of the infrared detector can be evaluated among other physical properties from the dark current.

The transmission coefficient of electrons is evaluated as a preliminary work to assess the bias-voltage-dependant dark current in our structure. After extracting the self-consistent solutions of the Schrödinger and Poisson equations providing the potential distribution, the distribution of charge carrier density along the structure is again defined. The described method of the Schrödinger equation solver is effective for a charge distribution considering all charge states with the energy above and below the Fermi energy at the injecting contact. However, this method does not allow finding out a probability of tunneling. For this reason, the description of carrier transport by a system of wave function equations have been applied considering the incident and reflected waves in every points of the discrete mesh and including the boundary conditions at the points 1 and N [45]. Thus, the transmission coefficient for tunneling is given by:

$$T(E_z, k) = \frac{k(N)}{k(1)} |B|^2$$

(8)

This transmission coefficient includes 3D to 2D from injecting contacts to the active part of this devise.

In dark conditions, the current density J flowing through the device is the one injected at the contacts: $J=J_{inj}=J_{dark}$. However, and while the above current-density is for steady state conditions, the continuity equations deal with time-dependent phenomena such as low-level injection, generation and recombination. The dark current in the photodiodes investigated can be given by the sum of several components, namely $J= J_{GR} +J_C+J_D$ where J_{GR} is the generation-recombination current the result of carrier generation inside the depletion region, J_C is the conduction current density and J_D is diffusion current due to thermal generation in the n regions. Thus, the net change of carrier concentration is the difference between generation and recombination, plus the net current flowing in and out of the region of interest. The continuity equation is [22]:

$$J = \frac{\partial n_i}{\partial t} = G_{ni} - \frac{n_i - n_I}{\tau_{ni}} + n_i \mu_{ni} \frac{\partial \xi_i}{\partial z_i} + \mu_{ni} \xi_i \frac{\partial n_i}{\partial z_i} + D_{ni} \frac{\partial^2 n_i}{\partial z_i^2} \qquad (9)$$

where n_i is the electron free carrier concentration in the ith mesh point (i=1…N, N is the number of mesh points), G_{ni} is the electron generation rate, n_I is the electron free carrier concentration at thermodynamic equilibrium, $\mu_{ni} = \dfrac{e\tau_{ni}}{m^*_{ie}}$ is the electron mobility, $D_{ni} = \dfrac{k_B T}{e} \dfrac{e\tau_{ni}}{m^*_{ie}}$ is the electron diffusion coefficient, τ_{ni} is the electron recombination life time and $\xi_i = \dfrac{\partial V_{H,i}}{\partial z}$ is the electric applied field along the z direction, V_H being the potential. In the absence of the optical excitation with photons, $G_{ni} = 0$.

The charge accumulation in the quantum wells is non-uniform and defines the potential distribution in the structure which is evaluated by the Poisson equation.

Crystals with cubic (zinc blende) structure are analyzed due to the easy fabrication by cleaving. In the analysis, the transition energy and the dipole moment are necessary for calculations of absorption coefficient. The modeled designs are analyzed with changing the number of the QW as well as their size. Each QW period contains compressively strained CdS/ZnSe. The active region of these structures is based on nine compressively strained CdS/ZnSe QWs with 5nm thick ZnSe layers, the thickness of CdS wells are parameters. BeTe barrier layers to optical confinement are 20 nm thick. For a single QW, the basic problems of carrier collection and quantum well band filling have made clear the limits on single quantum-well laser and how these can be overcome with multiple quantum-well active regions. In fact, for a multiple QW structure, the step-like density-of-states can improve the structure characteristics. The electron wave function shows an important carrier confinement in the active region. Such graded quantum wells reduces the lack of restrictions expansion of carriers out the active region, therefore, the wave function overlap is improved compared to the structure with a square ZnSe QW. By varying the CdS QW thickness in our structure, we can tailor the band alignment achieving the intersubband transition with 3.3μm. To this end, we have first calculated the subband energies for the structure by a self consistent calculation involving the Schrödinger function based on the effective mass approximation and the

Poisson equation. Figure 8 shows the intersubband transition E_2-E_1 for the 3.3µm detection as a function of CdS well size. The aimed wavelength is achieved with 2.7 nm CdS well width.

The numerical calculations provide the shame potential of the multi-QWs investigated system at room temperature (300 K) as well as the fundamental and the first electron energy levels with their relative wave function (Figure 9). The overlap $\left|\left\langle\Psi_1\middle|\Psi_2\right\rangle\right|^2$ equal to 43.8% makes possible intra-layer transitions accompanied by optical absorption.

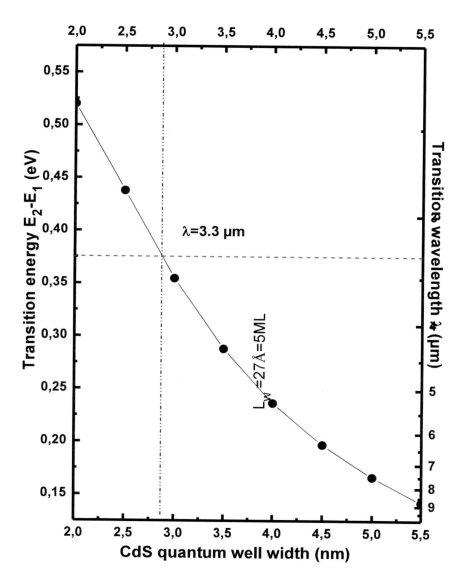

Figure 8. Energy and wavelength for intersubband transition E_2-E_1 versus CdS well width. The 3.3 µm (0.375 eV) required wavelength is achieved for a 2.7 nm well width.

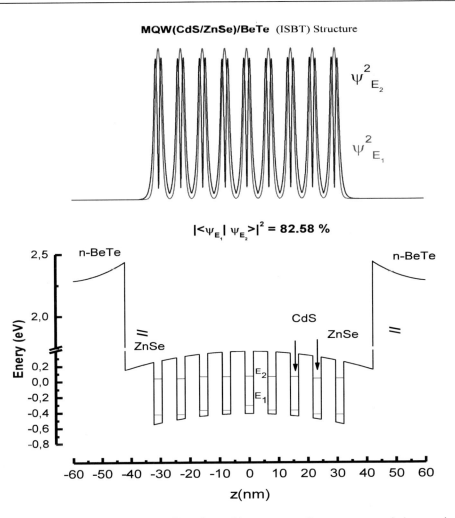

Figure 9. Calculated conduction band profiles of a multi quantum wells structure consisting on nine CdS QWs separated by ZnSe inner barriers. The whole is surrounded by two BeTe barriers under $N_D=10^{18}$ cm^{-3} carrier injection. Also are shown in the same figure the fundamental E_1 and the first level E_2 levels as well as the square wavefunctions $\psi_{E_1}^2$ and $\psi_{E_2}^2$.

Let us examine how these optical transition events contribute to the absorption in this multi-QWs system. We apply the equation (7) to calculate the linear absorption coefficient in our structure. Results for the system with $L_w=2.7$ nm are illustrated in Figure 10a) as a function of photon energy for different n-doping concentration of BeTe barriers. We can see a shift toward shorter wavelength for high doping concentrations ($N_D > 3.10^{18}$ cm^{-3}) and the value of the absorption peak is enhanced if the electron density increases.

This can be completely understood based on equation (7). The absorption pick is accentuate with a full width at half maximum (FWHM) of about 0.16 µm. When inspecting the electric field effect on the absorption coefficient, we mention from Figure 10 b), that the intensity of absorption spectra increases when increasing the applied electric field, no shift of the peak position is obtained. Two other transitions have been detected, with the same device parameters, around 4.5 and 6.5 µm but with very low absorption coefficient (in the inset).

In the following, we will discuss the performances of the modelled infrared detector plotted in Figure 9 and we will focus on the dark current. We have first evaluated the transmission coefficient of electrons as a preliminary work to assess the current; results are illustrated in Figure 11.

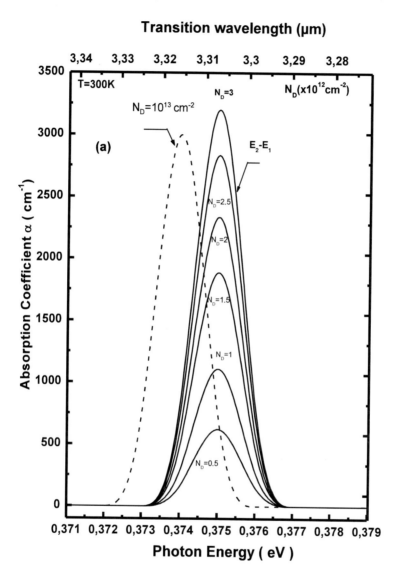

Figure 10a. The variation of the absorption coefficient (cm^{-1}) in BeTe/(multi-QWs CdS/ZnSe)/BeTe structure with the photon energy (a) for different values of BeTe barriers doping concentrations N_D.

The dark current density bias is illustrated in Figure 12 for different temperatures on 100 μm diameter detectors. The temperatures are ranging from 77 to 300 K. First, and as can be seen in Figure 12, we note an important decrease in current with the temperature. This reduction of the dark current at low temperatures is due to a decrease in thermionic emission from the doped conduction-band quantum wells. The dark current density scarcely exceeds 10^{-2} A/m^2 at 77 K. At room temperature, dark current density increased by about one order of magnitude and is around 10^{-1}A/m^2 confirming the good quality of the modelled detector ($I<10^{-9}$A). We note that the transmission coefficient and then the magnitude of this dark current strongly depend on the quantum well structure, i.e., barrier height and well width. And, at low bias, most of the carriers escape from the quantum well over the top of the barrier via thermionic emission; but, at higher bias, most of the carriers' tunnel through the barrier even for high values of barrier width due to the electric-field induced barrier lowering.

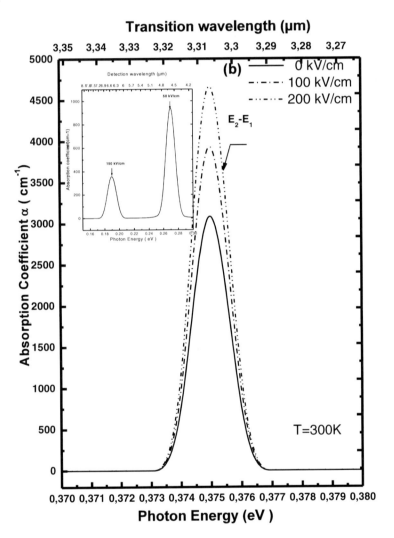

Figure 10. The variation of the absorption coefficient (cm^{-1}) in BeTe/(multi-QWs CdS/ZnSe)/BeTe structure with the photon energy (**b**) for different values of applied electric field. Other transitions have been detected, with the same device parameters, around 4.5 and 6.5 μm but with very low absorption coefficient (inset of figure b).

To conclude on electrical characterization, analysis of the dark current contributions was made on the previous J(V) curve in order to identify the bias dependent dominant current components. Three types of theoretical mechanisms were taken into account: conduction current, diffusion current, and generation-recombination current. The comparison between the different curves illustrated in Figure 13 proves that dark current density is dominated by GR current in the reverse bias; in the low forward bias, the diffusion component is dominant while conduction current contribution becomes predominant above 500 mV.

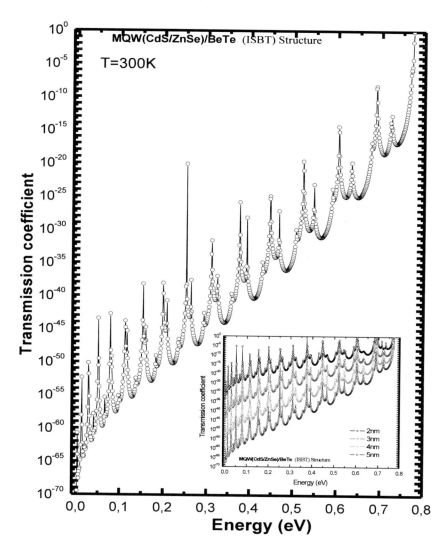

Figure 11. The transmission coefficient of electrons for the BeTe/ (MQWs ZnSe /CdS /ZnSe) / BeTe structure.

4. ABSORPTION COEFFICIENT AND REFRACTIVE INDEX CHANGES IN (CDS/ZNSE/BETE) QUANTUM WELLS (NONLINEAR PROPERTIES)

The linear absorption coefficient in the transition from a state 1 to a state 2 is already calculated. However, the non linear effects in asymmetric based QWs are at the base of a great number of potential applications for optical communication, sensing and computing. In fact, calculation provides that in addition to the incident optical frequencies, new frequencies corresponding to their sum or their difference can be generated if the photons energy is close to that of a transition of the material. For the low values of light intensity, the susceptibility and so the optical absorption is independent of the light intensity I. Nevertheless, when the electric field is intense, i.e. for high optical intensity, optical absorption becomes a function of the electric field. The third order nonlinear absorption coefficient $\alpha^{(3)}$ is given by the expression [26]:

$$\alpha^{(3)}(\omega, I) = -2\omega \sqrt{\frac{\mu_r}{\varepsilon_r}} \left(\frac{I}{\varepsilon_0 n_r c}\right) \frac{|M_{21}|^4 \rho_s \hbar \Gamma_0}{\left[(E_{21} - \hbar\omega)^2 + (\hbar\Gamma_0)^2\right]^2}$$

$$\times \left\{ 1 - \frac{|M_{22} - M_{11}|^2}{4|M_{21}|^2} \right. \tag{10}$$

$$\left. \frac{\left[(E_{21} - \hbar\omega)^2 - (\hbar\Gamma_0)^2 + 2(E_{21})(E_{21} - \hbar\omega)\right]}{(E_{21})^2 + (\hbar\Gamma_0)^2} \right\}$$

Note that in the case of a centro-symmetrical material there is not non-linearity of the second order. Thus, the total absorption coefficient can be expressed as:

$$\alpha(\omega, I) = \alpha^{(1)}(\omega) + \alpha^{(3)}(\omega, I) \tag{11}$$

where n_r denotes is the refractive index.

Another attractive optical property is the Kerr nonlinearity that is the refractive index dependence on the light intensity. For an optical Kerr material, the refractive index n can be written as $n = n(I) = n^1 + n^3(I)$; where n^1 is the linear refractive index, n^3 is the nonlinear refractive index (proportional to the real part of the third-order optical susceptibility and I is the light intensity. The CdS based asymmetric quantum wells exhibit a resonant third-order optical nonlinearity that can be enhanced by operating at wavelengths near the ISBT energy. Guided-wave switching and modulation can be achieved by perturbing the optical absorption coefficient and refractive index with an electric field. Carrier-induced refractive index changes are also important for laser design as well as new optical probing techniques.

The linear and third non linear refractive index changes are given by the following analytical forms [26]:

$$\frac{\Delta n^{(1)}(\omega)}{n_r} = \frac{1}{2n_r^2 \varepsilon_0} |M_{21}|^2 \rho_s \left[\frac{E_{21} - \hbar\omega}{(E_{21} - \hbar\omega)^2 + (\hbar\Gamma_0)^2}\right] \tag{12}$$

and

$$\frac{\Delta n^{(3)}(\omega)}{n_r} = \frac{\mu_r c}{4n_r^3 \varepsilon_0} |M_{21}|^2 \left[\frac{\rho_s I}{(E_{21} - \hbar\omega)^2 + (\hbar\Gamma_0)^2}\right] x$$

$$\left[4(E_{21} - \hbar\omega)|M_{21}|^2 - \frac{(M_{22} - M_{11})^2}{(E_{21})^2 + (\hbar\Gamma_0)^2} x \right. \tag{13}$$

$$\left. \left[(E_{21} - \hbar\omega)\left\{E_{21}(E_{21} - \hbar\omega) - (\hbar\Gamma_0)^2\right\} - (\hbar\Gamma_0)^2(2E_{21} - \hbar\omega)\right] \right]$$

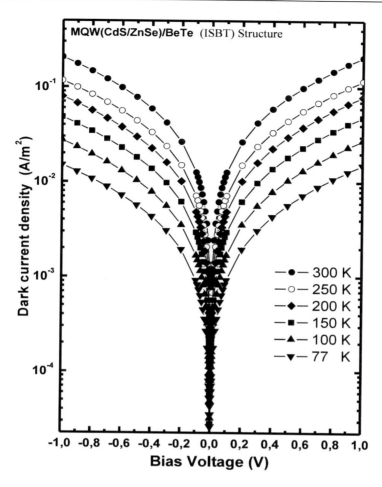

Figure 12. The bias-dependent dark current I(V) for different temperatures on 100 μm diameter BeTe/ (MQWs ZnSe/CdS/ZnSe)/BeTe detectors. The temperatures are ranging from 77 to 300K.

Therefore, the total refractive index change RIC can be written as:

$$\frac{\Delta n(\omega)}{n_r} = \frac{\Delta n^{(1)}(\omega)}{n_r} + \frac{\Delta n^{(3)}(\omega)}{n_r} \qquad (14)$$

The modeled structure (Figure 14) consists on a deep quantum well QW1 (ZnSe (0.6nm) /CdS/ZnSe (0.6nm)) and a quantum well ZnSe (QW2) not very deep coupled by a very thin BeTe barrier (0.6nm), the whole is surrounded by two BeTe barriers of 10nm width. The thicknesses of CdS layer and ZnSe QW2 are treated as parameters, their variations led to the adaptation of the band alignments and the realization of the desired transition for the communication by optical fibers with 1.55μm wavelength. In our calculations, the magnetic permittivity μ_r for CdS is deduced from the relation $n_r = \sqrt{\varepsilon_r \mu_r}$ taken equal to 2.38 and ρ_s is taken equal to 3.10^{16} cm^{-3}.

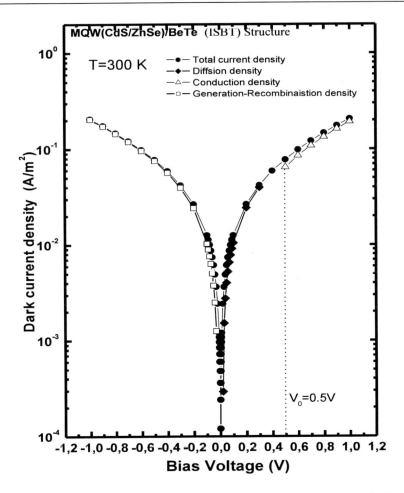

Figure 13. The bias-dependent different contribution of the dark current density J(V) for BeTe/(MQWs ZnSe/CdS/ZnSe)/BeTe detectors. The bias-dependent dark current I(V) (on linear scales) at 300K for 100 μm diameter detectors is illustrated in the inset.

Results are illustrated in Figure 14 for band alignment, the fundamental E_{11} and first excited one E_{13} in QW1 as well as the first level E_{12} in QW2. We note that the first X level in BeTe barrier is under E_{13}. The different transitions are schematized in the same figure. The origin is taken at the bottom of ZnSe conduction band. Energy variations E_{11}, E_{12} and E_{13} have been studied as a function of CdS well thickness for different values of QW2 width (not shown). The best confinement is given for ZnSe well width of 4 nm. Let us now examine the ISBTs, E_{11}-E_{12} the transition between the fundamental level from the QW1 and the level from the QW2 and E_{11}-E_{13} the direct transition between the fundamental level and the first excited level from the QW1. The variations of E_{11}-E_{13} and E_{11}-E_{12} transitions with CdS well width are illustrated in Figure 15 a). As shown, the 1.55 μm wavelength (0.8 eV) is approximately obtained with the transition E_{11}-E_{13} for CdS layer width of 1.6 nm. In order to reach the desired energy, we tried to exploit the doping parameter. The transition energy variations E_{11}-E_{13} and E_{11}-E_{12} with the doping concentration N_D are represented on Figure 15 b). It is straightforward to note that the variation of the energy E_{11}-E_{12} is more important than the of E_{11}-E_{13} transition. This last keep the same value of energy until doping concentration

$N_D=9.10^{18}$ then mark a light increase reaching the aimed energy 0.8 eV for approximately $N_D=10^{19} cm^{-3}$.

Now, we proceed to the theoretical determination of the linear absorption coefficient, the non-linear one and the total coefficient in our structure. The linear absorption coefficient $\alpha^{(1)}$ given by equation (7) is represented in Figure 16) as a function of photon energy for different doping concentration.

The peak for photon energy 0.8 eV is obtained with $N_D=10^{19} cm^{-3}$. A red shift is observed when N_D decreases while an increase in doping concentration leads to a blue shift. $\alpha^{(1)}$ variation according to the photon energy for the ISBT E_{11}-E_{13} with photon energy 0.8 eV is illustrated in Figure 17 at different temperatures. For a concentration of doping fixed $N_D=10^{19} cm^{-3}$, the absorption coefficient peak increases with the temperature, it is very accentuated (about 6500 cm^{-1}) at T=300K. The full width at half maximum (FWHM) slightly increases from 0.011eV at 150K up to 0.016 eV at T=300K. The linear absorption coefficients have been determined in the same way for the ISBT E_{11}-E_{12} and E_{12}-E_{13}. The absorption coefficients corresponding to transitions E_{11}-E_{12} and E_{12}-E_{13} are much lower than that corresponding to transition E_{11}-E_{13}.

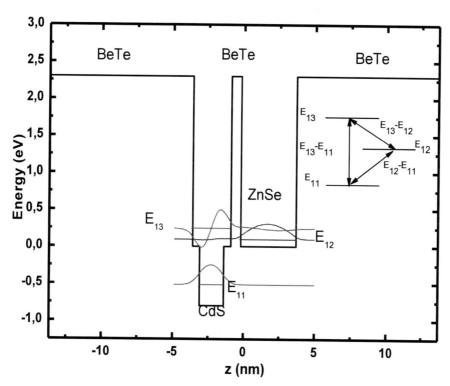

Figure 14. The conduction band profile as well as intersubband transitions in (BeTe: 10nm)/ (ZnSe: 0.6 nm/CdS/ZnSe: 0.6 nm)/ (BeTe: 0.6 nm)/ (ZnSe QW2)/ (BeTe: 10 nm) structure, the width of ZnSe QW2 is equal to 4 nm.

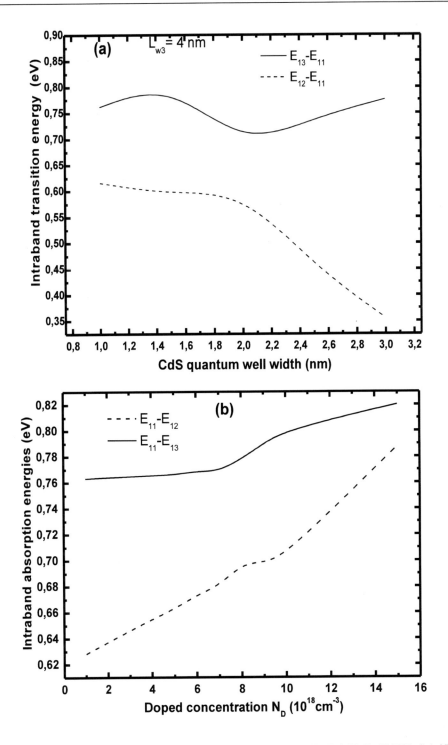

Figure 15. (a) Intersubband transition energies E_{11}-E_{13} and E_{11}-E_{12} in BeTe/ (ZnSe/CdS/ZnSe) / BeTe/ (ZnSe)/ BeTe asymmetric structure as a function of CdS well width (b) Intersubband transition energies E_{11}-E_{13} and E_{11}-E_{12} versus CdS well doping N_D, CdS and ZnSe well thicknesses are 1.6 nm and 4 nm respectively. The 1.5μm (0.8 eV) required wavelength is achieved with the transition E_{11}-E_{13} for CdS layer doping $N_D = 10^{19}$cm^{-3}.

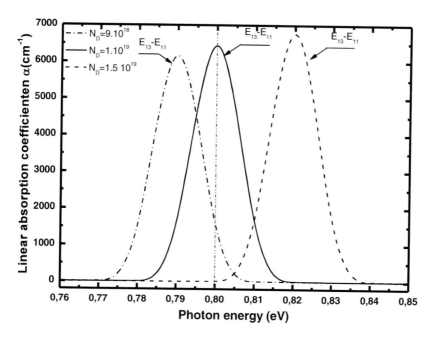

Figure 16. Linear absorption coefficient $\alpha^{(1)}$ (cm^{-1}) of the ISBT E_{11}-E_{13} in BeTe/ (ZnSe/CdS/ZnSe)/ BeTe/ (ZnSe)/ BeTe structure versus the photon energy for different values of CdS layer doping concentrations N_D.

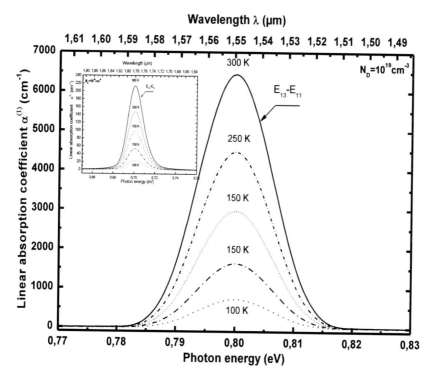

Figure 17. The variation of the linear absorption coefficient $\alpha^{(1)}$ (cm^{-1}) for the ISBT E_{11}-E_{13} with the photon energy for different values of temperature, the doping concentration is fixed N_D=10^{19}cm^{-3}. The linear absorption coefficient for the ISBT E_{11}-E_{12} is shown in the insert.

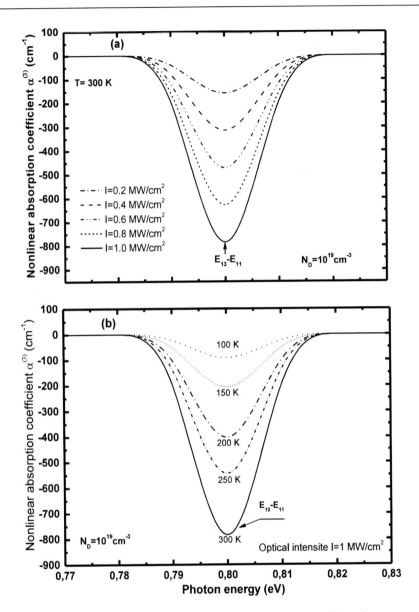

Figure 18. (a) The third absorption coefficient photon energy dependence $\alpha^{(3)}$ (cm^{-1}) for the ISBT E_{11}-E_{13} at different values of optical intensity, with fixed doping concentration $N_D=10^{19}$cm^{-3} and fixed temperature T=300K. (b) The third absorption coefficient photon energy dependence for the ISBT E_{11}-E_{13} at different temperatures with a fixed doping concentration and optical intensity I=1 MW/cm^2.

We proceed now to the evaluation of the third nonlinear absorption coefficient $\alpha^{(3)}$ as calculated from equation (10). The third absorption coefficient photon energy dependence is plotted in Figure 18 a) for the ISBT E_{11}-E_{13} with different values of optical intensity, at fixed doping concentration $N_D=10^{19}$cm^{-3} and fixed temperature T=300K. $\alpha^{(3)}$ is negative and the absorption peak deceases when the optical intensity increases, however, the values taken by this coefficient are much lower than those corresponding to the linear absorption coefficient. We have also studied the variations of the third order absorption coefficient with the photon energy at different temperatures with a fixed doping concentration and optical intensity I=1 MW/cm².

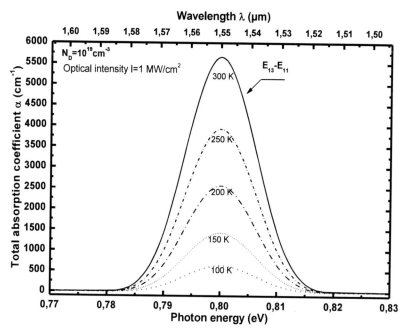

Figure 19. The total absorption coefficient photon energy dependence $\alpha\,(\omega, I)$ (cm⁻¹) for the ISBT E_{11}-E_{13} at different values of temperature with fixed optical intensity I=1 MW/cm² and doping concentration $N_D=10^{19}$ cm⁻³.

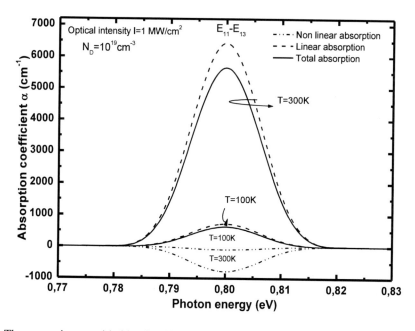

Figure 20. The correction provided by the third nonlinear absorption coefficient is as important as the temperature is high: the linear, the nonlinear, and the total absorption coefficients are shown versus the photon energy at T=100 K and 300 K.

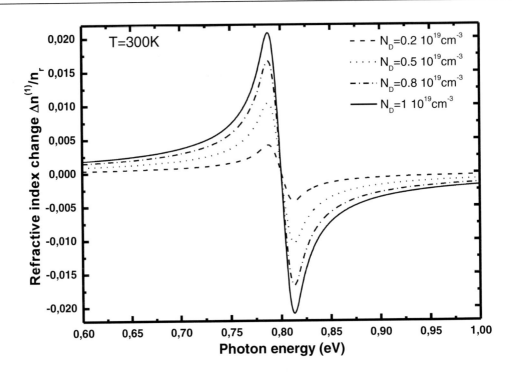

Figure 21. The linear refractive index change $\Delta n^{(1)}/n_r$ as a function of photon energy in the ISBT E_{11}-E_{13} for different doping concentrations at T= 300K.

As shown in Figure 18 b), the nonlinear absorption coefficient decreases as the temperature augments. Then, the total absorption coefficient is deduced, Figure 19 shows $\alpha(\omega,I)$ at different values of temperature with fixed optical intensity and doping concentration. The FWHM increases slightly, from 0.011 eV at T=200 K to 0.017 eV at 300K.

Correction provided by the third nonlinear absorption coefficient is as important as the temperature is high (Figure 20). Thus, at the ambient temperature, this nonlinear effect of absorption is important and must be taken into account.

We pass now to the investigation of the linear and the third-order nonlinear refractive index changes, $\Delta n^{(1)}/n_r$ and $\Delta n^{(3)}/n_r$ as calculated from equations (12) and (13). Figure 21 illustrates the linear refractive index change $\Delta n^{(1)}/n_r$ as a function of photon energy for different doping concentrations, our calculations are made at 300K.

This change in refractive index increases for energies near and below the transition energy E_{11}-E_{13}, and $\Delta n^{(1)}$ becomes negative for energies well above E_{11}-E_{13}. $\Delta n^{(1)}$ augments with doping content in absolute value, this increase is more sensitive in the vicinity of ISBT energy. The variations of the third non linear refractive index change $\Delta n^{(3)}/n_r$ versus the pump photon energy is shown in Figure 22 a) for different doping concentrations N_D and in Figure 22 b) for different light intensities I.

As expected, $\Delta n^{(3)}/n_r$ is negative below the transition energy E_{11}-E_{13} and becomes positive for energies above E_{11}-E_{13}, $\Delta n^{(3)}/n_r$ increases with N_D and with I below the transition energy. In Figure 23) the variations of the total refractive index change $\Delta n/n_r$ are plotted as a function of the photon energy compared with $\Delta n^{(1)}/n_r$ and $\Delta n^{(3)}/n_r$ for a light intensity I=1 MW/cm^2 and a doping concentration $N_D=10^{19}$cm^{-3}.

T= 300K and a doping concentration $N_D=10^{19}cm^{-3}$

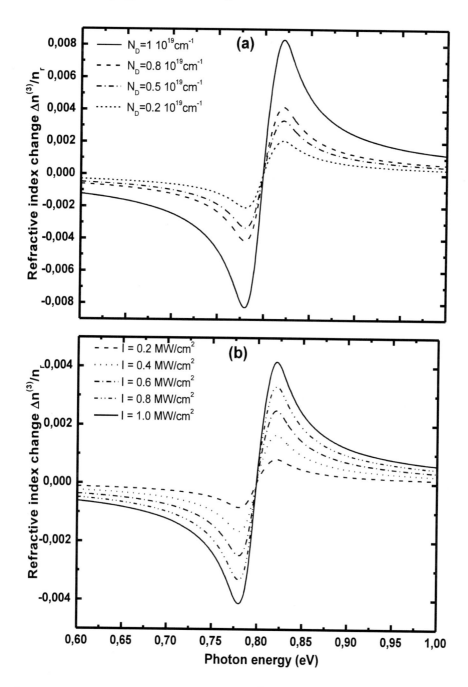

Figure 22. (a) The nonlinear refractive index change $\Delta n^{(3)}/n_r$ as a function of photon energy for different doping concentrations at T= 300K and I=1 MW/cm^2. (b) The nonlinear refractive index change $\Delta n^{(3)}/n_r$ as a function of photon energy for different light intensities I at T= 300K and $N_D=10^{19}cm^{-3}$.

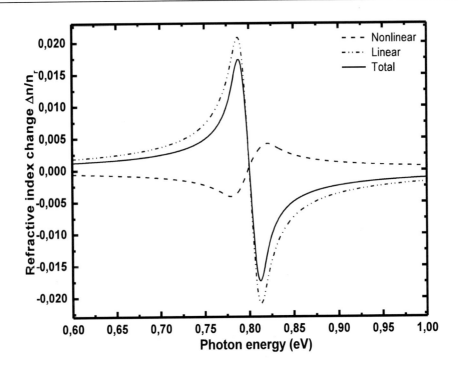

Figure 23. The nonlinear refractive index change $\Delta n/n_r$ as a function of photon energy compared with $\Delta n^{(1)}/n_r$ and $\Delta n^{(3)}/n_r$ for a light intensity I=1 MW/cm^2, a temperature

The largest changes in refractive index are not, like the peak of absorption coefficient at the ISBT energy E_{11}-E_{13} but at photon energies about 1.1meV of the line center. Δn is positive for energies below E_{11}-E_{13}, a result of the increase in absorption coefficient for fixed energies. It is straightforward to notice that $\Delta n/n_r$ just like the absorption coefficient, is mainly contributed by the linear terms. However, and while the linear and the third nonlinear RIC are of opposite signs, $\Delta n^{(3)}/n_r$ reduces the index changes which are as large as 10^{-2}.

CONCLUSION

While intersubband transitions in quantum wells and quantum dots are very important both from a fundamental physics perspective and for device applications, we have paid our attention to CdS/ZnSe based systems. Diverse devices are proposed and investigated. i) CdS/ZnSe/BeTe layered structures which lead to fabrication of light emitting diodes at 1.55 mm wavelength. We have presented a model of ISBT in n-type doped (CdS/ZnSe)/BeTe QWs. Strong ISBT has been demonstrated between the two lowest confined electron states. The optimum parameters lead to 1.55 μm light emission with good quantum efficiency; ii) The intersubband optical absorption in (CdS/ZnSe)/BeTe near-infrared light detectors under an external electric field have been studied. Simulated results including eigenvalues, absorption coefficient and electro-absorption properties have been discussed. The ISBT in ZnSe/CdS/ZnSe QW can be tuned in a wide range of the spectrum covering the optical fibre communication wavelengths (1.33-1.55 μm); iii) A multi quantum well ZnSe/CdS/ZnSe based structure is designed for the ISBT around 3.3 μm. The electro-optic performances of the infrared detector are estimated, the dark current dependence with the applied voltage and temperature are discussed. Our results have shown that the dark current doesn't exceed 10^{-9}A at 300 K, confirming the good quality of the modelled detector and showing its good sensitivity; iv) We note also that, the natural properties of the fast and nonlinear carriers' relaxation in the ISBT make it a promising candidate in the infrared field. We have demonstrated that ISBT at 1.55µm wavelength can be achieved in asymmetric (CdS/ZnSe/BeTe)/ (BeTe/ZnSe) QW with appropriate parameters at ambient temperature. The linear and third-order nonlinear optical properties have been theoretically deduced. The correction provided by the third nonlinear terms is as important as the temperature is high. So, at the ambient temperature, this nonlinear effect must be taken into account. The refractive index change is considerably reduced as the optical intensity increases, consequently, in practice, if we want to obtain a larger change in the refractive index, a relatively weaker incident optical intensity should be adopted. The changes in the index of refraction can be used for informational memory and to fabricate bistable components.

References

[1]   West, L. C; Eglash, S. J; *Appl Phys Lett* 1995, 46, 557.

[2]   Zhang, L; Xie, H. J; *Phys Rev* B 2003, 68, 235315.

[3]   Karabulut, I; Atav, Ü; Safak, H; *Phys Rev* B 2005, 72, 207301.

[4]   Karabulut, I; Safak, H; Tomak, M; *Solid State Commun* 2005, 135, 735.

[5]   Karabulut, I; Safak, H; *Physica* B 2005, 368, 82.

[6]   Faist, J; Capasso, F; Sivco, D; Sirtori, C; Hutchinson, A. L; Chu, S. N. G and Cho, A. Y; *Science* 1994, 264, 553.

[7]   Levine, B. F; Malik, R. J; Walker, J; Choi, K. K; Bethea, C. G; Kleinman, D. A and Vanderberg, M; *Appl Phys Lett* 1987, 50, 273.

[8]   *Intersubband Transitions in Quantum Wells: Physics and Devices,* edited by Li, S. S and Su Y.-K 1998, Kluwer Academic. Dodrecht.

[9]   Iizuka, N; Kaneko, K; Suzuki, N; Azano, T; Noda, S; Wada, O; *Appl Phys Lett* 2000, 77, 648.

[10] Iizuka, N; Kaneko, K; Suzuki, N; *Appl Phys Lett* 2002, 81, 1083.

[11] Akimoto, R; Akita, K; Sazaki, F; Hasama, T; *Appl Phys Lett* 2002, 81, 2998.

[12] Gmachi, C; Frolov, S. V; Ng, H. M; Chu, S-N. G and Cho, A. Y; *Electron Lett* 2001, 37, 378.

[13] Akimoto, R; Kinpara, Y; Akita, K; Sasaki, F and Kobayashi, S; *Appl Phys Lett* 2001, 78, 580.

[14] Abdi Ben Nasrallah, S. Sfina, N and Said, M; *Eur Phys J* B 2005, 47, 167.

[15] Smet, J. H; Peng, L. H; Hirayama,Y and Fonstad, C.G; *Appl Phys Lett* 1994, 64, 986.

[16] Li, B. S; Akimoto, R; Akita, K and Hasama, T; *Physica* E 2006, 34, 276.

[17] Li, B. S; Akimoto, R; Akita, K and Hasama, T; *Appl Phys Lett* 2006, 88, 221915.

[18] Waag A; Fisher, F; Lugauer, H.-J; Liz, Th; Zehnder, U; Ossau, W; Gerhardt, T; Möller, M and Landweher, G; *J Appl Phys* 1996, 80, 792.

[19] Göppert, M; Becker, R; Petillon, S; Grün, M; Maier, C; Dinger, A; Klingshirn, C; *Physica* E, 2000, 7, 89.

[20] Kim, D. J; Yu, Y.-M; Lee, J. W; Choi, Y. D; Applied Surface Science 2008, 254, 7522.
[21] Li, B. S; Akimoto, R and Shen, A; *Appl Phys Lett,* 2008, 92, 021123.
[22] Dinger, A; Petillon, S; Grün, M; Hetterich M and Klingshirn, C; *Semicond Sci Technol,* 1999, 14, 595.
[23] Li, X; Embden, J. V; Chon, J. W. M and Gu, M; *App Phys Lett,* 2009, 94, 103117.
[24] Wei, S.-H; Zhang, S. B and Zunger, A; *J Appl Phys,* 2000, 87, 1.
[25] Deligoz, E; Colakoglu, K; Ciftci, Y; *Physica* B, 2006, 373, 124.
[26] Sfina, N; Abdi-Ben Nasrallah, S; Mnasri, S and Said, M; J Phys D: *Appl Phys* 2009, 42, 000000.
[27] Cong, G. W; Akimoto, R; Akita, K; Hasama, T and Ishikawa, H; *Appl Phys Lett,* 2007, 90, 181919.
[28] Broser, I; Fricke, Ch; Lummer, B; Heitz, R; Perls, H and Hoffmann, A; *J Cryst Growth,* 1992, 117, 788.
[29] Ning, T; Gao, P; Wang, W; Lu, H; Fu, W; Zhou, Y; Zhang, D; Bai, X; Wang, E; Yang, G; *Physica* E 2009, 41, 715.
[30] Yoshikawa A and Sakai, Y; *Solid-State Electron,* 1977, 20, 133.
[31] Ghimpu, L; Ursaki, V. V; Potlog, T and Tiginyanu, I. M; *Semicond Sci Technol,* 2005, 20, 1127.
[32] Choi, I.H; Yu, P; *Phys Stat Sol* (B), 2005, 242, 1610.
[33] Kanemitsu,Y; Nagai, T; Kushida, T; Nakamura, S; Yamada, Y; Taguchi, T; *Appl Phys Lett,* 2002, 80, 267.
[34] Kanemitsu, Y; Nagai, T; Yamada, Y; Taguchi, T; *Appl Phys Lett,* 2003, 82, 388.
[35] Nagai, T; Kanemitsu, Y; Ando, M; Kushida, T; Nakamura, S; Yamada, Y; Taguchi, T; *Phys Stat Sol* (B), 2002, 229, 611.
[36] Hofmann, P; Horn, K; Bradshaw, M; Johnson, R.L; Fuchs, D; Cardona, M; *Phys Rev* B, 1993, 47, 1639.
[37] Senthil, K; Mangalaraj, D; Narayandass, S. K; Adachi, S; *Mater Sci Eng* B, 2000, 78, 53..
[38] Ortuno, M. B; Sotelo-Lerma, M; Mendoza-Galvan, A; Ramirez-Bon, R; *Thin Solid Films,* 2004, 457, 278.
[39] Ninomiya, S; Adachi, S; *J Appl Phys,* 1995, 78, 1183.
[40] Ben Zid, F; Bhouri, A; Mejri, H; Tlili, R and Said, M; *J Appl Phys,* 2002, 91, N 11.
[41] Brey, L; Dempsey, J; Johnson, N. F; and Halperin, B. I; *Phys Rev* B, 1990, 42, 1240.
[42] Akimoto, R; Akita, K; Sasaki, F; and Hasama, T; *Appl Phys Lett,* 2003, 81, N 16, 2998.
[43] Toporov, A. A; Nekrutkina, O. V; Nestoklon, M. O; Sorokin, S. V; Solnyshkov, D. D; Ivanov, S. V; Waag, A; Landwehr, G; *Phys Rev* B, 2003, 67, 113307.
[44] Abdi-Ben Nasrallah, S; Ben Afia, S; Belmabrouk, H; and Said, M; *Eur Phys J* B, 2005, 43, 3.
[45] Rama. M. V; *J Chem Phy,* 1991, 95, 8309.
[46] Madelung, O; *Semiconductors Basic Data,* 2nd revised edn. (Springer, 1996).
[47] Göppert, M; Becker, R; Petillon, S; Grün, M; Maier, C; Dinger, A; and Kligshirn, C; *J Cryst Growth,* 2000, 214/215, 625.
[48] Chen, W. Q; Andersson, T.G; *J Appl Phys,* 1993, 73, 4484.
[49] Ünlü, S; Karabulut, I; Safak, H; *Physica* E, 2006, 33, 319.
[50] Rostami, A; Baghban Asghari Nejad H; Rasooli Saghai, H; *Physica* B, 2008, 403, 2725.

[51]   Rostami, A; Rasooli Saghai, H; Baghban Asghari Nejad, H; *Physica* B, 2008, 403 2789.

[52]   Irmscher, S; Lewen, R; and Eriksson, U; *IEEE Photon Technol Lett,* 2002, 14 923.

[53]   Ahn, D; Chuang, S. L; *IEEE J Quantum Electron,* 1987, 23, 2196.

[54]   Han, X; Li, J; Wu, J; Cong, G; Liu, X; Zhu, Q; Wang, Z; *Physica* E, 2005, 28, 230.

[55]   Gmachl, C; Ng, H. M;. Chu, S.-N. G; and Cho, A. Y; Appl Phys Lett, 2000, 77, 3722 Wagner, J; Fuchs, F; Herres, N; Schmitz, J; Koidl, P; 1995 in: Li, S. S; Liu, H. C; Tidrow, M. Z; Beck, W. A; Singh, A; (Eds.), *Proceedings of the Third International Symposium on Long Wavelength Infrared Detectors and Arrays: Physics and Applications III,* 28–95 The Electrochemical Society, 201.

[56]   Razeghi, M; Mohseni, H; 2002 GaSb/InAs superlattices for infrared FPAs, in: Henini, M; Razeghi, M; (Eds.), *Handbook of Infrared Detection Technologies,* Elsevier, UK, 191.

[57]   Mehdi, I; Haddad, G. I; and Mains, R. K; *Superlatt Microstruc,* 1989, 5 443.

[58]   Loehr, J. P; Singh, J; Mains, R. K; and Haddad, G. I; *Appl Phys Lett,* 1991, 59, (17) 2070.

[59]   Esaki, L; and Sakaki, H; *IBM Tech Discl Bull,* 1977, 20, 2456.

[60]   Wada, O; *New J Phys,* 2004, 6, 183.

[61]   Lu, H; Shen, A; Tamargo, M. C; Song, C. Y; Liu, H. C; Zhang, S. K; and Alfano, R. R; *Appl Phys Lett,* 2006, 89, 131903.

[62]   Karabulut, İ; and Baskoutas, S; *J Appl Phys,* 2008, 103, 073512.

In: Light-Emitting Diodes and Optoelectronics: New Research   ISBN: 978-1-62100-448-6
Editors: Joshua T. Hall and Anton O. Koskinen   © 2012 Nova Science Publishers, Inc.

Chapter 5

INORGANIC-ORGANIC HYBRID EMITTING MATERIAL FABRICATED BY SOLVOTHERMAL SYNTHESIS

Takeshi Fukuda[*]

Department of Functional Materials Science, Graduate School of Science
and Engineering, Saitama University, Saitama, Japan

ABSTRACT

Eu-complex is one of the most interesting lanthanide with the organic ligand owing to its high photoluminescence (PL) quantum efficiency by the ultraviolet (UV)-excitation. Therefore, Eu-complex is expected as the red-emitting phosphor by the UV-light irradiation for the white light-emitting diode. However, the important problem for the practical application is that the poor long-term stability. Nowadays, we demonstrated the improved long-term stability against the UV-light irradiation by encapsulating the sol-gel derived glass network around Eu-complex. The vapor and oxygen are protected by the silica glass layer to react Eu-complex, resulting in the high long-term stability. Especially, the long-term stability was drastically improved by annealing in the autoclave container with high pressure, called as the solvothermal process. By optimizing the silane alkoxide and the solvent in the sol-gel starting solution, we successfully achieved the particle structure of the silica glass coated Eu-complex with the diameter of less than 100 nm. Among the tested combinations of the sol-gel process, the combination of TEOS and ethanol is the best condition to realize the high PL quantum efficiency and the long-term

---

[*] Author has been engaged in a research topic of the coating-technique for Eu-complex using the conventional sol-gel and solvothermal processes. Nowadays, several papers demonstrated about organic-inorganic emitting materials including Eu-complex and the sol-gel derived inorganic materials, such as the silica glass and the titanium dioxide. However, the reported long-term stability of Eu-complex is not sufficient for practical applications, and the controllability of the material structure is not enough. We tried to realize the improved long-term stability of Eu-complex using the solvothermal process. By optimizing the fabrication condition, we successfully demonstrated the improved the optical degradation characteristic of Eu-complex. Therefore, we think that this encapsulation technique via the solvothermal process is useful tool for in a near future. These results are not reached practical levels of many applications; however the further improvement is expected by investigation of many researchers.

stability. Therefore, this organic-inorganic emitting material will be applied as phosphors for practical applications.

1. INTRODUCTION

Since lanthanide compounds have several special advantages as optical and magnetic materials, many researchers have investigated for many applications in recent years. In the case of optical applications, Nd^{3+}-doped yttrium aluminum garnet (YAG) lasers [1], Er^{3+}-doped fiber amplifiers [2], and optical fiber lasers [3] have been already in a practical use. Nowadays, they are used for industrial/medical applications and optical fiber communication systems. In addition, many lanthanide compounds generate bright visible emission by the ultraviolet (UV) and vacuum UV excitations. Therefore, they have been widely investigated as phosphors for plasma displays [4,5], wavelength conversion films for silicon photovoltaic cell [6-8], bio sensors [9-12], and medical imaging detectors [13]. Especially, a practical application for the white light-emitting diode (LED) is interested due to the suitable emission spectrum at a visible wavelength region, the high photoluminescence (PL) quantum efficiency, the high thermal stability, and the excellent long-term stability [14-17]. In the case of the trivalent lanthanide ion, the most prevalent oxidation state in which those elements are present in nature corresponds to $4f^N$ (N=1-14). The shielding of 4f electrons from interactions with their surroundings (ligand-field interaction) by the filled $5s^2$ and $5p^6$ orbital is responsible for the interesting optical properties of the lanthanide ion [18-21].

Nowadays, several methods have been investigated to realize the white LED. A primitive white LED lamp consists of three-kinds of LED chips, which generate red, green, and blue lights, respectively [30]. In many cases, InGaN-based blue/green and AlGaInP-based red LEDs are used due to their high luminance efficiencies and suitable electroluminescence spectra for the white LED. In this method, an arbitrary color rendering index can be easily controlled by changing the emission intensity of the each LED. However, this type of the white LED has not been widely used for many lighting applications owing to the high fabrication cost. This is because both driver circuits and three kinds of LEDs with emission spectra of blue-, green-, and red-wavelength regions need to generate the white light emission. The drive voltages of three-kinds of LEDs are different each other; therefore, the stable white light cannot be realized using only one driver circuit. In addition, electro-luminescence spectra of each LEDs are too narrow for general lighting applications. This fact indicates that this type of the white LED is not suitable for the practical white LEDs.

One approach to solve the above-mentioned problem is a combination of an InAlGaN-based blue-LED and a phosphor, which generates complementary color (yellow) of the blue light from the LED. Currently, $Y_3Al_5O_{12}$ (YAG):Ce^{3+} is commonly used [31,32] for practical applications because of the broad PL excitation (PLE) spectrum at a blue wavelength region, the broad emission peak at a yellow wavelength region, the high PL quantum efficiency, and the high long-term stability against the blue light irradiation. The emission wavelength of YAG:Ce^{3+} phosphor is approximately 540 nm at a room temperature; therefore, the complementary white light can be generated in combination with the blue light from the InAlGaN-based blue LED. However, the disadvantage of this method is the variability of the color rendering index due to difficulties during the manufacturing of accurately controlling the ratio of the blue light transmitted through the phosphor. Also, the delicate control of the color temperature is linked to difficulties in realizing a white LED with the high color rendering index.

In recent years, several researchers demonstrated the enhanced luminance efficiency and the high color rendering index of the white LED by optimizing the device structure of the blue-LED and the crystal structure of the inorganic phosphor [14-17]. As a result, PL and PLE spectra can be controlled for several kinds of phosphors. The color rendering index indicates the ability of the white LED for lighting applications. The improved color rendering index has been demonstrated in combination with the several kinds of phosphors, and the arbitrary white emission from the LED can be obtained by controlling the emission spectrum and intensity from the phosphors. Nowadays, the luminous efficiency of the white LEDs has been already superior that of a conventional incandescent lamp in our daily life, such as general lamps, headlights of cars/motorcycles/bicycles, and backlights of liquid crystal displays. For further improved luminescence efficiency of the white LED, the high PL quantum efficiency with the large excitation spectrum at blue and UV wavelength regions is necessary. Therefore, several types of phosphors have been reported in previous papers [22-29]. The further improved luminance intensity and long-term stability have been required for the practical white LED instead of the fluorescent lamp, and many kinds of phosphors have been demonstrated by many researchers.

One effective method to realize the enhanced color rendering index of the white LED is the usage of blended phosphors with different PL spectra by the blue-light excitation. Recently, a combination of the blue LED chip and several kinds of phosphors was proposed to control the color rendering index of the white LED, and the high color rendering index was realized by optimizing the concentration of these phosphors in white LEDs [33-35]. The excitation light from the blue-LED is absorbed by green- and red-emitting phosphors, and the complementary white light is generated in combination with the blue light from the LED chip and the PL from phosphors. The PL spectrum of this white LED is easily controlled by changing the concentration of phosphors, resulting in the controllability of the color rendering index. However, the long-term stabilities of both phosphors are different by the blue-light irradiation in many cases. This fact cases the emitting white color changes gradually owing to the decreased PL intensity from green- and red-emitting phosphors [36,37].

Another approach is the combination with the UV-LED and several phosphors, which generate the visible emission by the UV-light irradiation. By now, the quad-chromatic white LED lamp consisting of a near UV-LED chip and several kinds of phosphors (orange-, yellow-, green-, and blue-emissions) was also demonstrated [38-40]. It is noted that the advantage of the white LED with the UV-LED excitation is that the arbitrary electro-luminescence spectrum can be realized by optimizing the emission wavelength and intensity of phosphors. This fact indicates that this type of the white LED will be used for the general lighting application in a near future. In general, optical characteristics of the UV-LED are lower than those of the blue-LED, such as the low luminance efficiency, the high driving voltage, the high fabrication cost, and short lifetime [41-43]. The device performances of the UV-LED have been progressed year by year; therefore, the many researchers have investigated novel phosphors for this type of white LED.

In this chapter, we demonstrated Eu-complex-based organic-inorganic hybrid phosphors via sol-gel and solvothermal processes. Eu-complex generates the sharp red emission by the UV light irradiation, and the silica glass layer protects to change the molecular structure of Eu-complex due to its low transmittance of vapor (water) and oxygen. As a result, the long-term and thermal stabilities improves by coating the silica glass layer around the Eu-complex. The silica glass layer is fabricated by the sol-gel and solvothermal processes, and we show the advantage of the solvothermal process compared to the conventional sol-gel process. By optimizing the fabrication condition, the long-term stability against the UV light irradiation was improved by coating the silica glass layer via the solvothermal process. Therefore, the inorganic-organic hybrid red phosphor with the practical long-term stability will be realized using the solvothermal process in a near future.

2. PRINCIPLE AND PROBLEM OF EUROPIUM-COMPLEX

For improved color rendering index and luminance efficiency of white LEDs, the red-emitting phosphor by the UV-light irradiation is necessary. This is because the reported optical characteristics of the red-emitting phosphors are lower than those of blue- and green-emitting phosphors [44,45]. In general, Eu-complex has the high absorption coefficient at the near UV wavelength region and the sharp PL spectrum at the red wavelength region [46-48]. Therefore, Eu-complex is one of the most interesting red-emitting phosphors for white LEDs with the high color rendering index. Eu^{3+} ion generates the rather weak red-emission corresponding to the $^5D_0 \rightarrow {}^7F_2$ transition; however, the combination with an appropriate organic ligand to form Eu-complex improves the absorption coefficient at the near UV wavelength region [50-52]. As a result, the PL intensity drastically improves by combining with the organic ligand. Furthermore, the organic ligand solves the problem of the concentration quenching due to the long inter-molecular distance. This is because that the molecular structure of the organic ligand is large enough to reduce the interaction between adjacent two Eu-complexes, resulting in the high PL quantum efficiency of Eu-complex. These superior optical characteristics indicate that Eu-complex is expected as the red-emitting phosphor excited by the near UV-light, and the color rendering index of the white LED will improve by adding Eu-complex. By now, Eu-complex has been investigated as lasing materials [53,54], wavelength conversion films for photovoltaic cells [6,55], and organic light-emitting diodes [56,57] due to the high PL quantum efficiency and the large excitation spectrum at the near UV-wavelength region.

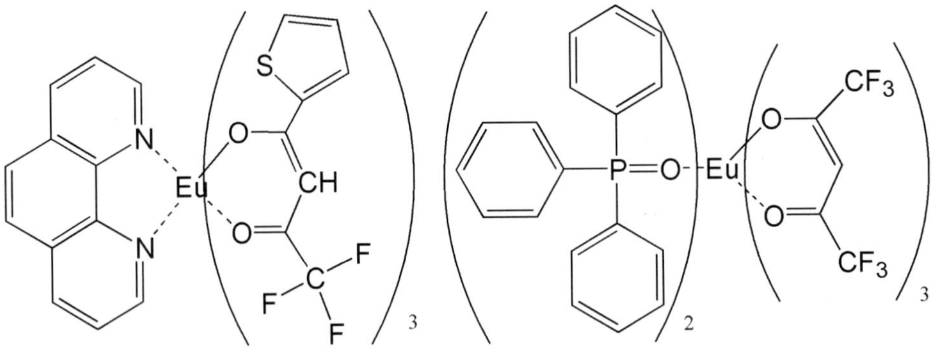

Figure 1. Molecular structures of Eu(TTA)₃phen and Eu(HFA)₃(TPPO)₂.

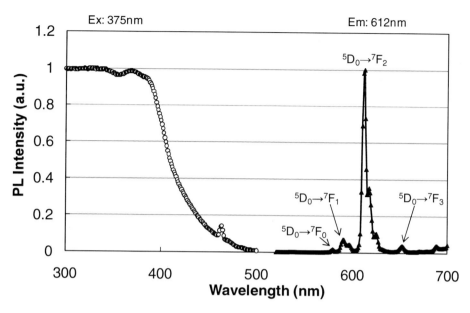

Figure 2. PL and PLE spectra of the Eu(TTA)$_3$phen powder.

Figure 1 shows molecular structures of two kinds of representative Eu-complexes, tris(2-thenoyltrifluoroacetonato)(1,10-phenanthroline)europium(III) (Eu(TTA)$_3$phen) and bis(tri phenylphosphine)tris(hexafluoroacetylacetonato)europium(III) (Eu(HFA)$_3$(TPPO)$_2$). In the case of Eu(HFA)$_3$(TPPO)$_2$, the HFA ligand acts as an accepter of the back energy. This is because the triplet state of the HFA ligand is close to the emitting state of Eu^{3+} ion [58], and the HFA ligand has the high absorption coefficient over 22,200 cm^{-1} at the UV-wavelength region. Therefore, the UV-induced exciton efficiently moves to the Eu^{3+} ion from the HFA ligand, resulting in the high PL quantum efficiency of Eu(HFA)$_3$(TPPO)$_2$. In addition, the asymmetric structure of Eu(HFA)$_3$(TPPO)$_2$ causes the high PL quantum efficiency. As a result, the reported PL quantum efficiency of Eu-complex reaches over 80 %. Figure 2 shows PL and PLE spectra of the Eu(TTA)$_3$phen powder. The excitation wavelength of the PL spectrum was 375 nm, and the monitor wavelength of the PLE spectrum was 612 nm. As clearly shown in Figure 2, the sharp red-emission with the center wavelength of 612 nm was observed by irradiating the near UV-light. The peak wavelength of the PL spectrum corresponds to the transition between $^5D_0$ and $^7F_2$ of Eu^{3+} ion [59]. In addition, the PLE spectrum shows the broad excitation at the wavelength region below 450 nm. Therefore, Eu(TTA)$_3$phen is the suitable red-emitting phosphor by the near UV irradiation.

The optical characteristics of Eu-complex are high enough for practical applications of white LEDs; however, one of the most serious problem is the long-term stability against the UV-light irradiation [60,61]. One experimental result about the weak resistivity of Eu-complex against the UV light irradiation is shown in Figure 3. Figure 3(a) shows the PL quantum efficiency of the Eu(TTA)$_3$phen powder as a function of the exposure time when the UV-light was continuously irradiated. The center wavelength of the excited light source was 254 nm, and the optical intensity was 0.6 mW/cm^2. The PL quantum efficiency continuously decreased with increasing irradiation time of the UV-light. In addition, photographs of Eu(TTA)$_3$phen with/without the UV-light irradiation are shown in Figure 3(b). The red-emission was reduced with increasing irradiation time of the UV-light. Furthermore,

Eu(TTA)₃phcn seems to be yellowish after irradiating the UV-light. These facts indicate the molecular structure of Eu(TTA)₃phen changed by the UV-light irradiation, resulting in the decreased PL quantum efficiency. This is because the bond strength between the Eu^{3+} ion and the organic ligand is lower than those of conventional inorganic materials. The organic ligand of Eu-complex absorbs the near UV-light, and the efficient energy transfer occurs from the organic ligand to the Eu^{3+} ion. Thus the PL quantum efficiency of Eu-complex decreased by changing the molecular structure of the organic ligand, and an unstable molecular structure is easily caused by the UV-light irradiation. On the other hand, it is noted that the optical degradation characteristics were reduced in the nitrogen atmosphere even though the UV-light was continuously irradiated [62]. This fact indicates that oxygen and/or vapor in the atmosphere causes the degradation of the Eu-complex during the UV-irradiation. The Eu-complex will be applied as the phosphor for white LEDs and the wavelength conversion film for photovoltaic cells; therefore, the high stability against the UV-light irradiation is necessary for practical applications.

Another problem of Eu-complex is the low thermal stability. The PL quantum efficiency of Eu(HFA)₃(TPPO)₂ is reduced by the thermal treatment in atmosphere [63]; however, the decrease trend of the PL quantum efficiency is suppressed by annealing in the nitrogen atmosphere. This result indicates the molecular structure of Eu-complex changes by the reaction between the Eu-complex and the oxygen and/or vapor during the thermal treatment, resulting in the low PL quantum efficiency [60,63,64]. In general, the high thermal stability of Eu-complex is necessary to mount on LED chips, circuit boards, and other electronic equipments. This is because the annealing temperature over 100 °C is required for a reflow soldering process. These facts indicate the coating technique around Eu-complex is important method to avoid the reaction between oxygen/vapor and Eu-complex.

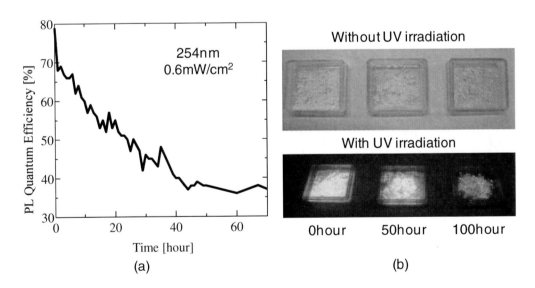

Figure 3. PL quantum efficiency as a function of the exposure time when the UV-light (254 nm, 0.6mW/cm²) was continuously irradiated. (b) Photograph of the Eu(TTA)₃phen powders with/without the UV-light irradiation.

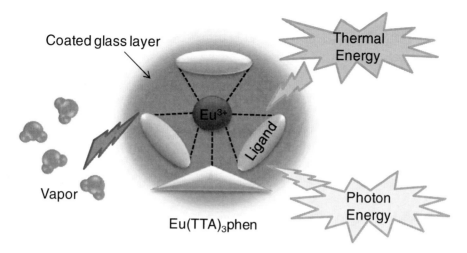

Figure 4. Concept image of the hybrid inorganic-organic phosphor containing Eu(TTA)₃phen and the coated glass layer fabricated by sol-gel and solvothermal processes.

Both optical and thermal degradations of Eu-complex are suppressed in the nitrogen atmosphere [62]; therefore, the coating technique using the sol-gel derived silica glass around Eu-complex has been investigated to improve the long-term stability against the UV-light irradiation [65-73]. Figure 4 shows the concept image of the hybrid inorganic-organic phosphor containing Eu-complex and the sol-gel derived silica glass. In a previous paper, Peng et al. reported a novel encapsulating method for Eu-complex as solutions in the solvent, and demonstrated an improved stability by encapsulating with base catalyzed hydrolysis of octyltrimetholysilane [74]. In addition, several previous papers have demonstrated that the sol-gel derived glass coated Eu-complex with the particle shape can be realized, and that this leads to the increase in the stability of the PL intensity against the UV-light irradiation [65-67]. Such a small particle size of the phosphor causes the high transparency against the visible light; therefore, nano-sized phosphors will be used for bio sensors [75,76], and wavelength conversion films of photovoltaic cells [6].

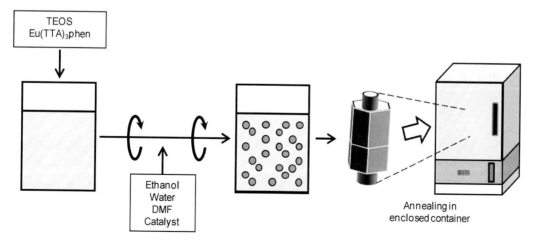

Figure 5. Flowchart of the sol-gel and solvothermal processes to coat the silica glass layer around the Eu(TTA)₃phen powder.

Our previous papers demonstrated the sol-gel derived silica glass film containing Eu-complex [63,64]. Since Eu-complex is easily dispersed in the silica glass network, the transparent emitting film with the high long-term and thermal stabilities can be realized. This fact indicates Eu-complex based inorganic-organic hybrid phosphors will become a key technology for further several applications due to the improved stability, which should be solved for the practical use. To fabricate the silica glass film by the conventional sol-gel process, phenyltrimethoxysilane (PTMS) and diethyldimethoxysilane (DEDMS) are used as silane alkoxide [63]. In addition, a mixture of DEDMS is chosen rather than PTMS only to form the transparent silica glass film. The function of DEDMS is to form a more flexible linear glass network than one obtained by the hydrolysis of PTMS only [77,78]. In this case, Eu-complex is incorporated in the sol-gel derived glass matrix at the same time as hydrolysis and condensation reactions of siloxane, and not after the sol-gel glass formation has finished. Therefore, the transparent and crack-free emitting thin film was obtained using DEDMS during the formation of the silica glass network around Eu-complex [63]. Another interesting progress of the thermal stability is the usage of deuterated methanol instead of ethanol used as the solvent in the sol-gel starting solution [64]. In a previous paper, the high frequency vibration of C-H and O-H bonds of Eu-complex is suppressed by the deuterated organic ligand [60]. As a result, the deuterated $Eu(HFA)_3(TPPO)_2$ has few high-frequency vibration modes of C-H and O-H bonds [79]. Therefore, we obtained the thermally stable Eu-complex in combination with the encapsulation method using the sol-gel derived silica glass and the organic ligand via a deuterated chemical structure. The most important finding is that the PL quantum efficiency remained equal to the original value of approximately 60 % at the annealing temperature below 180 °C [63].

3. SOL-GEL PROCESS TO ENCAPSULATE EUROPIUM-COMPLEX

A sol-gel synthesis consists of several processes to form an inorganic glass network around Eu-complex through the formation of a colloidal material (sol) and the solidification of the sol to form a network in a continuous liquid phase (gel). In general, the synthesis of colloids is occurred in a sol-gel starting solution including water, solvent, catalyst and silane alkoxide [80], such as tetrametoxysilane (TMOS) [81,82], tetraethoxysilane (TEOS) [83,84], poly(dimethylsiloxane) [85], PTMS [86,87], and DEDMS [88,89]. In addition, hydrolysis and condensation of TEOS describe the following equations [90].

$$n\ Si(OC_2H_5)_4 + 4n\ H_2O \rightarrow n\ Si(OH)_4 + 4n\ C_2H_5OH \tag{1}$$

$$n\ Si(OH)_4 \rightarrow n\ SiO_2 + 2n\ H_2O \tag{2}$$

In a previous paper, the time dependence of the Nuclear Magnetic Resonance (NMR) measurement for the solution containing sodium alkoxide is estimated at an ambient temperature [60]. The Si-OH signal shifts by adding the silane alkoxide, and it stops to shift after the enough hydrolysis time. The shift of the NMR signal is affected by concentrations of sodium and OH$^-$ ion in the sol-gel starting solution. Furthermore, the shift reflected the effect of surrounding Na$^+$ ion on Si-OH formed by the hydrolysis, as shown in Eq. (1).

It is noted that the silica glass network can be formed by the sol-gel process even though the annealing process is below 200 °C. Therefore, a most important advantage of the sol-gel process to encapsulate Eu-complex is the low temperature process including the final annealing process [63]. This is because Eu-complex is easily decomposed during the annealing process over 200 °C, and the high temperature process cannot be used to encapsulate around Eu-complex. Since the sol-gel synthesis can be performed at a temperature below 150 ° C [60], the thermal decomposition of Eu-complex does not occur during the sol-gel encapsulation process. In addition, a hybrid inorganic-organic emitting phosphors combining the organically modified silicate glass and Eu-complex has been demonstrated [65-67].

A most important disadvantage of the encapsulation around Eu-complex is the low-temperature sol-gel process, and the formed silica glass network consists of several organic components. Pure silica glass has low oxygen and vapor transmittances to prevent the degradation of Eu-complex while irradiating the UV-light; however, the organic component in the sol-gel derived glass network has the higher transmittance of oxygen and vapor compared to the pure inorganic silica glass network. This fact indicates that oxygen and vapor easily affect the molecular structure of Eu-complex when the UV-light is irradiated. Therefore, the poor long-term is obtained by encapsulating with the sol-gel derived silica glass network including the many organic components. In addition, the optical degradation characteristics of the Eu-complex coated by the sol-gel derived silica glass network are influenced by the annealing temperature [62]. Our approach to form the dense silica glass network with the less organic component is the solvothermal process [73]. We have investigated a novel method to fabricate the encapsulating glass layer around Eu(TTA)₃phen using a conventional sol-gel process and high-pressure annealing, referred to as the solvothermal process. The high-pressure annealing results in the efficient hydrolysis and condensation of silane alkoxide; therefore, the dense glass network is considered to be formed after the high-pressure annealing.

4. SILICA GLASS COATED EUROPIUM-COMPLEX BY SOLVOTHERMAL PROCESS

4.1. Principle of Solvothemal Process

A solvothermal process is one chemical method to synthesis of several materials, such as magnetic materials [91,92], oxides [93-95], and phosphors [96-100]. Since the sol-gel starting solution is annealed in a stainless autoclave container in this process [73], the pressure in the autoclave container is drastically improved due to the evaporated solvent and water. This fact causes the efficient hydrolysis of the silane alkoxide occurs during the annealing process compared to the conventional thermal annealing. In addition, the solvothermal synthesis allows controlling the size distribution and crystalline of metal oxide. These characteristics can be altered by changing experimental parameters, including the reaction temperature, the annealing time, the solvent type, the pressure in the autoclave container, and other several parameters.

4.2. Fabrication Process and Measurement

Figure 5 shows the flowchart of the sol-gel and solvothermal processes to coat the silica glass around Eu-complex. At first, the Eu(TTA)$_3$phen powder was added into the encapsulating agent, silane alkoxide. We used 10 kinds of silane alkoxide, such as TEOS, diethoxydiphenylsilane, dimethoxydiphenylsilane, trimethylchlorosilane, phenylethyl dichlorosilane, phenyltriethoxysilane, PTMS, diphenylsilane, triphenylsilane, and triphenylsilanol. The resulting solution was then injected into a mixture of distilled water, solvent and catalyst. In addition, the used solvents were ethanol, hexane, benzene, xylene, methanol, chlorobenzene, butanol, dichloromethane, chloroform, 1,2-dichloroethane, and N,N-dimethoxyformamide (DMF). To investigate of the solvent on the optical characteristics of Eu(TTA)$_3$phen encapsulated by the solvothermal process. In general, the catalyst is important material to control the reaction speed of hydrolysis and the water-alcohol induced condensation in the sol-gel starting solution and the formation of the silica glass network. And then, the solution was mixed at a room temperature for 10 minutes until Eu(TTA)$_3$phen was completely dissolved in a Teflon beaker. In the sol-gel starting solution, the glass layer was continuously deposited around Eu-complex during a mixing process. Therefore, a particle size can be controlled by changing the mixing time of the sol-gel starting solution. The rotation speed was maintained at 400 rpm by a magnetic stirrer for all the mixing processes. After the mixing process, the sol-gel solution with 1 mL was placed into the autoclave container with the capacity of 25 mL. Finally, the solution was dripped onto a glass substrate, and the sample was annealed in the autoclave container. The measured pressure was approximately 100 kPa. A reference sample was also prepared by annealing in an electric furnace under the same annealing conditions expect for the pressure. The concentration of the sol-gel starting solution and the mixing condition were the same as the sample to be encapsulated by the solvothermal process.

The PL quantum efficiency was measured by a luminance quantum efficiency measurement system (QEMS-2000, Systems Engineering Inc.), which consists of an integrated sphere and a UV-LED with a center wavelength of 375 nm as an excitation source [63,64]. The PL quantum efficiency was determined using a method based upon that originally developed by de Mello et al. [101]. In this approach, the quantum efficiency is given by the integrated PL intensity of the sample excited by the UV-LED divided by the decrease in excitation intensity caused by inserting the sample into the integrated sphere. In addition, the normalized PL intensity was also recorded as a function of time under UV-light irradiation using a spectrofluorometer [73]. The center wavelength and the optical intensity of the excitation UV light were 360 nm and 5 mW/cm^2, respectively. The PL intensity change was measured by a luminance spectrometer (FluoroMax-3, Horiba Jovin Yvon). In addition, PL and PLE spectra were measured using a spectrofluorometer. The excitation wavelength of the PL spectrum was 360 nm, and the PLE spectrum was monitored at 612 nm.

4.3. Dependence of Silane Alxoxide

We discuss the influence of the silane alkoxide in the sol-gel starting solution on optical and degradation characteristics of the Eu(TTA)$_3$phen coated by the silica glass via the solvothermal process. The molar ratio of silane alkoxide, deionized water, solvent (ethanol),

DMF, ammonia, and Eu(TTA)$_3$phen was 1:35:30:4:0.1:0.02, and the annealing temperature was fixed at 150 °C for the solvothermal process.

Tables 1 summarizes the PL quantum efficiency of the coated Eu(TTA)$_3$phen after the solvothermal process. In addition, the PL quantum efficiency of the original Eu(TTA)$_3$phen powder was 61 %. The PL quantum efficiency was influenced by the silane alkoxide due to the molecular structure change of Eu(TTA)$_3$phen. This is because the efficient energy transfer from the organic ligand to the Eu^{3+} ion cannot be occurred by changing the molecular structure of Eu(TTA)$_3$phen. Among the tested silane alkoxide, the highest PL quantum efficiency of 91 % was obtained from the Eu(TTA)$_3$phen after encapsulating with TEOS. This value was much higher than that of the original Eu(TTA)$_3$phen powder. The clearly mechanism of this result is not cleared; however, the one possible hypothesis is that the excess OH group was reduced by the thermal treatment in the organic solvent.

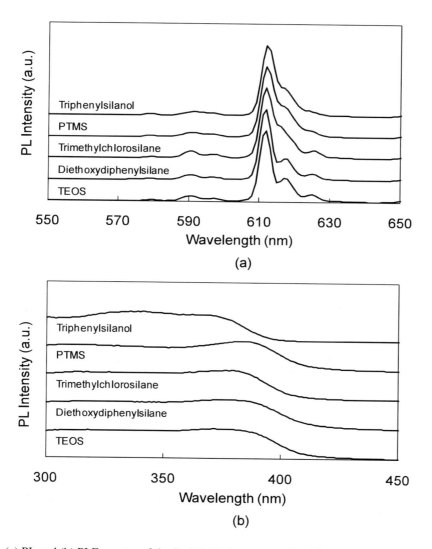

Figure 6. (a) PL and (b) PLE spectra of the Eu(TTA)$_3$phen encapsulated by the solvothermal process with the different silane alkoxide. The excitation wavelength for the PL spectrum was 360 nm, and the PLE spectrum was monitored at 612 nm.

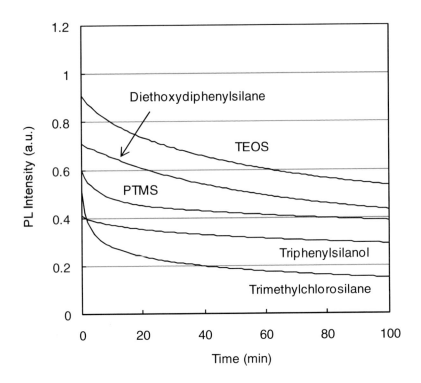

Figure 7. Relative PL intensity change against the UV-light irradiation time for Eu(TTA)$_3$phen encapsulated by the solvothermal process with the different silane alkoxide.

Table 1. PL quantum efficiency of the Eu(TTA)$_3$phen coated by the sol-gel derived silica glass with different silane alkoxide by the solvothermal process

Silane Alkoxide	PL Quantum Efficiency (%)
TEOS	91
diethoxydiphenylsilane	71
dimethoxydiphenylsilane	52
trimethylchlorosilane	51
phenylethyldichlorosilane	26
phenyltriethoxysilane	77
PTMS	60
diphenylsilane	44
triphenylsilane	79
triphenylsilanol	41

Figures 6(a) and 6(b) show PL and PLE spectra of the Eu(TTA)$_3$phen encapsulated by the solvothermal process with the different silane alkoxide, respectively. The excitation wavelength for the PL spectrum was 360 nm, and the PLE spectrum was monitored at 612 nm corresponding to the peak wavelength of the PL spectrum. As clearly shown in Figure 6(a), the PL spectra of all the samples showed the sharp red emission at the center wavelength of 612 nm. The emission peak corresponds to the $^5D_0 \rightarrow {}^7F_2$ transition in the Eu^{3+} ion [67,102]. On the other hand, the PLE spectrum of the encapsulated Eu(TTA)$_3$phen was shifted toward the shorter wavelength region compared to the Eu(TTA)$_3$phen powder [73]. However, a large excitation spectrum was observed in the near UV wavelength region, which indicates that the encapsulated Eu(TTA)$_3$phen could be applicable to red-emitting phosphors of UV-LED excited white LEDs. In addition, our previous experimental result showed that the PLE spectrum of the encapsulated Eu(TTA)$_3$phen was changed due to the change in the molecular structure during the sol-gel process [63]. Therefore, the difference in the PLE spectra can be explained by the change in the molecular structure of Eu(TTA)$_3$phen can be explained by the silane alkoxide.

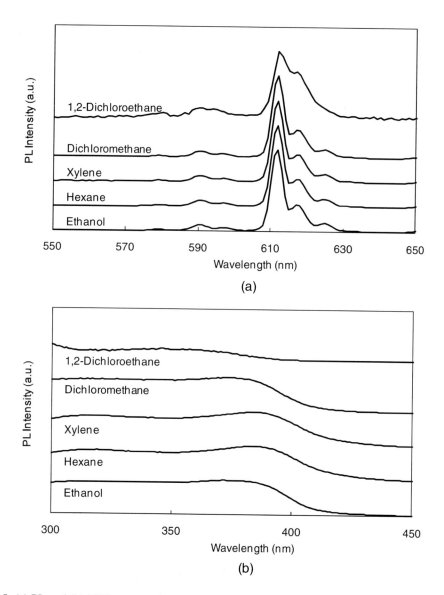

Figure 8. (a) PL and (b) PLE spectra of the Eu(TTA)$_3$phen encapsulated by the solvothermal process with the different solvents. The excitation wavelength for the PL spectrum was 360 nm, and the PLE spectrum was monitored at 612 nm.

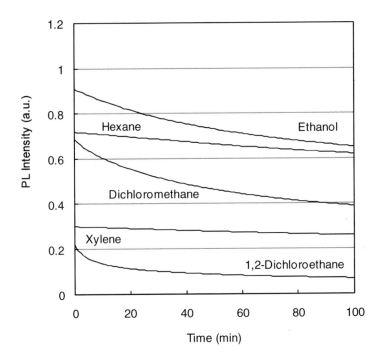

Figure 9. Relative PL intensity change against the UV-light irradiation time for Eu(TTA)$_3$phen encapsulated by the solvothermal process with the different solvents.

Figure 7 shows the relative PL intensity change against the UV-light irradiation time for Eu(TTA)$_3$phen encapsulated by the solvothermal process with the different silane alkoxide. The initial relative PL intensity was determined as the measured PL quantum efficiency, as summarized in Table 1. The PL intensity continuously decreased with increasing UV-light irradiation time due to the molecular structure change of Eu(TTA)$_3$phen. However, the optical degradation was suppressed by encapsulating the silica glass for all the cases of silane alkoxide. Especially, the highest performance was achieved when TEOS was used as the silane alkoxide.

4.4. Dependence of Organic Solvent

In the previous section, we showed that TEOS is one of the most suitable silane alkoxide to form the encapsulating glass layer in our experiment; therefore, we investigated the relationship between the solvent and optical characteristics of the glass coated Eu(TTA)$_3$phen. The molar ratio of TEOS, deionized water, solvent (ethanol, hexane, benzene, xylene, methanol, chlorobenzene, butanol, dichloromethane, chloroform, and 1,2-dichloroethane), DMF, ammonia, and Eu(TTA)$_3$phen was 1:35:30:4:0.1:0.02, and the annealing temperature was fixed at 150 °C for all the solvothermal process.

Table 2 shows the PL quantum efficiency of the fabricated sample by the solvothermal process with different solvents in the sol-gel starting solution. The PL quantum efficiency was changed by changing the solvent. Especially, the low PL quantum efficiency was obtained in the cases of benzene, xylene, chlorobenzene, and 1,2-dichloroethane. The highest PL quantum efficiency of 91 % was achieved when ethanol was used as the solvent in the sol-gel starting solution. This result also indicates that the molecular structure of Eu(TTA)$_3$phen is considered to affect the PL quantum efficiency. Therefore, the change in the molecular structure causes the low energy transfer from the organic ligand to the Eu3+ ion, resulting in the reduced PL quantum efficiency.

Figure 10. SEM image of the Eu(TTA)$_3$phen powder and the coated sample fabricated by the conventional annealing and the solvothermal process.

Table 2. PL quantum efficiency of the Eu(TTA)₃phen coated by the sol-gel derived silica glass with different solvents by the solvothermal process

Solvent	PL Quantum Efficiency (%)
Ethanol	91
Hexane	72
Benzene	11
Xylene	30
Methanol	68
Chlorobenzene	10
Butanol	80
Dichloromethane	69
Chloroform	64
1,2-Dichloroethane	22

Figures 8(a) and 8(b) show PL and PLE spectra of Eu(TTA)₃phen encapsulated by the solvothermal process with the different solvents, respectively. We obtained the sharp emission peak at 612 nm corresponding to the $^5D_0 \rightarrow {}^7F_2$ transition in the Eu^{3+} ion [67,102]. However, the exact PL/PLE spectra were not observed in the case of 1,2-dichloroethane due to the low PL quantum efficiency, as shown in Table 2. Furthermore, large excitation spectra were observed at the near UV wavelength region for all the cases of solvents expect for 1,2-dichloroethane. This result indicates that the encapsulated Eu(TTA)₃phen can be applicable for the white LED excited with the UV-LED by choosing the optimum solvent.

Figure 9 shows the relative PL intensity change against the UV-light irradiation time for Eu(TTA)₃phen encapsulated by the solvothermal process with different solvents. The initial relative PL intensity was also determined as the measured PL quantum efficiency, as shown in Table 2. As clearly shown in Figure 9, the relative PL intensity continuously decreased with increasing UV light irradiation time. This is because Eu(TTA)₃phen easily reacts to oxygen and/or vapor while irradiating the UV-light. Therefore, the efficiency of the energy transfer from the organic ligand to the Eu^{3+} ion is reduced by reacting Eu(TTA)₃phen, resulting in the low PL intensity. Among the tested solvents, ethanol is the best solvent for the high optical degradation characteristics of the glass coated Eu(TTA)₃phen.

Figure 10 shows SEM images of the Eu(TTA)₃phen powder and the glass coated Eu(TTA)₃phen particles, which were fabricated by the conventional annealing and the solvothermal processes. The Eu(TTA)₃phen powder has a rectangular shape with the length of several 10 □m long. On the other hand, the diameter of the glass coated Eu(TTA)₃phen showed a circle shape with the diameter of several 10 nm. The Eu(TTA)₃phen powder is easily dissolved in the sol-gel starting solution. Therefore, the small particle was obtained after the conventional sol-gel and solvothermal processes. Especially, the uniform particle structures were achieved in the case of the solvothermal process.

CONCLUSION

In conclusion, we demonstrated the Eu-complex with the high long-term stability against the UV-light irradiation by encapsulating the sol-gel derived silica glass using the solvothermal process. By optimizing silane alkoxide and solvent, we successfully demonstrated the improved PL quantum efficiency and the long-term stability. These experimental results indicate that this kind of the red-emitting phosphor will be used for the white LED exited with the UV-LED.

ACKNOWLEDGMENTS

The author thanks to Dr. N. Kijima of Mitsubishi Chemical Group, Science and Technology, Research Center, Inc. for providing several Eu-complexes and the discussion of experimental results. The author also thanks Miss S. Kato, Mr. S. Akiyama, Prof. Z. Honda and Prof. N. Kamata of Saitama University. This work was supported by a grant from the Japan Science and Technology Agency (JST; A-STEP AS2111253C).

REFERENCES

[1]   Noor, Y.M.; Tam, S.C.; Lim, L.E.N.; Jana, S. (1994). *J. Mater. Proc. Technol.*, 42, 95-133.

[2]   Ainslie, B.J. (1991). *IEEE J. Lightwave Tech.*, 9, 220-227.

[3]   Bellemare, A. (2003). *Prog. Quantum Electron.*, 27, 211-266.

[4]   Mahakhode, J.G.; Dhoble, S.J.; Joshi, C.P.; Moharil, S.V. (2007). *J. Alloys Compd.*, 438, 293-297.

[5]   Bizarri, G.; Moine, B. (2005). *J. Lumin.*, 113, 199-213.

[6]   Fukuda, T.; Kato, S.; Kin, E.; Okaniwa, K.; Morikawa, H.; Honda, Z.; Kamata, N. (2009). *Opt. Mater.*, 32, 22-25.

[7]   Richards, B.S. (2006). *Sol. Ener. Mater. Sol. Cells,* 90, 2329-2337.

[8]   Strümpel, C.; McCann, M.; Beaucarne, G.; Arkhipov, V.; Slaoui, A.; Švrček, V.; Caiñzo, C. del; Tobias, I. (2007). *Sol. Ener. Mater. Sol. Cells,* 91, 238-249.

[9]   Niedbala, R.S.; Feindt, H.; Kardos, K.; Vail, T.; Burton, J.; Bieska, B.; Li, S.; Milunic, D.; Bourdelle, P., Vallejo, R. (2001). *Anal. Biochem.,* 293, 22-30

[10] Byrappa, K.; Devaraju, M.K.; Paramesh, J.R., Basavalingu, B.; Soga, K. (2008). *J. Mater. Sci.,* 43, 2229-2233.

[11] 11. Yan, Z.; Zhou, L.; Zhao, Y.; Wang, J.; Huang, L.; Hu, K.; Liu, H.; Wang, H.; Guo, Z.; Song, Y.; Huang, H.; Yang, R. (2006). *Sens. Actuators B Chem.,* 119, 656-663.

[12] Wu, M.; Long, S.; Frutos, A.G.; Eichelberger, M.; Li, M.; Fang, Y. (2009). J. *Recept Signal Transduct. Res.,* 29, 202-210.

[13] Cavouras, D.; Kandarakis, I.; Panaylotakis, G.S.; Kanellopoulos, E.; Triantis, D.; Nomicos, C.D. (1998). *Appl. Rad. Isoto.,* 49, 931-937.

[14] Kimura, N.; Sakuma, K.; Hirafune, S.; Asano, K.; Hirosaki, N.; Xie, R.J. (2007) *Appl. Phys. Lett.,* 90, 051109.

[15] Miyamoto, Y.; Kato, H.; Honna, Y.; Yamamoto, H.; Ohmi, K. (2009). *J. Electrochem. Soc.*, 156, J235-J241.

[16] Nazarov, M.; Yoon, C. (2006). *J. Sol. Stat. Chem.*, 179, 2529-2533.

[17] Yang, H.; Lee, D.-K.; Kim, Y.-S. (2009). *Mater. Chem. Phys.*, 114, 665-669.

[18] Dorenbos, P. (2002). *J. Lumin.*, 99, 283-299.

[19] Chen, Y.; Gong, M.; Cheah, K.W. (2010). *Mater. Sci. Eng.* B, 166, 24-27.

[20] Carlos, L.D.; Ferreira, R.A.S.; Bermudez, V.Z.; Ribeiro, S.J.L. (2009*). Adv. Mater.*, 21, 509-534.

[21] Bos, A.J.J.; Dorenbos, P.; Bessière, A.;Viana, B. (2008). *Radiat. Meas.*, 43, 222-226.

[22] Fukuda, Y.; Ishida, K.; Mitsuishi, I.; Nunoue, S. (2009). *Appl. Phys. Exp.*, 3, 012401.

[23] Shikako, S.; Jiye, W. (2001). *J. Alloys Compd.*, 327, 82-86.

[24] De la Rosa, E.; Dodriguez, R.A.; Díaz-Torres, L.A.; Salas, P.; Meléndrez, R.; Barboza-Flores, M. (2005). *Opt. Mater.*, 27, 1245-1249.

[25] Shimomura, Y.; Honma, T.; Shigeiwa, M.; Akai, T.; Okamoto, K.; Kijima, N. (2007). *J. Electrochem. Soc.*, 154, J35-J38.

[26] Zhang, X.; He, H.; Li, Z.; Yu, T.; Zou, Z. (2008). *J. Lumin.*, 128, 1876-1879.

[27] Watanabe, H.; Wada, H.; Seki, K.; Itou, M.; K.; Kijima, N. (2009). *J. Electrochem. Soc.*, 155, F31-F36.

[28] Yamaga, M.; Masui, Y.; Sakuta, S.; Kodama, N.; Kaminaga, K. (2005). *Phys. Rev.* B, 71, 205102.

[29] Xie, R.-J.; Mitomo, M.; Uheda, K.; Xu, F.-F.; Akimune, Y. (2002). *J. Am. Ceram. Soc.*, 85, 1229-1234.

[30] Yamada, M.; Narukawa, Y.; Tamaki, H.; Murazaki, Y.; Mukai, T. (2005). *IEICE Trans. Electron.*, E88-C, 1860-1871.

[31] Tsao, J.Y. (2004). *IEEE Circuits Devices,* 20, 28-37.

[32] Nishimura, S.; Tanabe, S.; Fujioka, K.; Fujimoto, Y. (2011). Opt. Mater., 33, 688-691.

[33] Narukawa, Y.; Niki, I.; Izuno, K.; Yamada, M.; Murazaki, Y.; Mukai, T. (2002). *Jpn. J. Appl. Phys.*, 41, L371-L373.

[34] Chung, W.; Yu, H.J.; Park, S.H.; Chun, B.-H.; Kim, S.H. (2011). *Mater. Chem. Phys.*, 125, 162-166.

[35] Shin, J.-S.; Kim, H.-J.; Jeong, Y.-K.; Kim, K.-B.; Kang, J.-G. (2011). *Mater. Chem. Phys.*, 126, 591-595.

[36] Akasaki, I. (2007). *J. Cry. Growth,* 300, 2-10.

[37] Yamada, M.; Naitou, T.; Izuno, K.; Tamaki, H.; Murazaki, Y.; Kameshima, M.; Mukai, T. (2003). *Jpn. J. Appl. Phys.*, 42, L20-L23.

[38] Cao, F.-B.; Tian, Y.-W.; Chen, Y.J.; Xiao, L.-J.; Wu, Q. (2009). *J, Lumin.*, 129, 585-588.

[39] Kaufmann, U.; Kunzer, M.; Köhker, K.; Obloh, H.; Pletschen, W.; Schlotter, P.; Wagner, J.; Ellens, A.; Rossner, W.; Kobusch, M. (2002). *Phys. Stat. Sol.* (a), 192, 246-253.

[40] Lm, W.B.; Fourré. Y.; Brinkley, S.; Sonoda, J.; Nakamura, S.; DenBaars, S.P.; Seshadri, R. (2009). *Opt. Exp.*, 17, 22673-22679.

[41] Akasaki, I. (2007). *J. Crystal Growth,* 300, 2-10.

[42] Razeghi, M.; McClintock, R. (2009). *J. Crystal Growth,* 311, 3067-3074.

[43] Mukai, T.; Nagahama, S.; Kozaki, T.; Sano, M.; Morita, D.; Yanamoto, T.; Yamamoto, M.; Akashi, K.; Masui, S. (2004). *Phys. Stat. Sol.* (a), 12, 2712-2716.

[44] Dai, P.; Zhang, X.; Li, X.; Wang, G.; Zhao, C.; Liu, Y. (2011). *J. Lumin.*, 131, 653-656.

[45] Hu, Y.; Zhuang, W.; Ye, H.; Wang, D.; Zhang, S.; Huang, X. (2005). *J. Alloys Compd.*, 390, 226-229.

[46] Schwendemann, T.C.; May, P.S.; Berry, M.T.; Hou, Y.; Meyers, C.Y. (1998). *J. Phys. Chem.* A, 102, 8690-8694.

[47] Wang, L.-H.; Wang, W.; Zhang, W.-G.; Kang, E.-T.; Huang, W. (2000). *Chem. Mater.*, 12, 2212-2218.

[48] Lis, S.; Hnatejko, Z.; Barczynski, P.; Elbanowski, M. (2002). J. *Alloys Compd.*, 344, 70-74.

[49] Gallardo, H.; Conte, G.; Tuzimoto, P.; Bortoluzzi, A.; Peralta, R.A.; *Neves*, A. (2008). Inorg. Chem. Commun., 11, 1292-1296.

[50] Iwanaga, H.; Amano, A.; Furuya, F.; Yamasaki, Y. (2006). *Jpn. J. Appl. Phys.*, 45, 558-562.

[51] Hasegawa, Y.; Yamamuro, M.; Wada, Y.; Kanehisa, N.; Kai, Y.; Yanagida, S. (2003). *J. Phys. Chem.* A, 107, 1697-1702.

[52] Frey, S.T.; Horrocks Jr., W.D. (1995). *Inorg. Chim. Acta*, 229, 383-390.

[53] Nakamura, K.; Hasegawa, Y.; Wada, Y.; Yanagida, S. (2004). *Chem. Phys. Lett.*, 398, 500-504.

[54] Nakamura, K.; Hasegawa, Y.; Kawai, H.; Yasuda, N.; Wada, Y.; Yanagida, S. (2006). *J. Alloys Compd.*, 408-412, 771-775.

[55] Donne, A.L.; Dilda, M.; Crippa, M.; Acciarri, M.; Binetti, S. (2011). *Opt. Mater.*, 33, 1012-1014.

[56] Adachi, C.; Baldo, M.A.; Forrest, S.R. (2000). *J. Appl. Phys.*, 87, 8049-8055.

[57] Liua, L.; Xua, Z.; Loua, Z; Zhanga, F.; Sunb, B; Pei, J. (2007). J. *Lumin.*, 122-123, 961-963.

[58] Sato, S.; Wada, M. (1970). *Bull. Chem. Soc. Jpn.*, 43, 1955-1962.

[59] Hasegawa, Y.; Kawai, H.; Nakamura, K.; Yasuda, N.; Wada, Y.; Yanagida, S. (2006). *J. Alloys Compd.*, 408-412, 669-674.

[60] Kin, E.; Fukuda, T.; Yamauchi, S.; Honda, Z.; Ohara, H.; Yokoo, T.; Kijima, N., Kamata, N. (2009). *J. Alloys Compd.*, 480, 908-911.

[61] Binnemanss, K. (2009). Chem. Rev., 109, 4283-4374.

[62] Kin, E.; Fukuda, T.; Kamata, N.; Ohara, H.; Yokoo, T.; Kijima, N. (2009). *J. Light & Vis. Env.*, 33, 82-87.

[63] Fukuda, T.; Yamauchi, S.; Honda, Z.; Kijima, N.; Kamata, N. (2009). *Opt. Mater.*, 32, 207-211.

[64] Fukuda, T.; Yamauchi, S.; Honda, Z.; Kamata, N.; Kijima, N. (2009). *Phys. Stat. Sol. Rap. Res. Lett.*, 2, 296-298.

[65] Qian, G.; Wang, M. (2000). *J. Am. Ceram. Soc.*, 83, 703-708.

[66] Jin, T.; Inoue, S.; Machida, K.; Adachi, G. (1998). *J. Alloys Compd.*, 265, 234-239.

[67] Strek, W.; Sokolnicki, J.; Lengendziewicz, J.; Maruszewski, K.; Reisfeld, R.; Pavich, T. (1999). Opt. Mater., 13, 41-48.

[68] Lenaerts, P.; Ryckebosch, E.; Driesen, K.; Deun, R.V.; Nockemann, P.; Görller-Walrand, C.; Binnemans, K. (2005). *J. Lumin.*, 114, 77-84.

[69] Hao, X.; Fan, X.; Wang, M. (1999). T*hin Solid Films,* 353, 223-226.

[70] Zaitoun, M.A.; Kim, T.; Jaradat, Q.M.; Momani, K.; Qaseer, H.A.; El-Qisairi, A.K.; Qudah, A.; Radwan, N.E. (2008). *J. Lumin.,* 128, 227-231.

[71] Fan, X.; Wu, X.; Wang, M.; Qiu, J.; Kawamoto, Y. (2004). *Mater. Lett.,* 58, 2217-2221.

[72] Kin, E.; Fukuda, T.; Kato, S.; Honda, Z.; Kamata, N. (2009*). J. Sol-Gel Sci. Technol.,* 50, 409-414.

[73] Kato, S.; Fukuda, T.; Akiyama, S.; Honda, Z.; Kamata, N.; Kijima, N. (2011). *Jpn. J. Appl. Phys.,* 50, 01BF02.

[74] Peng, H.; Wu, C.; Jiang, Y.; Huang, S.; McNeill, J. (2007). *Langmuir,* 23, 1591-1595.

[75] Sharma, P.; Brown, S.; Walter, G.; Santra, S.; Moudgil, B. (2006). *Adv. Coll. Interf. Sci.,* 123-126, 371-485.

[76] Zhang, C.; Sun, L.; Zhang, Y.; Yan, C. (2010*). J. Rare Earth,* 28, 807-819.

[77] Łączka, M.; Cholewa-Kowalska, K.; Kogut, M. (2001). *J. Non-Cry. Sol.,* 287, 10-14.

[78] Binnemans, K.; Lenaerts, P.; Driesen, K.; Görller-Walrand, C. (2004*). J. Mater. Chem.,* 14, 191-195.

[79] Schwendemann, T.C.; May, P.S.; Berry, M.T.; Hou, Y.H.; Meyers, C.Y. (1998*). J. Phys. Chem.* A, 102, 8690-8694.

[80] Wang, D.; Bierwagen, G.P. (2009). *Prog. Org. Coat.,* 64, 327-338.

[81] Jung, K.T.; Chu, Y.-H.; Haam, S.; Shul, Y.G.; (2002). *J. Non-Cryst. Solids,* 298, 193-201.

[82] Vollet, D.R.; Donatti, D.A.; Ruiz A.I.; Maceti, H. (2003). *J. Non-Cryst. Solids,* 324, 201-207.

[83] Einarsrud, M.-A.; Nilsen, E. (1998). *J. Non-Cryst. Solids,* 226, 122-128.

[84] Fahrenholtz, W.G.; Smith, D.M.; Hua, D.-W. (1992). *J. Non-Cryst. Solids,* 144, 45-52.

[85] Karwa, M.; Hahn, D.; Mitra, S. (2005). *Anal. Chim. Acta,* 546, 22-29.

[86] Förster, T.; Scholz, S.; Zhu, Y.; Lercher, J.A. (2011). *Microporous Mesoporous Mater.,* 142, 464-472.

[87] Jürgen-Lohmann, D.L.; Nacke, C.; Simon, L.S.; Legge, R.L. (2009). *J. Sol-Gel Sci. Technol.,* 52, 370-381.

[88] Ardhyananta, H.; Wahid, M.H.; Sasaki, M.; Agag, T.; Kawaguchi, T.; Ismail, H.; Takeichi, T. (2008). *Polymer,* 49, 4585-4591.

[89] Suda, S.; Iwaida, M.; Yamashita, K.; Umegaki, T. (1996*). J. Non-Cry. Sol.,* 197, 65-72.

[90] Wu, C.-H.; Jeng, J.-S.; Chia, J.-L.; Ding, S. (2011). *J. Coll. Interf. Sci.,* 353, 124-130.

[91] Tian, Y.; Yu, B.; Li, X.; Li, K. (2011). *J. Mater. Chem.,* 21, 2476-2481.

[92] Sun, X.; Xing, Y.; Liu, X.; Liu, B.; Hou, S. (2010). *Dalton Trans.,* 39, 1985–1988.

[93] Wahi, R.K.; Liu, Y.; Falkner, J.C.; Colvi, V.L. (2006). *J. Colloid Interface* Sci., 302, 530–536.

[94] Matos, J.; García, A.; Zhao, L.; Titirici, M.M. (2010). Appl. Catal. A: *General,* 390, 175-182.

[95] Xu, L.; Hu, Y.-L.; Pelligra, C.; Chen, C.-H.; Jin, L.; Huang, H.; Sithambaram, S.; Aindow, M.; Joesten, R.; Suib, S.L. (2009). *Chem. Mater.,* 21, 2875-2885.

[96] Xie, Y.; Yin, S.; Hashimoto, T.; Tokano, Y.; Sasaki, A.; Sato, T. (2010*). J. Mater. Sci.,* 45, 725-732.

[97] Byun, H.-J.; Lee, J.C.; Yang, H.; (2011). *J. Colloid Interface Sci.,* 355, 35-41.

[98]   Sulawan, K.; Kouichi, N.; Valery, P.; Somchai, T.; Masato, K.; *J. Am. Ceram. Soc.*, 92, S16-S20.

[99]   Li, X.; Liu, H.; Wang, J.; Cui, H.; Han, F. (2004). *Mater. Res. Bull.*, 39, 1923-1930.

[100] Nishimura, F.; Isobe, T. (2005). *J. Rare Earths*, 2005, 46, 146-147.

[101] Mello, J.C.de; Wittmann, H.F.; Friend, R.H. (1997). *Adv. Mater.*, 9, 230-232.

[102] da Silva, L.C.C.; Martins, T.S.; Filho, S.; Teotônio, E.E.S.; Isolani, P.C.; Brito, H.F.; Tabacniks, M.H.; Fantini, M.C.A.; Matos, J.R. (2006). *Micro. Meso. Mater.*, 92, 94-100.

In: Light-Emitting Diodes and Optoelectronics: New Research   ISBN: 978-1-62100-448-6
Editors: Joshua T. Hall and Anton O. Koskinen   © 2012 Nova Science Publishers, Inc.

Chapter 6

ENHANCED EFFICIENCY OF ZNTE-BASED GREEN LIGHT-EMITTING DIODES

*Tooru Tanaka[1,2], Katsuhiko Saito[3], Qixin Guo[3]
and Mitsuhiro Nishio[1]*

[1]Department of Electrical and Electronic Engineering,
Graduate School of Science and Engineering, Saga University, Saga, Japan
[2]PRESTO, Japan Science and Technology Agency (JST),
Kawaguchi, Japan
[3]Synchrotron Light Application Center,
Saga University, Japan

ABSTRACT

ZnTe is expected to use in a variety of optoelectronic devices such as pure green light-emitting diodes (LEDs) because of its direct transition band gap of 2.26eV at room temperature (RT). Although p-type ZnTe was obtained easily, the growth of an n-type material was difficult due to a self-compensation effect and an incorporation of residual impurities. Among several attempts to realize the n-type conductivity in ZnTe, a thermal diffusion technique showed high potential for obtaining n-type ZnTe, resulting in a development of ZnTe-based green LEDs. In this chapter, we review our recent research results on the development of ZnTe LEDs with a special emphasis on the thermal diffusion technique through the oxide layer, which plays an important role as a diffusion limiting layer. Using the Al-diffused layer with a controlled Al-concentration, the output power of ZnTe LED has been increased significantly.

INTRODUCTION

ZnTe-related materials are expected to use in a variety of optoelectronic devices including light emitting diodes (LEDs), solar cells, and terahertz devices. Among them, the application of ZnTe to pure green LEDs is promising because of its excellent feature for emitting pure green light represented by its direct transition band gap of 2.26 eV at room temperature (RT). This compound does not contain expensive rare metals such as In and Ga, leading to the fabrication of LEDs with low cost.

So far, a lot of efforts have been performed to realize electroluminescence (EL) devices using ZnTe[1-25]. The first successful ZnTe EL devices were reported in 1964 with a metal-insulator-semiconductor (MIS) and a metal-semiconductor (MS) structure. [1,2] Then, green or yellow-green ELs were demonstrated at 77 K[3-5] and also at RT.[5-8] Although the wall-plug efficiencies of these devices were not so high, these results obviously prove the potential application of ZnTe to green light emitting devices. Also, relatively strong red ELs originated from the localized states of oxygen incorporated in ZnTe were demonstrated in MIS diodes[4,5,9-11]. The reason for the use of MIS or MS structures was caused by a difficulty in realizing n-type conductivity in ZnTe, while p-type ZnTe can be easily obtained.

In 1994, the successful n-type dopings in ZnTe were reported by using metalorganic vapor phase epitaxy (MOVPE) and molecular beam epitaxy (MBE) using Al and Cl as donors, respectively [26, 27]. After that, several techniques were reported for the fabrication of n-type ZnTe, including laser doping[12,13], MBE[28], and thermal diffusion [14-17]. These successes have opened up pn-junction LEDs based on ZnTe [12-25]. Among these techniques, the thermal diffusion has been extensively studied as a potential method for low-cost LEDs [17, 21-25].

In 2008, we found that the light output power can be improved by adapting our original thermal diffusion technique that uses an oxide layer as a diffusion-limiting layer inserted between Al diffusion source and p-ZnTe substrate [23, 25]. This enables to control an Al concentration in an n-type diffused layer, resulting in the formation of good pn junction. Furthermore, we fabricated the ZnTe LED with a thin film structure and achieved a wall plug efficiency of 0.15%, which is comparable to commercially available GaP LED [29].

In the first section of this chapter, we describe an increase of output power of ZnTe LED by fabricating a high quality n-type ZnTe layer using the thermal diffusion technique through an oxide layer prepared by an O radical irradiation to the ZnTe clean surface. In the second section, the results using the Al oxide layer directly deposited on the ZnTe clean surface by irradiating O radical and Al flux simultaneously are reviewed. In this case, an Al concentration at the surface of ZnTe can be adjusted by changing the Al oxide layer thickness. The obtained maximum output power was as high as 18 μW under a forward current density of 8 A/cm^2 in spite of a strong self-absorption effect. In the final section, the fabrication of ZnTe LEDs with a thin film structure is reported. As a result of a reduced self-absorption, the output power of LED was greatly increased to approximately 45 μW at the wavelength of 557 nm under an injection current of 10 mA at room temperature in spite of a simple homo-junction structure. The wall-plug efficiency of the LED is improved to that of the commercially available GaP LED.

PROPERTIES OF ZNTE LED FABRICATED BY THERMAL DIFFUSION OF AL THROUGH A SURFACE OXIDATION LAYER

In this section, we describe the results of ZnTe LED fabricated by thermal diffusion of Al through a surface oxidation layer. The surface oxidation layer was prepared by direct irradiation of O radical to p-ZnTe clean surface, and the O radical irradiation times were varied between 0 and 60 min. An Al thin film of 200 nm thickness was deposited on the surface oxidation layer. Thermal diffusion was carried out in an electrical furnace at 420 °C for 5 h. After the thermal diffusion, a Pd film was deposited on the back surface of the substrate as a *p*-type ohmic contact.

Figures 1(a), 1(b), 1(c), and 1(d) show RHEED patterns of the ZnTe substrate after the H radical cleaning and with subsequent O radical irradiation for 5, 15, and 30 min, respectively. While some ring patterns were observed immediately after the wet etching, the ring patterns disappeared and a clear spot pattern was observed after the H radical cleaning, as shown in Figure 1(a), indicating that the native oxides and/or contaminations were removed by the cleaning process using H radicals. Then, O radicals were irradiated onto the clean surface of ZnTe. The RHEED patterns after the O radical irradiation showed halo like patterns, indicating that some amorphous oxides were formed on the surface of ZnTe owing to the O irradiation. Because the spot pattern disappeared with increasing irradiation time, the thickness of the oxidation layer increased with increasing the O radical irradiation time.

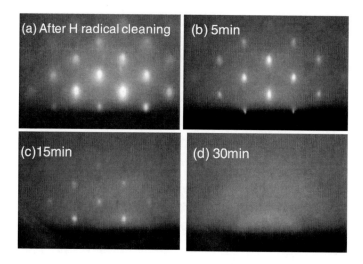

Figure 1. RHEED patterns of ZnTe after H radical cleaning (a), and following O radical irradiation for 5 (b), 15 (c), and 30 min (d) [23]. Copyright 2008 The Japan Society of Applied Physics.

Figure 2. High-resolution TEM image of a cross section of sample prepared by O radical irradiation for 30 min [23]. Copyright 2008 The Japan Society of Applied Physics.

Figure 2 shows an HR-TEM image of a cross section of the sample prepared by O radical irradiation for 30 min. The oxidation layer of approximately 2.3 nm thickness was observed between ZnTe and the Al film. Considering that the oxidation layer was prepared by O radical irradiation, the uniformity of the thickness of the oxidation layer was fairly good. Al atoms diffused through this oxidation layer during the thermal diffusion.

Figure 3 shows the depth profiles of the Al concentration of the Al-diffused samples prepared by O radical irradiation. Here, the low concentration of Al in the Al layer is an artifact due to the detector saturation. The Al concentration was found to decrease with increasing O radical irradiation time. Because the thickness of the oxidation layer increased with increasing O radical irradiation time, the Al concentration on the surface of ZnTe decreased with increasing the O radical irradiation time, leading to a decrease in the Al concentration in the diffused layer. In thermal diffusion, the surface concentration of an impurity is equal to the solid solubility of the impurity in a material at a given temperature. To reduce the surface concentration, the diffusion temperature must be lowered, which results in a decrease in the diffusion coefficient of the impurity, leading to a shallower diffusion depth. Therefore, it is worth noting that control of the surface concentration of Al in ZnTe is possible by using a thin oxidation layer prepared by O radical irradiation. A rapid decrease in the Al concentration was observed at the concentration of approximately 10^{18} cm^{-3} in all the samples, yielding a fairly steep interface between the diffused layer and the substrate. This results from the fact that the diffusion coefficient of Al is dependent on the Al concentration in ZnTe, as has been reported previously [30].

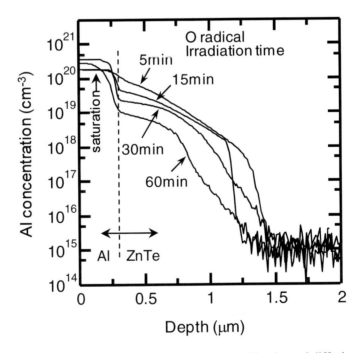

Figure 3. Depth profiles of Al concentration of samples fabricated by thermal diffusion through oxidation layer prepared by O radical irradiation [23]. Copyright 2008 The Japan Society of Applied Physics.

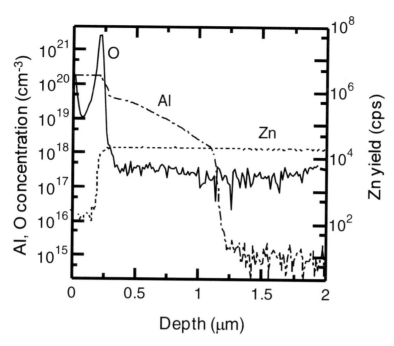

Figure 4. Depth profiles of O and Al concentrations measured by SIMS [23]. Copyright 2008 The Japan Society of Applied Physics.

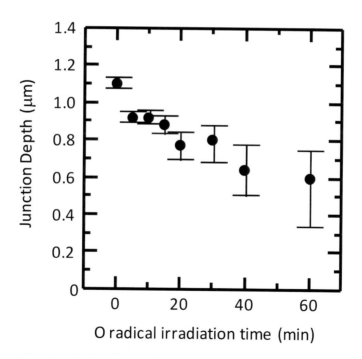

Figure 5. Variation of junction depth with O radical irradiation time [23]. Copyright 2008 The Japan Society of Applied Physics.

To determine whether the O atoms diffused simultaneously, the depth profile of the O concentration was measured by SIMS, and the result is shown in Figure 4. The depth profile of the Al concentration is also shown in Figure 4 for comparison. It was confirmed that the O atoms barely diffused into the ZnTe at the diffusion temperature of 420 °C. The diffusion coefficient of O appears to be relatively low in ZnTe at this diffusion temperature.

The depths of the pn-junction of the samples with an LED structure were evaluated by EBIC measurements. Figure 5 shows the variation of the junction depth with O radical irradiation time. Consistent with the change in the depth profile of the Al concentration shown in Figure 3, the junction depth became shallower with increasing O radical irradiation time. However, the uniformity of the junction depth also decreased with increasing O radical irradiation time. This is probably due to increased in-plane variation of the thickness of the oxidation layer, which causes a difference in the Al concentration on the surface of ZnTe.

Figure 6 shows the current-voltage (*I-V*) curves of the ZnTe LEDs fabricated with samples irradiated with O radicals with different durations. A bias voltage was applied from - 2 to 7 V. Rectification behavior was observed in all the samples in the initial stage. The series resistance and turn-on voltage were found to decrease as the O radical irradiation time increased. Because the difference in these samples was the Al concentration near the surface of ZnTe, as shown in Figure 3, the presence of the heavily Al-doped layer resulted in a high resistance. According to the photoluminescence spectra of an Al-diffused ZnTe layer[21], a high compensation was observed in the region with a high Al concentration near the surface of ZnTe. Therefore, excessive doping of Al was found to cause the formation of a highly resistive layer owing to the compensation effect. On the other hand, it was found that the presence of the oxidation layer did not affect the current transport in the LEDs. Since the oxidation layer was quite thin, as shown in Figure 2, the current may pass through the oxidation layer by tunneling.

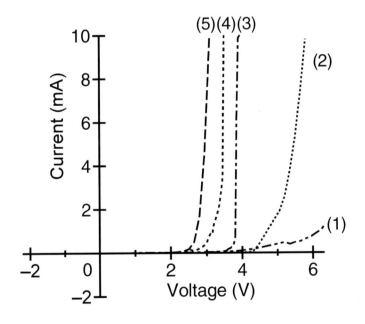

Figure 6. *I-V* curves of the ZnTe LED fabricated using samples without (1) and with O radical irradiation for (2) 5, (3) 20, (4) 30, and (5) 40 min [23]. Copyright 2008 The Japan Society of Applied Physics.

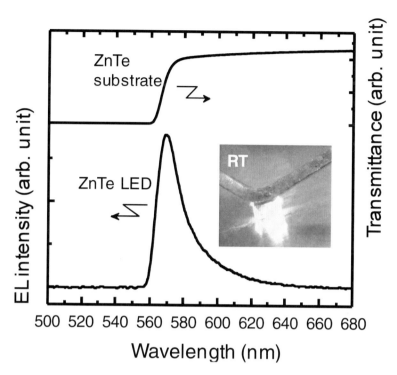

Figure 7. Electroluminescence spectrum of ZnTe LED and transmission spectrum of ZnTe substrate. The inset is a photograph of a ZnTe LED operating at room temperature [23]. Copyright 2008 The Japan Society of Applied Physics.

Figure 8. RHEED patterns of ZnTe after Al oxide layer deposition for 10 min on clean ZnTe surface [25]. Copyright 2009 The Japan Society of Applied Physics.

LEDs fabricated at an O radical irradiation time at less than 15 min did not exhibit EL at room temperature. In these samples, the rectification behavior deteriorated and the leakage current increased with every *I-V* measurement. This results from the breakdown of the diode structure owing to the increased electric field at the highly resistive region.

On the other hand, LEDs fabricated at O radical irradiation time between 20 and 40 min exhibited strong EL at room temperature, and the rectification behavior was observed repeatedly. When the O radical irradiation time exceeded 40 min, the EL intensity became weak. This is probably due to the lowered doping level of Al or the increase in the in-plane variation of the junction depth with increasing O radical irradiation time. Therefore, it is important to control the O radical irradiation time in a suitable range to obtain strong EL intensity.

Figure 7 shows an EL spectrum at room temperature of a ZnTe LED fabricated at an O radical irradiation time of 40 min, and the inset is a photograph of the LED. The EL spectrum was measured from the side wall of the LED tip since both the front and back sides of the LED were covered with contacts. The emission peak was observed at 570 nm. Because the LED had a homojunction structure, the near-band-edge emission at approximately 550 nm was completely absorbed by the ZnTe itself, as was observed by comparing the EL spectrum with the transmission spectrum of ZnTe. Despite such a self-absorption effect due to ZnTe, a light output power as high as 6 μW was obtained under a current density of 8 A/cm^2 at room temperature.

IMPROVED LED PERFORMANCE BY USING AL OXIDE LAYER AS A DIFFUSION LIMITING LAYER

In the former section, we described the increase of the output power of ZnTe LED by the thermal diffusion technique using the diffusion limiting layer. The presence of a surface oxidation layer improves the LED performance by the formation of an Al-doped layer with a moderate concentration of Al. However, the uniformity of the diffusion depth (the distance between the ZnTe and pn-junction) became worse with increasing irradiation time of O radicals, probably because of a lack of uniformity in the irradiation of O radicals. Therefore, the further improvement of LED performance was expected by another approach to reduce the in-plane variation of the diffusion depth. In this section, we describe Al thermal diffusion in ZnTe through an Al oxide layer deposited on the clean surface of ZnTe, instead of the surface oxide layer prepared by O radical irradiation.

Al diffusion was carried out in the same manner as reported in the previous section, except for the deposition of Al oxide layer. The Al oxide layer was deposited by simultaneous irradiation with both an Al beam evaporated using a Knudsen cell and O radicals onto the clean surface of ZnTe. The Al oxide layer deposition time was varied between 0 and 15 min. Then, an Al thin film with a thickness of 200 nm was deposited on the Al oxide layer. Thermal diffusion was carried out in an electrical furnace at 420 °C for 5 and 20 h. After the thermal diffusion, the Al thin film remaining on the surface of ZnTe was used as an n-type contact, and a Pd film was deposited on the back of the substrate to form a p-type contact.

Figure 8 shows a RHEED pattern of the ZnTe substrate after the deposition of the Al oxide layer for 10 min on a clean surface of ZnTe. The RHEED pattern is composed of an amorphous halo pattern and weak spots from the ZnTe substrate, indicating that the surface of the ZnTe substrate is covered with the amorphous Al oxide layer. Since the spots from the

ZnTe substrate disappeared with increasing Al oxide layer deposition time, the thickness of the Al oxide layer was concluded to increase with the deposition time.

Figures 9(a) and 9(b) show HR-TEM images of the cross sections of samples with an Al oxide layer deposited for 5 min before and after the thermal diffusion process, respectively. The thickness of the Al oxide layer was approximately 2.3 nm before the thermal diffusion, and its uniformity was found to be good. After the thermal diffusion, the thickness of the Al oxide layer increased to approximately 3.0 nm. This increase in the thickness is suspected to be due to the formation of an Al-rich Al oxide layer resulting from an increase in the Al concentration in the Al oxide layer by the thermal diffusion of Al. It is important that the Al oxide layer did not disappear and maintained a sharp interface between the Al oxide and ZnTe even after the thermal diffusion process, indicating that the use of an Al oxide layer is effective in the thermal diffusion process.

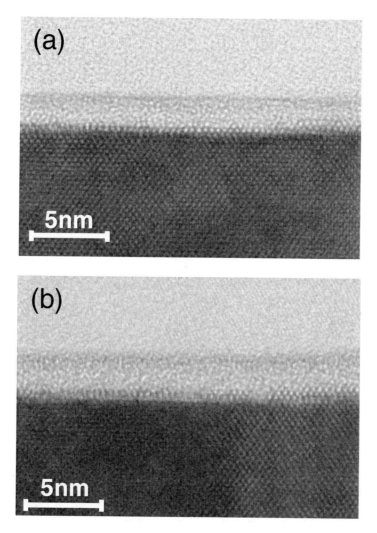

Figure 9. High-resolution TEM images of cross section of sample with an Al-oxide layer deposited for 5 min (a) before and (b) after thermal diffusion process[25]. Copyright 2009 The Japan Society of Applied Physics.

Figures 10(a) and 10(b) show Al 2p and O 1s XPS spectra of the Al oxide layer deposited for 60 min on ZnTe, respectively. In Figs. 10(c) and 10(d), the spectra of α-Al₂O₃ measured under the same condition are given for comparison. Since the XPS measurements were performed *ex situ*, native oxides on the surface of both the Al oxide layer and α-Al₂O₃ were removed by Ar ion sputtering just before the measurements. The chemical composition of the Al oxide layer was determined to be Al:O=39:61 within an accuracy of ±1%. Since the chemical shifts of Al 2p and O 1s observed in the Al oxide layer on ZnTe were in good agreement with those observed in α-Al₂O₃, the Al and O atoms in the Al oxide layer were bonded as Al₂O₃. From the above XPS and RHEED results, it was concluded that the Al oxide layer on ZnTe was amorphous Al₂O₃.

In order to evaluate the electrical conductivity of the Al₂O₃ layer on ZnTe, test samples with structures of (a) Al/p-ZnTe/Pd and (b) Al/Al₂O₃/p-ZnTe/Pd were prepared. Here, Pd is an ohmic contact for p-ZnTe. The thickness of the Al₂O₃ layer in sample (b), which was deposited for 10 min, was estimated to be 4.6 nm. Figure 11 shows the current-voltage (*I-V*) curves of the samples. Sample (b) showed approximately 20 Ω higher series resistance than sample (a) owing to the presence of the Al₂O₃ layer. From the difference in series resistance, the resistivity of Al₂O₃ layer was roughly estimated to be on the order of 10^5 Ω cm.

Figure 12 shows the depth profiles of the Al concentration in the samples fabricated by thermal diffusion at 420 °C for 20 h through the Al₂O₃ layer prepared with three different deposition times. Here, the low concentration of Al in the Al layer is an artifact owing to detector saturation. The Al concentration in the diffused layer was found to decrease with increasing deposition time of the Al₂O₃ layer

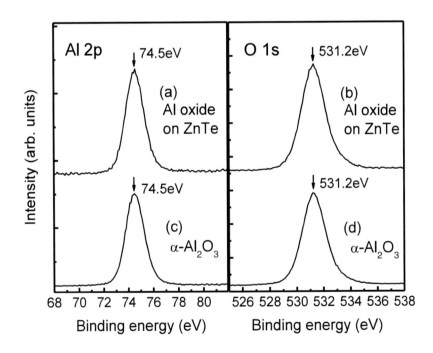

Figure 10. Al 2p and O 1s XPS spectra for Al-oxide layer deposited for 60 min on ZnTe. Those for α-Al₂O₃ measured under the same conditions are shown for comparison[25]. Copyright 2009 The Japan Society of Applied Physics.

Because the thickness of the Al_2O_3 layer increased with increasing deposition times, the concentration of Al atoms arriving at the ZnTe surface decreased, which results in the decrease of Al concentration in the diffused layer.

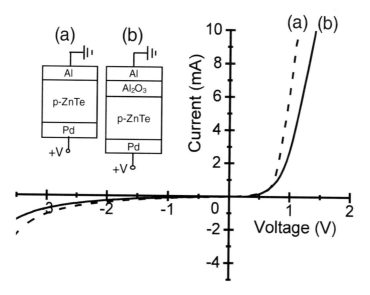

Figure 11. *I-V* curves of samples with structures of (a) Al/p-ZnTe/Pd and (b) Al/Al_2O_3/p-ZnTe/Pd [25]. Copyright 2009 The Japan Society of Applied Physics.

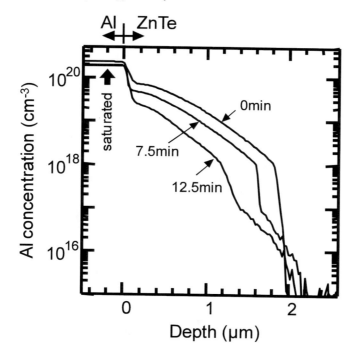

Figure 12. Depth profiles of Al concentration in samples fabricated by thermal diffusion at 420 °C for 20 h through Al_2O_3 layer prepared at three different deposition times[25]. Copyright 2009 The Japan Society of Applied Physics.

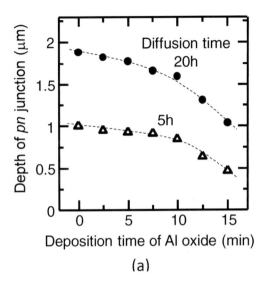

(a)

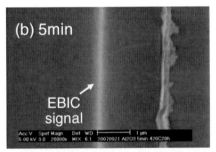

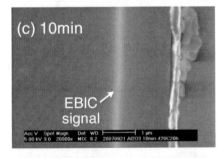

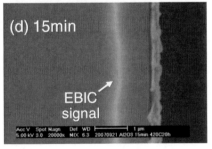

Figure 13. (a) Variations in the depth of pn-junction in samples with LED structure fabricated at diffusion time of 5 or 20 h as a function of Al_2O_3 deposition time. The combined SEM images and EBIC signals of the samples fabricated by thermal diffusion at 420 °C for 20 h using Al_2O_3 layers deposited for (b) 5, (c) 10, and (d) 15 min [25]. Copyright 2009 The Japan Society of Applied Physics.

Thus, the Al_2O_3 layer clearly plays an important role as "a diffusion-limiting layer" in the thermal diffusion of Al into ZnTe. The fact that the control of the surface concentration of Al in ZnTe is possible when the Al_2O_3 layer is inserted between Al and ZnTe without changing the diffusion temperature indicates that this technique is a useful approach for fabricating ZnTe-based LEDs.

In good agreement with our previous study[30] and the results described in the former section, a rapid decrease in the Al concentration was observed at an Al concentration of approximately 10^{18} cm^{-3} in all the cases in Figure 12, yielding a fairly steep interface between the diffused layer and the substrate. This results from the fact that the diffusion coefficient of Al is concentration-dependent in ZnTe. The depth profile of O concentration was also measured by SIMS, and it was confirmed that O atoms only nominally diffused into ZnTe at the diffusion temperature of 420 °C used in the experiments.

Figure 13(a) shows the variations in the depth of the pn-junction in the samples with an LED structure fabricated with a diffusion time of 5 or 20 h, as a function of the deposition time of the Al_2O_3 layer. In both cases, the depth of the pn-junction became shallower with increasing deposition time of the Al_2O_3 layer, consistent with the change in the depth profiles of the Al concentration shown in Figure 5. Figures 13(b)-13(d) show combined images of the SEM micrograph and the EBIC signal for the samples fabricated by thermal diffusion at 420 °C for 20 h using Al_2O_3 layers deposited for 5, 10, and 15 min. The uniformity of the junction depth was quite good in the samples with the Al_2O_3 layer deposited for less than 10 min, but it became slightly worse when the deposition time of Al_2O_3 exceeded 10 min. However, the in-plane variation of the junction depth was only less than ±0.1 μm, even in the sample with the Al_2O_3 deposition time of 15 min. Therefore, it can be concluded that the in-plane variation of the junction depth can be reduced by using the Al_2O_3 layer as a diffusion-limiting layer, compared with the case of using the oxide layer prepared by O radical irradiation (more than ±0.2 μm) as described in the former section.

Figure 14 shows the I-V curves of ZnTe LEDs fabricated at various deposition times of

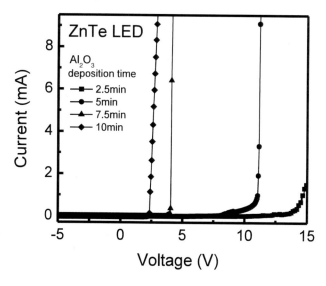

Figure 14. *I-V* curves of ZnTe LEDs fabricated at various deposition times of Al_2O_3 layer. Bias voltage from -5 to 15 V was applied [25]. Copyright 2009 The Japan Society of Applied Physics.

the Al_2O_3 layer. A bias voltage was applied from -5 to 15 V. When the deposition time of Al_2O_3 was as short as 2.5 or 5 min, the turn-on voltage and series resistance were quite high. The Al concentration was so high when the deposition time was less than 5 min that the existence of the heavily Al-doped layer led to a high resistance. According to the photoluminescence spectra of an Al-diffused ZnTe layer[21], a high compensation was induced in the region with high Al-concentration near the surface of ZnTe. Therefore, excessive doping of Al was found to cause the formation of a highly resistive layer as a result of the compensation effect.

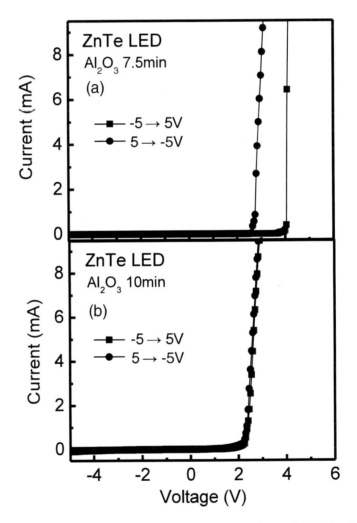

Figure 15. *I-V* curves of ZnTe LEDs fabricated at Al_2O_3 deposition times of (a) 7.5 and (b) 10min. The bias voltage from -5 to 5 V or from 5 to -5V was applied [25]. Copyright 2009 The Japan Society of Applied Physics.

When the deposition time of Al_2O_3 increased to 7.5 min, the turn-on voltage was reduced to approximately 4 V. However, the current increased almost vertically above the turn-on voltage, indicating that breakdown took place. As shown in Figure 15(a), when the bias voltage was applied from 5 to -5 V, the turn-on voltage was reduced to 2 V. This behavior indicates that an avalanche breakdown was involved in the current transport in the LED fabricated at the Al_2O_3 deposition time of 7.5 min. Although the Al concentration at the surface of the diffused layer decreased with increasing thickness of the Al_2O_3 layer, a highly resistive thin layer might still remain in this sample, causing the avalanche breakdown. On the other hand, a rectification behavior with a turn-on voltage of approximately 2 V was observed in the sample fabricated at the deposition time of Al_2O_3 for 10 min. Also, the behavior related to the avalanche breakdown was never observed in the samples fabricated at the Al_2O_3 deposition time above 10 min, as shown in Figure 15 (b). Therefore, it is found that the highly resistive layer is not present in these samples.

Figure 16 shows variations in the light output from ZnTe LEDs as a function of the deposition time of the Al_2O_3 layer. Since the top and back surfaces of the LED were covered with the contact metal, the luminescence was extracted only from the side wall of the LED tip. The light output was measured using an integrating sphere with a multichannel photodetector under a forward current density of 8 A/cm^2. The LEDs fabricated using the Al_2O_3 layer deposited for less than 2.5 min did not show any EL at room temperature. When the deposition time of the Al_2O_3 layer was increased to 5 or 7.5 min, strong EL was observed in the LEDs fabricated at the diffusion time of 5 or 20 h, respectively. A photograph of the ZnTe LED operating at room temperature is shown in the inset in Figure 16. In these LEDs, the luminescence was observed from a small area in the LED chip, indicating that current flow was localized in a particular area. This is related to the avalanche breakdown observed in the I-V curves. The reason for the high light output in these samples is not yet clear, but the presence of an avalanche plasma might exert a positive effect on luminescence. Further increase in deposition time of the Al_2O_3 layer resulted in a decrease in light output.

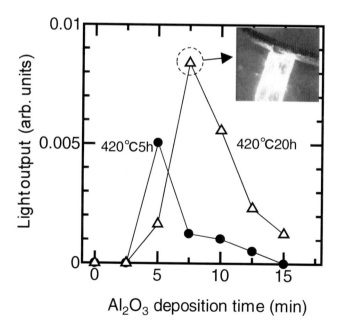

Figure 16. Variations in light output from ZnTe LEDs as a function of Al_2O_3 deposition time. Photograph of ZnTe LED operating at room temperature is shown in the inset [25]. Copyright 2009 The Japan Society of Applied Physics.

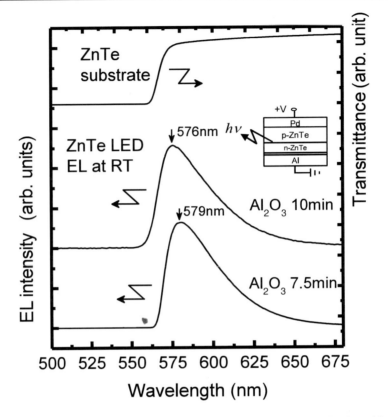

Figure 17. EL spectra measured from side wall of ZnTe LEDs fabricated at Al_2O_3 deposition times of 7.5 and 10 min under a forward current density of 4 A/cm^2, and transmission spectrum for ZnTe substrate with a thickness of 500 μm [25]. Copyright 2009 The Japan Society of Applied Physics.

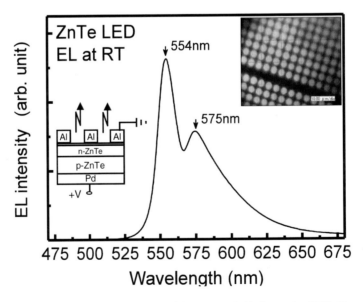

Figure 18. EL spectrum and photograph of LED with patterned Al electrodes [25]. Copyright 2009 The Japan Society of Applied Physics.

In these LEDs, the luminescence was observed from the entire junction plane, consistent with the disappearance of the avalanche behavior. Therefore, we consider that the luminescence takes place through the recombination of injected minority carriers in these LEDs. The decrease in light output may be partially due to the increase in the in-plane variation of the junction depth with increasing Al_2O_3 deposition time. The maximum light output of the LED fabricated in this experiment was as high as 18 µW under a forward current density of 8 A/cm^2, which is three times higher than those of LEDs described in the former section.

Figure 17 shows the EL spectra measured from the side wall of ZnTe LEDs fabricated at the Al_2O_3 deposition times of 7.5 and 10 min under a forward current density of 4 A/cm^2, as well as the transmission spectrum for the ZnTe substrate with a thickness of 500 µm. For both LEDs, the emission peak was observed at approximately 580 nm in spite of the different current transport behavior. Therefore, the origin of the emission peak was inferred to be the same. Upon comparing the EL spectra with the transmission spectrum of ZnTe shown in Figure 17, it is obvious that the emission of wavelength shorter than 580 nm is completely absorbed by ZnTe itself because LEDs have a homojunction structure. In order to clarify the origin of the emission, the Al film was patterned by photolithography. The EL spectrum and a photograph of the LED are shown in Figure 18. The band-edge emission peak was clearly observed at 554 nm, while the emission occurred only under the Al contact because of the large spreading resistance. Therefore, the origin of the emission can be assigned to band-to-band transition, and the emission peak observed at approximately 580 nm results from the self-absorption effect by ZnTe.

ENHANCED OUTPUT POWER OF ZNTE LED BY ADAPTING THIN FILM STRUCTURE

As described in the former section, the influence of the self-absorption was significant in the homo-junction LED. The light output is expected to be improved greatly by suppressing the self-absorption effect. In this section, we describe the enhancement of the performance in the ZnTe LED by utilizing a thin film structure. The output power of the LED is considerably increased and the red-shift is improved significantly, leading to the wall-plug efficiency more than 0.15 %.

The pn-junction was fabricated in the same manner as described in the former section. In order to reduce the thickness of p-ZnTe side, e.g., down to several µm, after the thermal diffusion, the sample was mounted on the highly conductive p-ZnTe wafer having Pd ohmic contacts on both sides, and p-ZnTe side of the sample was lapped by mechanical polishing and chemical etching. Then, Pd ohmic contact was deposited on the top surface of p-ZnTe using photolithography, and each LED tip was isolated with the area of 430 × 430 µm^2 by reactive ion etching (RIE) using CH_4 and H_2 gases. The detail of the RIE process was reported in ref [31]. Finally, the LED dies with a dimension of 450 × 450 × 300 µm^3 were separated using a conventional dicing machine, and were encapsulated into the epoxy.

Figure 19 shows the scanning electron microscope (SEM) image of ZnTe LED die with the structure having the reduced thickness at p-ZnTe side, i.e., the thin film structure. The inset shows the bird-view image near the edge of thin film LED. It is confirmed that relatively steep edge is formed by RIE process and that the thickness of LED is approximately 4 µm.

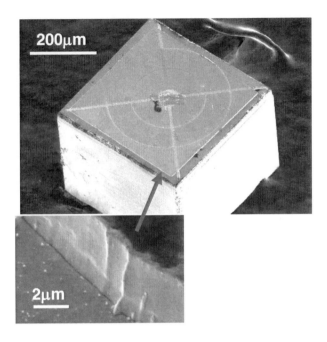

Figure 19. SEM image of ZnTe LED die with thin film structure. The inset shows the bird-view image near the edge of LED [29]. Copyright 2009 The Japan Society of Applied Physics.

The *I-V* characteristic of ZnTe LE D at RT is reported on Figure 20. The device displays a low operating voltage of 2.95 V at 10 mA and a series resistance of 56 Ω. The high series resistance is partly attributed to the existence of Al_2O_3 layer at the interface between Al and ZnTe. As written in the former section, the resistivity of Al_2O_3 layer is estimated to be 20 Ω. As a result, it can be said that fairly low conductivity of the Al-diffused layer is formed. Further reduction could be possible by replacing Al_2O_3 with other conductive materials and by optimizing the thermal diffusion condition. From the inset of Figure 20, a leakage current of 20 μA at -5 V can be determined. The somewhat large leakage current might be due to the surface current since the pn-junction is exposed to the air at the LED edges, as shown in Figure 19.

Figure 21 shows an optical microscope image of LED under the forward bias. The optical microscope image shows that the full area is free of cracks, even after the LED processing. This fact may be important, since ZnTe is so brittle that cracks can be easily induced by mechanical polishing. Therefore, the device process must be optimized to avoid cracking. This is a key point for the realization of good working devices, since cracks can cause short-circuits of the pn-junction and decrease the output power.

Figure 22 shows the EL spectrum of the thin film LED. The EL spectrum of LED with a thickness of 500 μm, in which the lapping process was not carried out, is also shown in Figure 22 for comparison. The injection current density was fixed at 5.4 A/cm^2. The thin film LED emits green light at 557 nm whereas the thick LED has a weak emission peak at 582 nm. As a result of the thickness reduction in p-ZnTe side, the emission peak can be observed at shorter wavelength close to the band edge emission. The difference in the EL spectrum between thin film LED and the thick LED will be due to the strong self-absorption effect. It can be found that the self-absorption is dramatically suppressed and the light extraction is greatly enhanced. Thus, we obtain the LED with an increase in output power more than 10 times and a shortened emission peak.

The EL spectra of ZnTe LED with injection currents of 1, 3, 5, and 10 mA at RT are shown in Figure 23. The emission peak is observed at 557 nm independent of the injection current. The asymmetric EL spectra are attributed to the fact that the band-edge emission will be still absorbed by ZnTe itself. The output power of the LED as a function of the injection current is shown in the inset of Figure 23.

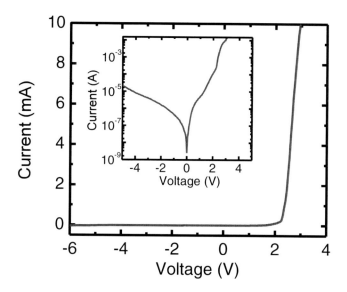

Figure 20. I-V characteristic of ZnTe LED at room temperature. The inset shows the same characteristic with a log scale [29]. Copyright 2009 The Japan Society of Applied Physics.

Figure 21. Photo of LED under the forward bias [29]. Copyright 2009 The Japan Society of Applied Physics.

The increase in output power with current is nearly linear, indicating minimal heating effects. At 10mA, the ZnTe LED emitted approximately 45 µW. Therefore, the wall-plug efficiency of the ZnTe LED is estimated to be more than 0.15 %. This efficiency is comparable to that obtained in commercially available GaP LEDs with an emission peak at 557 nm.

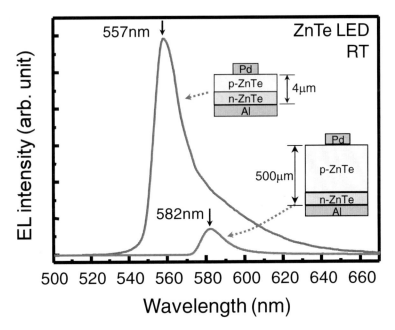

Figure 22. EL spectrum of LED with thin film structure. EL spectrum of LED with a thickness of 500 µm is also shown for comparison. Current density is fixed at 5.4 A/cm². [29] Copyright 2009 The Japan Society of Applied Physics.

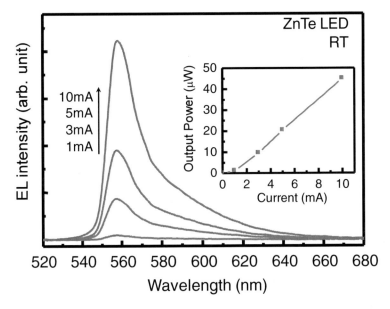

Figure 23. EL spectra of ZnTe LED with injection currents of 1, 3, 5, and 10 mA at room temperature. Output power as a function of the injection current is shown in the inset. [29] Copyright 2009 The Japan Society of Applied Physics.

CONCLUSION

In this chapter, we reviewed our recent research results on the development of ZnTe LEDs with a special emphasis on the thermal diffusion technique through the oxide layer, which plays an important role as a diffusion limiting layer. Using the Al-diffused layer with a controlled Al-concentration, the output power of ZnTe LED was increased significantly. In order to avoid the self-absorption effect, ZnTe LED with a thin film structure was developed. Finally, the output power of ZnTe LED was achieved to approximately 45 μW with a peak wavelength of 557 nm under the injection current of 10 mA at RT, resulting in the wall-plug efficiency of more than 0.15 %. This efficiency is comparable to that obtained in commercially available GaP green LED. Further increase in the efficiency is expected by realizing LED with double hetero structure, and therefore, ZnTe is promising for future high-efficiency green LED.

REFERENCES

[1]   M. G. Miksic, G. Mandel, F. F. Morehead, A. A. Onton, and E. S. *Schlig, Phys. Lett.* 11 (1964) 202.

[2]   P. C. Eastman, R. R. Haering, and P. A. Barnes, *Solid-State Electron.* 7 (1964) 879.

[3]   B. L. Crowder, F. F. Morehead, and P. R. Wagner, *Appl. Phys. Lett.* 8 (1966) 148.

[4]   D. Bortfeld and H. Kleinknecht, *J. Appl. Phys.* 39 (1968) 6104.

[5]   J. P. Donnelly, A. G. Foyt, W. T. Lindley, and G. W. Iseler, *Solid-State Electron.* 13 (1970) 755.

[6]   J. Gu, K. Tonomura, N. Yoshikawa, and T. Sakaguchi, *J. Appl. Phys.* 44 (1973) 4692.

[7]   J. L. Tissot, G. Labrunie, and J. Marine, *IEEE Trans. Electron Devices* 26 (1979) 1202.

[8]   D. Bensahel, J. L. Tissot, L. Revoil, and J. C. Pfister, *J. Lumin.* 21 (1980) 259.

[9]   D. I. Kennedy and M. J. Russ, *J. Appl. Phys.* 38 (1967) 4387.

[10] D. I. Kennedy and M. J. Russ, *Solid-State Electron.* 11 (1968) 513.

[11] D. I. Kennedy and M. J. Russ, *J. Phys. Chem. Solids* 32 (1971) 847.

[12] V. N. Iodko and A. K. Belyaeva, *Mater. Sci. Forum* 182-184 (1995) 353.

[13] V. N. Iodko, V. P. Gribkovskii, A. K. Belyaeva, Y. R. Suprun-Belevich, and Z. A. Ketko, *J. Cryst. Growth* 184-185 (1998) 1170.

[14] K. Sato, T. Asahi, M. Hanafusa, A. Noda, A. Arakawa, M. Uchida, O. Oda, Y. Yamada, and T. Taguchi, *Phys. Status Solidi* A 180 (2000)267.

[15] K. Sato, M. Hanafusa, A. Noda, A. Arakawa, M. Uchida, T. Asahi, and O. Oda, *J. Cryst. Growth,* 214-215 (2000) 1080.

[16] K. Sato, M. Hanafusa, A. Noda, A. Arakawa, T. Asahi, M. Uchida, and O. Oda, *IEICE Trans. Electron.* E83-C (2000) 579.

[17] T. Tanaka, Y. Kume, M. Nishio, Q. Guo, H. Ogawa, and A. Yoshida, *Jpn. J. Appl. Phys.* 42 (2003) L362.

[18] J. H. Chang, T. Takai, K. Godo, J. S. Song, B. H. Koo, T. Hanada, and T. Yao, *Phys. Status Solidi* B 229 (2002) 995.

[19] K. Kishino, I. Nomura, Y. Ochiai, and S. B. Che, *Phys. Status Solidi* B 229 (2002) 991.

[20] I. Nomura, Y. Ochiai, N. Toyomura, A. Manoshiro, A. Kikuchi, K. Kishino, *Phys. Stat. Solidi* B 241 (2004) 483.

[21] T. Tanaka, Y. Matsuno, Y. Kume, M. Nishio, Q. Guo, and H. Ogawa, *Phys. Status Solidi* C 1 (2004) 1026.

[22] T. Tanaka, K. Hayashida, K. Saito, M. Nishio, Q. Guo, and H. Ogawa, *Phys. Status Solidi* B 243 (2006) 959.

[23] T. Tanaka, M. Nishio, Q. Guo, and H. Ogawa, *Jpn. J. Appl. Phys.* 47 (2008) 8408.

[24] T. Tanaka, K. Saito, M. Nishio, Q. Guo, and H. Ogawa, *J. Mater. Sci.: Mater. Electron.* 20 (2009) S505.

[25] T. Tanaka, M. Nishio, Q. Guo, and H. Ogawa, *Jpn. J. Appl. Phys.* 48 (2009) 022203.

[26] H. Ogawa, S. I. Gheyas, H. Nakamura, M. Nishio, and A. Yoshida, *Jpn. J. Appl. Phys.* 33 (1994) L980.

[27] I. W. Tao, M. Jurkovic, and W. I. Wang, *Appl. Phys. Lett.* 64 (1994) 1848.

[28] J. H. Chang, T. Takai, B. H. Koo, J. S. Song, T. Handa, and T. Yao, *Appl. Phys. Lett.* 79 (2001) 785.

[29] T. Tanaka, K. Saito, M. Nishio, Q. Guo, and H. Ogawa, *Appl. Phys. Express* 2, 122101 (2009).

[30] T. Tanaka, N. Murata, K. Saito, Q. Guo, M. Nishio, and H. Ogawa, *Phys. Status Solidi* B 244 (2007) 1685.

[31] Q. X. Guo, M. Matsuse, T. Tanaka, M. Nishio, H. Ogawa, Y. Chang, J. Wang, and S. L. Wang, *J. Vac. Sci. Technol.* A 19 (2001) 2232.

In: Light-Emitting Diodes and Optoelectronics: New Research   ISBN: 978-1-62100-448-6
Editors: Joshua T. Hall and Anton O. Koskinen   © 2012 Nova Science Publishers, Inc.

Chapter 7

SELF-INTRODUCED LATTICE DISTORTION, INVISIBLE CAVITY AND HIDDEN COLLECTIVE BEHAVIOR OF A POLYMERIC NANOFIBER LASER

Sheng Li[1,2] Wei-Feng Jiang[1] and Thomas F. George[2]

[1]Department of Physics, Zhejiang Normal University, Jinhua, Zhejiang, China
[2]Office of the Chancellor and Center for Nanoscience,
Departments of Chemistry & Biochemistry and Physics & Astronomy,
University of Missouri-St. Louis, St. Louis, Missouri, US

ABSTRACT

A polymeric laser generally undergoes two unconventional processes: (1) microscopic lattice and electronic evolution and (2) macroscopic localization of light emission. After an external pulsed laser beam is applied to pump a conjugated polymer comprised of parallel polymer fibers, such as polythiophene, the external excitation immediately destroys the periodic structure of the polymer chain, self-introducing localized lattice distortion along the polymer fiber chain. Along with the continuous optical pumping, the electron populations of the exciton in a single polymer fiber are reversed. With multiple light scattering in the fibers, the external gain then counteracts the leakage that causes non-coherent light emission, and finally localizes the light in the middle of the fiber bunch. Concurrent with the localization of the polymer fiber laser, an associated hidden collective behavior is also uncovered. It is revealed that the multiple scattering, instead of phase tuning in the traditional resonator, results in the cavity of the fiber laser to be comprised of only randomly-distributed polymer fibers, and accordingly, to be "invisible."

I. INTRODUCTION

Solid-state lasers with organic materials can date back to the 1960s when Soffer and McFarland fabricated a laser based on dye-doped polymers. [1] In 1974, Avanesjan *et al.* demonstrated the first laser based on a single crystal of fluorine-doped anthracene.[2] In spite of considerable research and effort during the 1960s and 1970s, [1-3] the shortage of high-quality organic single crystals limited the development of lasers based on organic materials. By the end of the 1980s, the inventions of organic transistors [4, 5] and polymer light-emitting diodes (PLEDs) [6] finally broke this limitation.

In 1992, the first organic semiconductor laser was developed in a solution consisting of a conjugated polymer. [7] In 1996, this was extended to conjugated polymers [8-11] where, in particular, Tessler *et al.* demonstrated an optically-pumped polymer microcavity laser. [8] Since then, it has been a topic of vigorous research. Conjugated polymer lasers with a variety of resonators, such as distributed feedback [12] and photonic bandgap structures, [13] have been extensively fabricated by using a range of coating and imprinting techniques, and launching a new field known as "plastic lasers." The one-dimensional nature of a conjugated polymer gives it flexibility. By taking advantage of this, the first single conjugated polymer nanowire laser was realized. [16] Further, the polymeric laser is also a bit dependent on temperature. [15] These prominent virtues of conjugated polymers have thereby paved the way to fabricate nanolasers.

When Quachi *et al* isolated individual *p*-sexiphenyl nanofiber lasers, it was observed that the lasing emission is localized along the nanofibers. [17-19] Such emission is attributed to coherent light propagation in a one-dimensional random *p*-sexiphenyl nanofiber, where the disorder structure is introduced into polymer chain by the air gaps breaking among the single nanofiber. Here, a question is raised as to whether there exists new channel to self-introduce the disorder to the one-dimensional polymer chain besides the air gap. To answer this, we have to pull the focus back to the "flexibility" of a conjugated polymer. The so-called "flexibility" simply reflects that the lattice of a conjugated polymer sensitively depends on the electronic state. The laser, as a typical excitation of electrons, is triggered by electron population inversion. Considering just this, an assumption is naturally proposed that it is highly possible for the polymeric laser to destroy the periodic bond structure of the polymer, thus inducing the occurrence of disorder structure. To clarify the assumption, the description both of electronic states and the lattice structure become key steps.

For the electronic behavior of a polymer laser, through comparing it with the contribution of an exciton to the luminescence of a polymer light-emitting diode (PLED), scientists have generally regarded organic lasing effect as an excitonic behavior. [15] Actually, an exciton (electron-hole pair) has no electron population inversion, which means it is ill-suited to describe the lasing effect. Thus, the first challenge facing us is to describe the process of an electron transition in the polymeric laser, and then to search out the new states contributing to the polymeric lasing effect. Further, due to the strong electron-lattice coupling as mentioned before, the lattice structure is sensitive to a change of electronic states. Thus, it is necessary to develop a viable method that is not only with respect to the evolution of electron transition, but also links the lattice structure with the electronic behavior.

Different from an inorganic laser, the one-dimensional nature of a conjugated polymer causes the polymeric lattice to sensitively depend on the electronic state, namely self-trapping, which gives the conjugated polymer special flexibility. Taking advantage of this, Quachi and D. O'Carroll fabricated conjugated polymer nanowire lasers based on polythiophene and polyfluorine (PFO) nanofibers. [16-18] Although both are made of conjugated polymer nanofibers, their cavities are different, where the PFO nanofiber laser is comprised of a Fabry-Perot resonator, while the polythiophene nanofiber laser is made up of only some randomly-distributed parallel polythiophene nanofibers. Without the help of the resonator, the laser emission of the polythiophene laser is localized on the nanofiber, forming a luminescent spot. The simple structure of the polythiopene nanofiber laser thereby paves the way to fabricate the nanolaser, although it also raises questions as to where to find the "invisible" resonator of the laser and what is the mechanism of the localization of the laser emission. One could assume the localized lasing emission to be attributed to the coherent light propagation in a one-dimensional nanofiber. Yet, on the basis of this assumption, we are unable to exhibit the "hidden" resonator of the laser.

Localized light emission is generally regarded as a collective phenomenon. The electronic behavior, however, is a typical microscopic dynamical process. Compared with the contribution of an exciton to the luminescence of a PLED, scientists have taken it for granted that organic lasing is a similar excitonic behavior.[15] A laser, as a typical excitation of electrons, is triggered by electron population inversion. Actually, an exciton (electron-hole pair) has no electron population inversion, which is ill-suited to describe the lasing effect. Furthermore, thanks to the self-trapping effect, the occurrence of the lasing effect is also able to induce the change of the lattice structure of the polymer.

Therefore, the investigation of the conjugated polymer nanofiber laser has to involve three strategic effects: self-trapping, electron population inversion and multiple scattering. How to unite the microscopic behavior with the collective behavior becomes the first key challenge. In this chapter, the molecular dynamics and multiple-scattering process are fused together to bridge the gap between the microscopic process and collective behavior, not only describing the electron transition process along with the evolution of lattice structure, but also exhibiting the localized laser emission phenomenon. Further, with a newly-developed phase distribution diagram, the invisible resonant cavity is clearly identified. The electron transition process is embedded inside the traditional method. The developed method combines the electron transition process with the self-trapping effect, and helps to extract the new state in the conjugated polymer that contributes to the lasing effect. Due to the self-trapping in the conjugated polymer, the new excitation contributing to the polymer laser is also a localized state, and its lattice structure even looks like an exciton. However, different from the exciton, the new excitation not only generates the electron population reversion, but also strengthens the localized distortion, inducing the occurrence of localized lattice disorder that spans many unit clusters/groups along the polymer chain.

II. MODEL

Firstly, for the prominent self-trapping of these conjugated polymers, the extended Hubbard-Su-Schreiffer-Heeger Hamiltonian provides a convenient and accurate description as follows: [21]

$$H = -\sum_{l,s}\{t_0 + \alpha(u_{l+1} - u_l) + (-1)^l t_e\} \times [c_{l+1,s}^+ c_{l,s} + H.c.] + \frac{K}{2}\sum_l (u_{l+1} - u_l)^2 + H' + H_E \tag{1}$$

,

$$H' = U\sum_{l,s} n_{l,\uparrow} n_{l,\downarrow} + V\sum_{l,s,s'} n_{l,s} n_{l+1,s'} . \tag{2}$$

Here, t_0 is a hopping constant; α is an electron-lattice coupling constant; $c_{l,s}^+(c_{l,s})$ denotes the electron creation (annihilation) operator with spin s at the unit cluster/group l and corresponding occupation number $n_{l,s} = c_{l,s}^+ c_{l,s}$; u_l is the displacement of unit cluster/group l; K is an elastic constant; t_e is the Brazovskii-Kirova term; and U and V are the on-site and nearest-neighbor Coulomb repulsion strengths, respectively.

In order to describe the electron's behavior, we have to know its energy spectrum ε_μ and wave function Φ_μ, which are functionals of the lattice displacement u_l, as determined by the eigenequation

$$H\Phi_\mu = \varepsilon_\mu \Phi_\mu . \tag{3}$$

Since atoms are much heavier than electrons, based on the Feynman-Hellmann theorem, the atomic movement of the lattice can be described through classical dynamics by the equation

$$M\frac{d^2 u_l}{dt^2} = -\sum_{\mu}^{occ}\frac{\partial \varepsilon_\mu}{\partial u_l} + K(2u_l - u_{l+1} - u_{l-1}) . \tag{4}$$

Further, since the polymer is not a strongly-correlated system, the electron-electron interaction term can be treated by the Hartree-Fock approximation.[20-22] Thus, assuming an electronic wave function $\Phi_\mu = \{Z_{n,\mu}^s\}$, the electron-electron interaction term H' take the form

$$H' = \sum_{l,s}\left\{U\left(\sum_{\mu}^{occ}\left|Z_{l,\mu}^{-s}\right|^2 - \frac{1}{2}\right) + V\left[\sum_{s'}\left(\sum_{\mu}^{occ}\left|Z_{l-1,\mu}^{-s'}\right|^2 + \sum_{\mu}^{occ}\left|Z_{l+1,\mu}^{-s'}\right|^2 - 2\right)\right]\right\}c_{l,s}^\dagger c_{l,s}$$
$$- \sum_{l,s}\left(V\sum_{\mu}^{occ}Z_{l,\mu}^s Z_{l+1,\mu}^s\right)\left(c_{l,s}^\dagger c_{l+1,s} + H.c.\right) \tag{5}$$

where occ stands for the occupation or population of electrons.

The traditional molecular dynamics combines all the previous equations. During the relaxation process associated with the lasing effect, the electron occupations are changed due to transitions. However, a conventional molecular dynamics approach fixes the electron occupations, which makes it difficult to describe the electronic transitions of the lasing effect. To resolve this dilemma, we introduce electron population rate equations into the molecular dynamics treatment.

To begin, if there are three electron occupied energy levels marked by a, b and c, and the gain is designated by g that pumps an electron in c to a, the evolutions of their related electron populations, P_a, P_b and P_c, are given as

$$\frac{dP_a}{dt} = gP_c - \gamma_{ab}P_a$$

$$\frac{dP_b}{dt} = \gamma_{ab}P_a - \gamma_{bc}P_b$$

$$P_c = n - P_a - P_b \,, \qquad (6)$$

where γ_{ab} (γ_{bc}) is the transition rate between energy levels a and b (b and c), and n is the total electron number.

When an external laser is focused on a wide-gap conjugated polymer nanofiber consisting of many f randomly-distributed parallel conjugated polymer chains, there is light emission from a chain just like when a dipole oscillates and emits an electromagnetic wave under an external field. Thus, if we represent a conjugated polymer chain by a line dipole, the conjugated polymer fiber comprised of parallel conjugated polymer chain becomes an ensemble of line dipoles. Accordingly, the mechanism can be imagined as follows: With the external pumping, every line dipole accepts the external gain. Meanwhile, the electric field generated by all others also drive dipoles to oscillate, thus emitting electromagnetic radiation. Upon repetition, this forms so-called multiple scattering. Thus, we shall assume N identical line dipoles, which are aligned along the z-axis and randomly embedded in a two-dimensional uniform medium, to stand for N parallel conjugated polymer chains. All lengths are scaled by the average distance between chains. The total electric field driving the ith dipole is comprised of the electric field radiating from the other dipoles, given as

$$E_i(\vec{r}) = \sum_{j \neq i}^{N} G_j(\vec{r} - \vec{r}_j) \,. \qquad (7)$$

Here, the field can be taken as $E_i(\vec{r}) = \vec{e}|\vec{E}|e^{i\phi}$. If the phase ϕ of the field in the random medium is constant, the whole system will behave as a coherent state, exhibiting the property of the lasing effect. Especially, based on the spatial distribution of coherence, the size and shape of the cavity can be clearly illustrated.

Combining the dipole oscillation, threshold behavior of the laser, [23] fusing rate equations and conventional molecular dynamics, we can quantitatively describe the dynamical evolution of not only the microscopic dynamical process of the polymeric nanofiber, but also the collective behavior of multiple light-scattering in the conjugated polymer chains. Using these coupled rate equations and conventional molecular dynamics, we can quantitatively describe the dynamical evolution of not only the electronic states but also the lattice structure in a conjugated polymer chain.

III. Results and Discussion

As mentioned in Section II, u_l is defined as the displacement of the unit cluster/group l along the polymer chain. For the convenience of presentation, in Fig. 1 a sphere is used to depict a unit cluster/group of polymer, such as polythiophenes, poly(phenylene vinylene)s or polyfluorenes. Without electron-lattice and electron-electron interactions, it can be expected that there is no displacement for any unit cluster of the polymer, and thus the clusters periodically distribute along the polymeric chain with lattice constant a, as shown in Fig. 1(a). Actually, the electron-lattice coupling of the conjugated polymer makes the state unstable and breaks the original symmetry of lattice structure. The symmetry breaking drives the unit clusters/groups to move in different directions. Finally, the unit cluster/groups of the conjugated polymer alternate between single and double bonds, giving lattice dimerization. As illustrated in Fig. 1(b), the displacement of the lth unit cluster/group is u_l, and the displacement of $(l+1)^{\text{th}}$ unit cluster/group is $-u_{l+1}$. For a clear demonstration, we choose a conjugated polymer consisting of 200 unit clusters/groups and use the lattice configuration $(-1)^l u_l$ to describe the lattice structure. Therefore, the lattice dimerization not only makes lattice configuration constant, as depicted in Fig. 2, but also produces an energy gap between the highest-occupied molecular orbital (HOMO) and lowest-occupied molecular orbital (LUMO), where the valance band, i.e., HOMO, is fully occupied while the conduction band is empty.

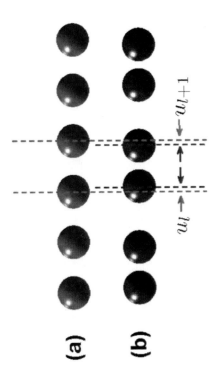

Figure 1. Periodic lattice structure. (a) periodic distribution of the units of clusters of the polymer chain with lattice constant a; and (b) lattice dimerization of the polymer chain with lattice constant a.

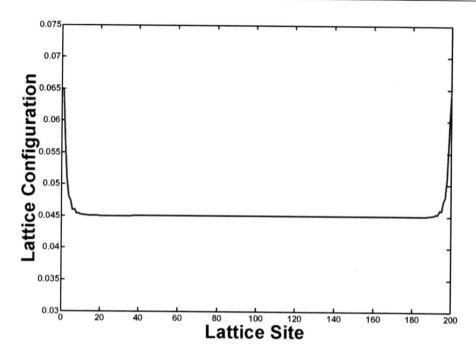

Figure 2. Lattice configuration of lattice dimerization.

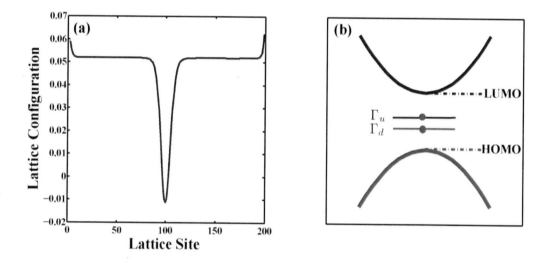

Figure 3. Profile of the singlet exciton: (a) lattice configuration; and (b) electronic spectrum. The value of the gap is 5.12 eV, and the energy values of Γ_u and Γ_d are 1.60 eV and -1.48 eV, respectively.

When an external laser beam or pulse is applied to the conjugated polymer, an electron is easily excited from the HOMO to the LUMO. Near 50 fs, it is found that the original periodic lattice is rapidly destroyed. After 300 fs, a localized lattice distortion appears over the background of the homogeneous dimerization, as illustrated in Fig. 3(a), leading ultimately to a stable lattice distortion along the polymer chain. Associated with the localized lattice distortion, the prominent self-trapping effect of the polymer moves the HOMO and LUMO states toward the center of the gap, giving two localized energy levels Γ_u and Γ_d at the center of gap, as shown in Figure 3(b). Thus, both lattice distortion and two localized states jointly contribute to forming a singlet exciton.

During the formation of the singlet exciton, the light emitted by each conjugated polymer chain also propagates inside the nanofiber and leads to multiple scattering between polymer chains. Here, the focus is on its energy spatial distribution. As seen in Fig. 4, the electric field intensity of light emission is spatially extensive. We randomly choose the detector positions in the fiber to describe the phase distribution of the light field with which the phase ϕ is associated. Interestingly, as shown in Figure 5, the phase of the emitting light is not prone to a certain value, which means that when the singlet exciton contributes to light emission, the polythiophene nanofiber radiates non-coherent light which, apparently, does not reflect the behavior of the lasing effect. According to the electronic structure of the exciton in Fig. 3(b), it is predictable that lasing does not exist because there is no electron population inversion, where each of the localized states is occupied by one electron. The singlet exciton only contributes to light emission, not to the lasing effect, which is consistent with the phase spatial distribution of the light emission.

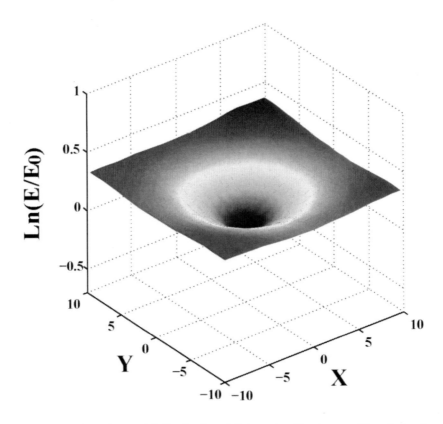

Figure 4. Electric field intensity spatial distribution of light emitted by the nanofiber during the formation of the singlet exciton, where E_0 is the electric field intensity in the center of the material. All lengths are scaled by the average distance between polymer chains.

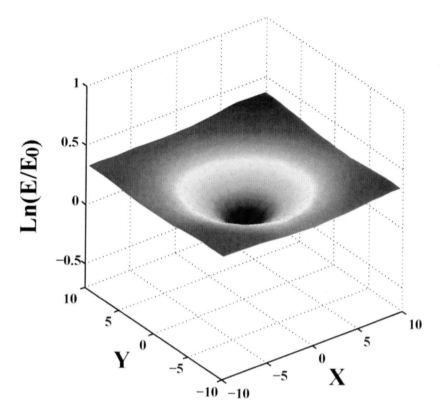

Figure 5. Phase spatial distribution of the light field during the formation of the singlet exciton in the nanofiber. All lengths are scaled by the average distance between polymer chains.

The whole process of the excitation of the singlet exciton, once the gain approaches and exceeds a threshold value, is depicted in Figs. 6 and 7. At beginning, the singlet exciton is comprised of an electron-hole pair, where each of localized states is occupied by one electron as illustrated in Fig. 3(b). When the time reaches 100 ps, the electron population of Γ_u is 1.48, while that of Γ_d becomes 0.52, and the electron population of HOMO is 2, as depicted in Fig. 6. At 300 ps, the electron population of Γ_u with higher energy is larger than that of Γ_d. This means that the electron population undergoes inversion, which also indicates lasing of the polymeric molecule. This is occurring still within the relaxation process of the excitation. Up to 500 ps, the electron populations of Γ_u and Γ_d become 0.09 and 1.91, respectively, while the electron population of HOMO still remains at 2. At 500 ps, the electron population becomes stable compared with that seen before, which is the end of the excitation of the singlet exciton. In other words, the relaxation time of the excitation is about 500 ps. After then, the electron population inversion of the two localized states, as the signature of lasing, sustains laser emission of the conjugated polymeric molecule. Before the excitation, the lattice configuration of the singlet exciton is shown in Fig. 3(a). After the optical pumping, the whole relaxation of the lattice configuration is described in detail in Fig. 7, and the associated lattice distortion of the conjugated polymer chain becomes more severe than that of the singlet exciton.

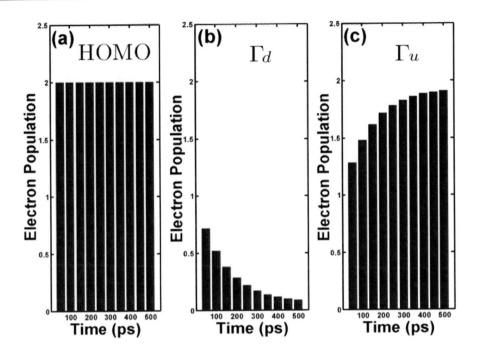

Figure 6. Time-dependent electron populations of three energy levels – HOMO, Γ_u and Γ_d – of a conjugated polymer semiconductor after it undergoes external optical pumping until population inversion between Γ_u and Γ_d.

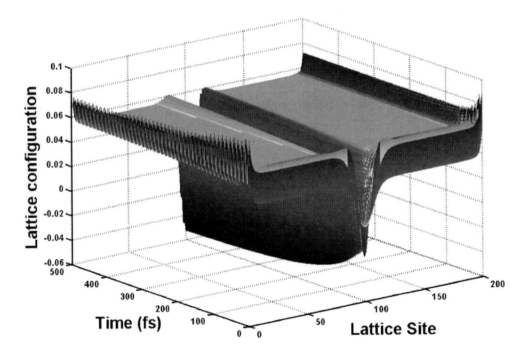

Figure 7. Three-dimensional picture for the time-dependent lattice configuration induced by external optical pumping over the entire time range.

As mentioned previously, when an external optical pulse pumps an exciton of the conjugated polymer, the lattice undergoes relaxation, leading to the electron population inversion. At the end of the relaxation of the excitation, i.e., 500 ps, the associated lattice becomes stable and also induces the localized lattice distortion, which not only totally destroys the initial periodic structure of dimerization, but also forms the new lattice structure. For a clear demonstration, we use a sphere to depict each unit cluster/group of conjugated polymer (Fig. 8), with three series of spheres demonstrating three different lattice structures. The first series of red spheres shown in Fig. 8(a) describes the periodic distribution of the unit clusters/groups of conjugated polymer with constant lattice length a. Figure 8(b) illustrates the lattice dimerization whose lattice length also becomes $2a$. For convenience of presentation, 20 unit clusters/groups of the polymer from the 81st to the 100th lattice site are chosen to show the final lattice structure of the polymer after external pumping (Fig. 8(c)). Based on Fig. 8(c) as compared with Figs. 8(a,b), it is found that the external pumping finally not only destroys the initial periodic structure of the polymer chain, but also induces the occurrence of the disordered lattice structure. Considering the symmetry of the polymer chain, it is appropriate to utilize the lattice configuration $(-1)^{l} u_{l}$ to present the size of the newly formed lattice distortion. As displayed in Fig. 9, the finally formed lattice is locally distorted and spans over 40 unit clusters/groups. If the size of each unit cluster/group is about 5Å, then the width of the self-introduced lattice distortion is 200Å. Further, due to the self-trapping and electron population inversion, the newly-localized lattice distortion moves the localized states in the middle gap Γ_{u} and Γ_{d} to 1.452 eV and -1.260 eV, giving lasing emission with the wavelength 457.227 nm (2.712 eV), which is consistent with the experimental result that the lasing emission occurs at 450-460nm. [16]

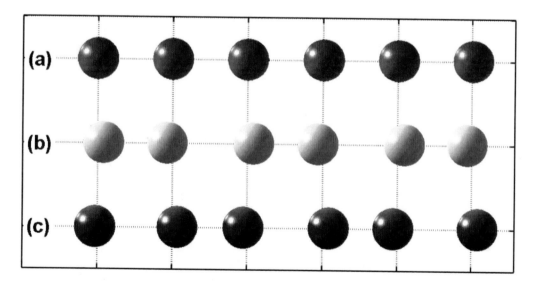

Figure 8. Lattice structure where a sphere depicts each unit cluster of polymer. (a) periodic lattice structure of the polymer chain with lattice constant a ; (b) lattice dimerization of the polymer chain with lattice constant a ; and (c) final lattice structure of the polymer after external pumping.

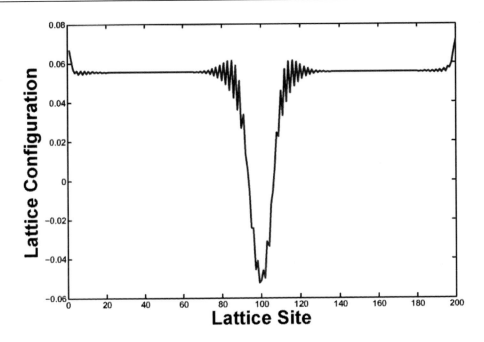

Figure 9. Fhe final lattice configuration of the polymer after external pumping.

Thus, the whole dynamics of the polymeric laser can be described as follows: Once an external optical pulse continuously pumps the conjugated polymer fibers, the induced electron population inversion not only causes laser emission, but also triggers lattice relaxation, finally forming a local disordered lattice distortion that spans over 40 unit clusters/groups along the polymer chain. Then, during the coherent light propagation along the polymeric fiber, the self-introduced local lattice distortion also strengthens multiple scattering of coherent light besides the air gap breaking in the polymer fiber, which in turn induces localization of laser emission known as the random laser. [24, 25] This explains the observation of the localized lasing emission intensity of an isolated individual p-sexiphenyl nanofiber. [17-19]

After surpassing the threshold, the external gain not only induces the electron population inversion, but also counteracts the leakage caused by the non-coherent extensive modes. Naturally, when the laser light emitted by each nanothiophene chain propagates inside the fiber, the collective behavior of light emission in the nanothiophene fiber also should be changed. Thus, the strong multiple scattering has to self-select a new mode without energy leakage to emission. As characterized in Fig. 6, the original extensive spatial distribution of the light field disappears, and the light is well-confined insider the fiber, which is consisten with the experimental observation. [17, 18]

Turning to the phase distribution of the light emission, we show the coherence behavior of the light emitted by the fiber in Fig. 10: Most of the phase of the entire light field is inclined to the specific value of 1.5, indicating the whole system to have a novel collective behavior. In other words, once the gain exceeds a critical value, besides light localization, the multiple scattering forces most of the thiophene chain to emit light with the same phase of 1.5. Accordingly, the "invisible" cavity for the lasing is comprised of randomly-distributed parallel nanothiophene chains. The red zone with the same phase, illustrated in Fig. 11, depicts the size of the resonator.

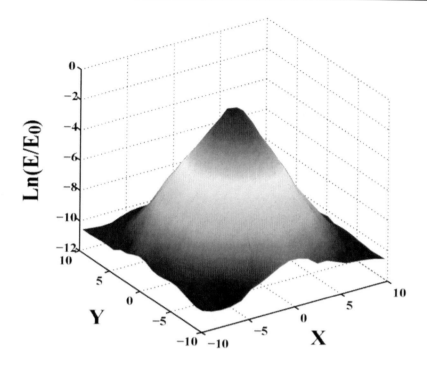

Figure 10. Localized electric field intensity spatial distribution of the light field inside the randomly-distributed polymer chains after electron population inversion occurs, where E_0 is the electric field intensity at the center of the nanofiber. All lengths are scaled by the average distance between polymer chains.

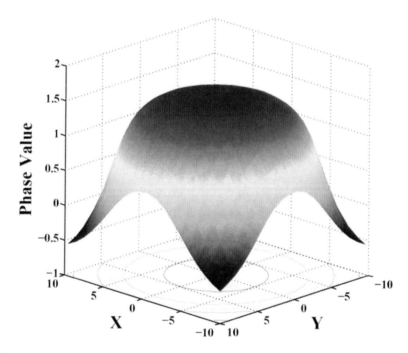

Figure 11. Phase spatial distribution of the light field after electron population inversion occurs in the nanofiber. All lengths are scaled by the average distance between polymer chains.

In conclusion, based on the newly-developed molecule dynamics approach that includes electronic transitions, it is found that after a conjugated polymer (like PLED0 undergoes photoexcitation, a singlet exciton is initially formed. The light propagates inside the nanofiber and leads to an extensive energy spatial distribution. Then, due to the continuous optical pumping, the electron populations of two localized states of the singlet exciton undergo inversion, along with generating localized lattice distortion. Through multiple light scattering, all of the polymer chains act as a "clever" mesh to select a coherent and localized mode to emit laser radiation, where the resonator consists of randomly-distributed parallel conjugated polymer chains, which is consistent with experimental observations.

We thank S. Y. Liu and X. Sun for helpful discussions. This work was supported by the National Science Foundation of China under Grants 20804039 and 10932010, and the Zhejiang Provincial Natural Science Foundation of China under Grant Y4080300.

REFERENCES

[1]   B. H. Soffer and B. B. McFarland, *Appl. Phys. Lett.* 10, 266 (1967).

[2]   N. Karl, *Phys. Stat. Sol.* (a) 13, 651 (1972).

[3]   O. S. Avanesjan, V. A. Benderskii, V. K. Brikenstein, V. L. Broude, L. I. Korshunov, A. G. Lavrushko, and II Tartakovskii, *Mol. Cryst. Liq. Cryst.* 29, 165 (1974).

[4]   A. Tsumura, H. Koezuka, and T. Ando, *Appl. Phys. Lett.* 49, 1210 (1986).

[5]   C. W. Tang and S. A. Vanslyke, *Appl. Phys. Lett.* 51, 913 (1987).

[6]   J. H. Burroughes, D. D. C. Bradley, A. R. Brown, R. N. Marks, K. Mackay, R. H. Friend, P. L. Burns, and A. B. Holmes, *Nature* 347, 539(1990).

[7]   D. Moses, *Appl. Phys. Lett.* 60, 3215 (1992).

[8]   N.Tessler, G. J. Denton, and R. H. Friend, *Nature* 382, 695 (1996).

[9]   F. Hide, M. A. DiazGarcia, B. J. Schwartz, M. R. Andersson, Q. B. Pei, and A. J. Heeger, *Science* 273, 1833(1996).

[10]   W. Holzer, A. Penzkofer, S. H. Gong,A. Bleyer, and D. D. C. Bradley, *Adv. Mater.* 8, 974 (1996).

[11]   S. V. Frolov, M. Ozaki, W. Gellermann, Z. V. Vardeny, and K.Yoshino, Jpn. *J. Appl. Phys.*, Part 2, 35, L1371 (1996).

[12]   G. Heliotis, R. D. Xia, G. A. Turnbull, P. Andrew, W. L. Barnes, I. D. W. Samuel, and D. D. C. Bradley, *Adv. Funct. Mater.* 14, 91 (2004).

[13]   S. Riechel, C. Kallinger, U. Lemmer, J. Feldmann, A. Gombert, V. Wittwer, and U. Scherf, *Appl. Phys. Lett.* 77, 2310 (2000).

[14]   V. G. Kozlov, V. Bulovic, and S. R. Forrest, *Appl. Phys. Lett.* 71, 2575 (1997).

[15]   I. D. W. Samuel and G. A. Turnbull, *Chem. Rev.* 107, 1272 (2007).

[16]   D O'Carroll, I. Lieberwirth, and G. Redmond, *Nature Nanotech.* 2, 180 (2007).

[17]   F. Quochi, F. Cordella, A. Mura, G. Bongiovanni, F. Balzer, and H. G. Rubahn, *J. Phys. Chem. B,* 109, 21690 (2005).

[18]   F. Quochi, F. Cordella, A. Mura, G. Bongiovanni, F. Balzer, and H. G. Rubahn, *Appl. Phys. Lett.* 88, 041106 (2006).

[19]   A. Andreev, F. Quochi, F. Cordella, A. Mura, G. Bongiovanni, and H. Sitter, *J. Appl. Phys.* 99, 034305 (2006).

[20] Quochi F, Cordella F, Mura A, Bongiovanni G, Balzer F and Rubahn H-G 2005 *J. Phys. Chem.* B 109 21690.

[21] S. Li, T. F. George, and X. Sun, *J. Phys. Chem. B* 111, 1067 (2007).

[22] S. Li, L. S. Chen, T. F. George, and X. *Sun, Phys. Rev.* B 70, 075201 (2004).

[23] S. Li, X. L. He, T. F. George, B. P. Xie, and X. Sun, *J. Phys. Chem. B,* 113, 400 (2009).

[24] A. L Burin, M. A. Ratner, H. Cao, and R. P. H. Chang, *Phys. Rev. Lett.* 87, 215503 (2001); 88, 093904 (2002).

[25] H. Cao, *J. Phys. A: Math. Gen.* 38, 10497 (2005).

[26] S. Li, Z. J. Wang, L. S. Chen, T. F. George, and X. Sun, Appl. Phys. Lett. 86, 171109 (2005); S. Li, S. Wang, Z. J. Wang, T. F. George, and X. Sun, *Appl. Phys. Lett.* 88, 111103 (2006).

In: Light-Emitting Diodes and Optoelectronics: New Research   ISBN: 978-1-62100-448-6
Editors: Joshua T. Hall and Anton O. Koskinen   © 2012 Nova Science Publishers, Inc.

Chapter 8

Photonic Bandgap Defect Structure on IV-VI Semiconductor: Resonating Cavity without Cleaving

Shaibal Mukherjee

Discipline of Electrical Engineering, Indian Institute of Technology, Indore, India

1.1. Background

For a few decades, lead chalcogenide diodes have been one of the most popular and commercially available semiconductor mid-infrared (MIR) lasers. However, due to the low operating temperatures and low external efficiencies, the performances of lead salt lasers could not reach the desired level. Even at low cryogenic heat-sink temperature, lasing output powers for single mode operation are less than few mWs. Moreover, an affinity towards multimodal operation and mode hopping is quite familiar. The threshold level of laser operation for IV-VI materials is enhanced due to the existence of four-fold L-valley degeneracy near band extrema. The degeneracy does not get lifted off for [100] orientation which is the most common growth orientation for QW laser structures. This, in turn, limits exploiting the supreme advantages of lead chalcogenide materials for high temperature and long wavelength operation i.e., reduced threshold level originating from a low nonradiative recombination rate. Also [100] growth orientation leads to inferior epitaxial material quality because it does not allow dislocation gliding phenomenon.

It is already well-known that the QW gain threshold is minimum for [111] lead salt crystal orientation compared to its other two counterparts. Therefore, this orientation suits for higher temperature laser operation of lead salt materials. Also experimentally we already have exhibited superior performances from lead salt vertical cavity surface emitting laser (VCSEL) grown on (111) BaF_2 substrate [1]. Those VCSELs have demonstrated pulsed-mode operation near room temperature. Lasing output power as high as 300 mW and threshold density as low as 10.5 kW/cm^2 are achieved [2]. Pulsed laser emission at elevated temperature from VCSELs on (111) substrate is obtained [3] [4]. The continuous wave (CW) lasing at 4.03 μm is observed up to 230 K. The minimum value of threshold excitation density of 2.6 kW/cm^2 has been measured at 190 K, 65 °C lower than that of the pulsed operation.

However the design complexity of distributed Bragg reflector (DBR) during MBE fabrication and low power operation severely prevent the popularity of VCSELs while realizing long wavelength and high power lasing devices. The best growth substrate for realizing high temperature devices is silicon (Si). Si (thermal conductivity = 1.56 W/cm K) has almost 78 times higher thermal conductivity than lead salt materials (thermal conductivity = 0.02 W/cm K) at 300 K. Moreover, by using Si one can reduce manufacturing cost of device fabrication as well as avail of the established device fabrication technologies on this substrate. Although unlike lead salt materials, the natural cleavage plane for Si is (111). Therefore in order to fabricate lasing device, there is a necessity to adopt some valid technology which does not undergo a cleaving process while making a Fabry-Perot resonating cavity.

The best solution in this regard is achieved by implementing the idea of well-known photonic bandgap (PBG) methodology. The physics of PBG does not depend on crystal growth orientation. PBG scheme demonstrates a perfect optical analogy of semiconductor bandgap where the electric potential distribution is governed by a periodicity of regularly distributed lattice dielectric medium (mainly air) in another highly contrasting dielectric medium. This helps in the confinement of light modes by creating a specific frequency bandgap which in other way opens a new horizon of tailoring the light-matter interaction in the crystal lattice. Due to the existence of bandgap effect in photonic crystals, optical waves corresponding to the bandgap frequency are forbidden [5] to propagate whereas waves outside this range are allowed to propagate. This basically provides a provision for suitably engineering the crystal design and localizing light modes within a semiconductor slab. The size of the bandgap is determined by two factors: the dielectric contrast of the constituent materials that build up the photonic crystal and the higher dielectric material filling fraction [6]. Several research studies [7, 8 ,9, 10, 11] regarding photonic crystal fabrication and characterizations in the semiconductor structure are cited in the literature.

1.2. INTRODUCTION

A periodic arrangement of atoms and molecules is known as crystal. If a tiny basic building block of atoms or molecules is repeated in space, a crystal lattice structure would be formed. Quite similar to the crystal structure, if the dielectric constant of a material system is periodically altered in space t is known as photonic crystal (PC). It is well known that all semiconductor materials have a specific energy bandgap in between conduction and valence band. Electrons and holes are "forbidden" to occupy any energy level within the bandgap. In a similar way, photonic crystal possesses a "forbidden" frequency band in which no propagation of electromagnetic waves is possible. Based on the number of directions the dielectric periodicity is exhibited, one-, two-, or three- dimensional photonic crystal or photonic bandgap structure is generated. Light scattering occurring inside a PBG structure can produce many photonic phenomena analogous to the electronic behavior inside crystals.

PBG crystal structure is analogous to normal crystal lattice in terms of atomic periodicity, except the fact that the periodicity is of the order of a micrometer rather than a fraction of a nanometer. The implementation of PBG effect to achieve a semiconductor laser by controlling spontaneous emission was demonstrated by Yablonovitch in 1987 [12]. The fundamental idea behind the concept was to design a photonic structure in order to originate a frequency bandgap that coincides with the semiconductor spontaneous emission. Since that time, the concept of PBG phenomenon is well adopted and practiced all over the world. Various significant applications [13, 14, 15] of PBG phenomenon are noticed to obtain millimeter and microwave frequencies in order to achieve desired performances for antennas and waveguides.

1.3. FINITE DIFFERENCE TIME DOMAIN (FDTD) METHOD

The FDTD method is a well established mathematical analysis tool which is extensively implemented for integrated and diffractive optical device simulations. Through rigorous and accurate numerical solution of electromagnetic Helmholtz's equations, this methodology helps in modeling light propagation, diffraction and scattering phenomena, and polarization and reflection effects inside a crystal. It also assists in modeling material anisotropy and dispersion without the requirement of any presumed field behavior. FDTD method is a very powerful simulation tool that provides effective analysis of sub-micron devices with very finite structural details. The miniaturization of device dimension helps in light confinement and consequently, the large dielectric constant difference of the semiconductor materials to be implemented in a typical device design.

1.4. TWO DIMENSIONAL FDTD EQUATIONS

Opti-FDTD software is implemented to perform the FDTD simulation for the lead salt PBG structure. The FDTD methodology is fundamentally based on numerical solutions to electromagnetic Helmholtz's equations. As shown in Figure 1.1, the photonic structure is arranged in the X-Z plane [16]. The light propagation inside the device is considered to be along Z-direction. The Y-direction is considered to be infinite. This fundamental approximation removes all the $\partial/\partial y$ partial derivatives from electromagnetic Maxwell's equations and divides them into two independent sets of numerical equations. These are known as transverse electric (TE) and transverse magnetic (TM) equations.

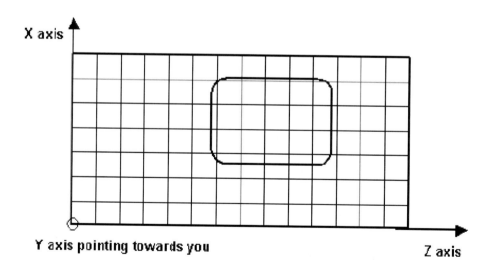

Figure 1.1. Numerical representation of the 2D computational domain.

The spatial steps in the X- and Z-directions arc designated as Δx and Δz respectively. Each mesh point in the crystal lattice structure is linked with a specific type of dielectric material (i.e., lead salt substrate or air) and contains information about its basic properties such as dielectric constant and dispersion parameters.

1.4.1. TE Waves

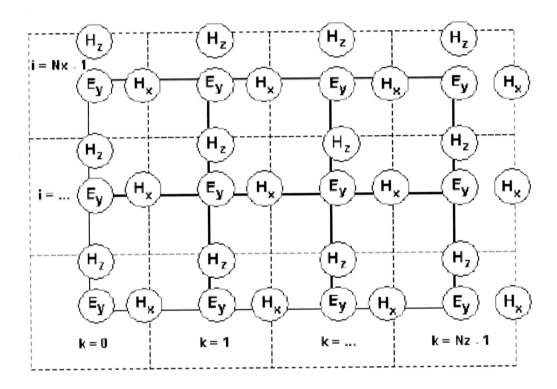

Figure 1.2. Position of the TE fields in the computational domain.

For two-dimensional TE case in X-Z plane, electric and magnetic field components like E_y, H_x and H_z would exist. The electromagnetic equations in lossless media are given as:

$$\frac{\partial E_y}{\partial t} = \frac{1}{\varepsilon}\left(\frac{\partial H_x}{\partial z} - \frac{\partial H_z}{\partial x}\right)$$

$$\frac{\partial H_x}{\partial t} = \frac{1}{\mu_0}\frac{\partial E_y}{\partial z}$$

$$\frac{\partial H_z}{\partial t} = -\frac{1}{\mu_0}\frac{\partial E_y}{\partial x} \tag{1.1}$$

where $\varepsilon = \varepsilon_0 \varepsilon_r$ is the dielectric permittivity of the crystal lattice medium and μ_0 is the vacuum magnetic permeability. ε_0 and ε_r are the vacuum permittivity and the relative permittivity of the crystal medium respectively. The refractive index of the medium is given by $n = \sqrt{\varepsilon_r}$.

Each component of the electromagnetic field inside the crystal lattice structure is represented by a two-dimensional (2D) array - $E_y(i,k)$, $H_x(i,k)$ and $H_z(i,k)$ - equivalent to the 2D mesh grid demonstrated in Figure 1.1. The indices i and k designate step size in space along the X- and Z- directions, respectively. Figure 1.2 illustrates the distribution of the TE field components in the mesh structure in account. The distribution model of the TE field pattern is explained as follows. The E_y field, as shown in Figure 1.2, coincides in space with the mesh node location in Figure 1.1. The solid lines in Figure 1.2 symbolize the mesh in Figure 1.1. The E_y field is designated to be the center of the FDTD cell in space. Each FDTD shell is indicated by dashed lines in Figure 1.2. The magnetic field components, H_x and H_z, form the cell edges. The positions of the E_y field is associated with the integer values of indices i and k. The locations of H_x field is associated with integer values of i and $(k + 0.5)$ indices. Similarly, the H_z field is associated with integer $(i + 0.5)$ and k indices. Implementing numerical discrimination of the field components and applying them in the first derivative of equation (1.1), we would have the following relation:

$$\frac{E_y^n(i,k) - E_y^{n-1}(i,k)}{\Delta t} = \frac{1}{\varepsilon \Delta z}\left[H_x^{n-\frac{1}{2}}\left(i, k + \frac{1}{2}\right) - H_x^{n-\frac{1}{2}}\left(i, k - \frac{1}{2}\right)\right]$$
$$- \frac{1}{\varepsilon \Delta x}\left[H_z^{n-\frac{1}{2}}\left(i + \frac{1}{2}, k\right) - H_z^{n-\frac{1}{2}}\left(i - \frac{1}{2}, k\right)\right] \qquad (1.2)$$

The total set of numerical equation (1.1) takes the following form in Cartesian co-ordinate system:

$$E_y^n(i,k) = E_y^{n-1}(i,k) + \frac{\Delta t}{\varepsilon \Delta z}\left[H_x^{n-\frac{1}{2}}\left(i, k + \frac{1}{2}\right) - H_x^{n-\frac{1}{2}}\left(i, k - \frac{1}{2}\right)\right]$$
$$- \frac{\Delta t}{\varepsilon \Delta x}\left[H_z^{n-\frac{1}{2}}\left(i + \frac{1}{2}, k\right) - H_z^{n-\frac{1}{2}}\left(i - \frac{1}{2}, k\right)\right]$$
$$H_x^{n+\frac{1}{2}}\left(i, k + \frac{1}{2}\right) = H_x^{n-\frac{1}{2}}\left(i, k + \frac{1}{2}\right) + \frac{\Delta t}{\mu_0 \Delta z}\left[E_y^n(i, k+1) - E_y^n(i,k)\right]$$
$$H_z^{n+\frac{1}{2}}\left(i + \frac{1}{2}, k\right) = H_x^{n-\frac{1}{2}}\left(i + \frac{1}{2}, k\right) - \frac{\Delta t}{\mu_0 \Delta x}\left[E_y^n(i+1, k) - E_y^n(i,k)\right] \qquad (1.3)$$

The scheme is actually based on implementing Yee's [17] numerical approach to establish central difference approximations for both temporal and spatial derivatives. The superscript n denotes the time steps used in the numerical algorithm. The space sampling for the calculation is done on a sub-micron scale. To ensure numerical stability of the mathematical calculation in the absorbing medium, we have utilized Courant-Friedrichs-Levy (CFL) condition [18] as follows:

$$\Delta t \le 1 \Big/ \left(c\sqrt{1/(\Delta x)^2 + 1/(\Delta z)^2} \right) \qquad (1.4)$$

1.4.2. TM Waves

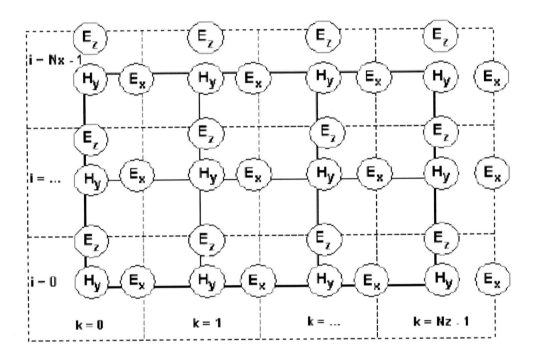

Figure 1.3. Position of the TM fields in the computational domain.

Similarly, for the 2D TM case in X-Z plane, electric and magnetic field components like E_x, H_y and E_z would exist. The Helmholtz's equations in lossless media are given as:

$$\frac{\partial H_y}{\partial t} = -\frac{1}{\mu_0}\left(\frac{\partial E_x}{\partial z} - \frac{\partial E_z}{\partial x}\right)$$

$$\frac{\partial E_x}{\partial t} = -\frac{1}{\varepsilon}\frac{\partial H_y}{\partial z}$$

$$\frac{\partial E_z}{\partial t} = \frac{1}{\varepsilon}\frac{\partial H_y}{\partial x} \qquad (1.5)$$

The distribution of the TM field components in the computational domain is shown in Figure 1.3. In this case, the electric field components E_x and E_z form the FDTD cell edges. The magnetic field component H_y is associated with the center of the FDTD cell. The numerical equations in the TM algorithm can be demonstrated in a similar way to equation (1.3).

1.5. PLANE WAVE EXPANSION (PWE) METHOD

The band structure of the lead salt PBG lattice for both TE and TM modes are plotted with the help of very popular PWE method. By this methodology, the Maxwell's equation in a transparent, time invariant, source-less, and non-magnetic medium can be described as:

$$\nabla \times \frac{1}{\varepsilon(\vec{r})} \nabla \times \vec{H}(\vec{r}) = \frac{\omega^2}{c^2} \vec{H}(\vec{r}) \tag{1.6}$$

where $\varepsilon(\vec{r})$ is the dielectric permittivity which is a function of space. c is the speed of light in vacuum, and $\vec{H}(\vec{r})$ is the optical magnetic field vector having a definite angular frequency ω. The time dependence of $\vec{H}(\vec{r})$ is designated by a term $e^{i\omega t}$. Equation (1.6) is known as the "master" equation [19]. This equation represents a Hermitian eigen value problem.

According to the Bloch theorem, the magnetic field can be represented in a medium with infinite periodicity as:

$$\vec{H}(\vec{r}) = e^{ik\vec{r}} \vec{h}_k(\vec{r}) \tag{1.7}$$

where $\vec{h}(\vec{r}) = \vec{h}(\vec{r} + \vec{R})$ for all possible arrangements of crystal lattice vectors $\vec{R}$. Thus, combining equations (1.6) and (1.7), we can obtain:

$$\left(\nabla + i\vec{k}\right) \times \left[\frac{1}{\varepsilon_{\vec{r}}}\left(\nabla + i\vec{k}\right)\right] \times \vec{h}_k = \frac{\omega^2}{c^2} \vec{h}_k \tag{1.8}$$

Equation (1.8) is the fundamental equation by solving which one can get the dispersion relation inside a particular crystal lattice structure. This equation is generally transformed into a finite domain by simply expanding the magnetic field in a finite basis of simple plane waves. Several methodologies [20, 21] have already been adopted in solving the final discretized problem. The final solution of the discretized problem would result in a dispersion relation between modal frequencies and wave vector $\vec{k}$, generally plotted in the form of a band diagram.

1.6. FINITE DIFFERENCE METHOD (FDM)

There are a number of numerical approaches available for the electromagnetic analysis of structures having complicated geometries and designs. Numerical schemes such as finite element, finite difference, and boundary element methods are all well known and well established. However, most of these popular mathematical and analytical tools involve complex algebra and therefore in order to find a numerical solution they all prove to be time killer. Therefore, a fast and precise method for the modal analysis of a complicated structure such as the PBG crystal lattice has been implemented in this study.

The method fundamentally incorporates finite difference (FD) scheme. It is based on the perturbation correction technique [22, 23], devised with a field convergence algorithm. This method removes the limitation of FDTD scheme in obtaining individual modal field distribution. This limitation of FDTD arises because the source used in the method is an impulse function in the time domain covering an infinite spectrum, thus field solutions are superposition of all possible modes. But FDM, through a small correction in field and mode index evaluates individual mode field solutions in the crystal lattice structure.

The concerned lead salt PBG structure is constructed with the usual discretization notation and index profile, $n^2(x, y) = n_{ij}^2$ where n_{ij} corresponds to the index profile along the cross-section of the lattice. Since the field at the boundaries must vanish, assuming an initial value of the field, e next to one boundary, approximate field distribution and mode index are calculated. After this primary calculation, taking these values as the initial conditions, the verification of the discretized Helmoltz's equation at all grid points by the basic mode convergence equation is carried out. The field value is represented by $e_{i,j}$ and can be given as:

$$e_{i,j} = \frac{e_{i,j+1} + e_{i,j-1} + e_{i+1,j} + e_{i-1,j}}{4 - (\Delta x)^2 k_0^2 \left(n_{i,j}^2 - n_{eff}^2\right)} \tag{1.9}$$

In equation (1.9) k_0 is free space wave vector, $n_{i,j}$ is the refractive index at the computational point and n_{eff} is the effective refractive index of the semiconductor material. In this case, we assume $\Delta x = \Delta y$. Through a series of convergence scans, the resulting field distribution yields a more accurate value of mode index [24], n_{effm} through the equation

$$n_{effm}^2 = \frac{\int_{-\infty}^{+\infty}\int_{-\infty}^{+\infty} \left[e_{i+1,j} + e_{i-1,j} + \frac{2\varepsilon_{i,j-1}}{\varepsilon_{i,j-1} + \varepsilon_{i,j}} e_{i,j-1} + \frac{2\varepsilon_{i,j+1}}{\varepsilon_{i,j+1} + \varepsilon_{i,j}} e_{i,j+1} - \left\{ 4 + \frac{\varepsilon_{i,j} - \varepsilon_{i,j-1}}{\varepsilon_{i,j} + \varepsilon_{i,j-1}} + \frac{\varepsilon_{i,j} - \varepsilon_{i,j+1}}{\varepsilon_{i,j} + \varepsilon_{i,j+1}} - k_0^2 (\Delta y)^2 \varepsilon_{i,j} \right\} e_{i,j} \right] e_{i,j}\, dx\, dy}{k_0^2 (\Delta y)^2 \int_{-\infty}^{+\infty}\int_{-\infty}^{+\infty} e_{i,j}^2\, dx\, dy} \tag{1.10}$$

The aforesaid process of recurring convergence scan and consequent assessment of the mode effective index is continued till a desired precision in the value of n_{effm} is achieved. For very fast convergence, the algorithm is altered with a relaxation factor by which the change in the field before and after convergence scans is added in the next step of convergence. The method described above results in the scalar fields of the waveguide modes. In order to get the solution for the polarized modes, the approximate modes can be calculated through the TE and TM modal solution of the constituent slab waveguides. For mode field correction, the FD discretization can simply be adapted by incorporating the semivectorial Helmoltz's equation. In case of the TE modes, the continuity condition leaves the scalar Helmoltz's equation unchanged to represent the semi-vectorial form. For the TM modes, the corresponding equation [25] is given as:

$$\left(\frac{\partial^2}{\partial y^2} + k_0^2 n^2 - \beta^2 \right) E_y + \frac{\partial}{\partial y}\left(E_y \frac{\partial}{\partial y} \log n^2 \right) = 0 \tag{1.11}$$

Considering a three-point centered difference approximation for the operator $\partial^2/\partial y^2$ and implementing a uniform sampling grid, the field in case of semi-vectorial form, is converged to an x-polarized mode as given by the following convergence scan equation.

$$e_{i,j} = \frac{e_{i+1,j} + e_{i-1,j} + \dfrac{2\varepsilon_{i,j-1}}{\varepsilon_{i,j-1} + \varepsilon_{i,j}} e_{i,j-1} + \dfrac{2\varepsilon_{i,j+1}}{\varepsilon_{i,j+1} + \varepsilon_{i,j}} e_{i,j+1}}{4 - k_0^2 (\Delta y)^2 \left(\varepsilon_{i,j} - n_{eff}^2 \right) + \dfrac{\varepsilon_{i,j} - \varepsilon_{i,j-1}}{\varepsilon_{i,j} + \varepsilon_{i,j-1}} + \dfrac{\varepsilon_{i,j} - \varepsilon_{i,j+1}}{\varepsilon_{i,j} + \varepsilon_{i,j+1}}} \tag{1.12}$$

From the field distribution, the n_{effm} is then calculated using the integral as in equation (1.13).

$$n_{effm}^2 = \frac{\displaystyle\int_{-\infty}^{+\infty}\int_{-\infty}^{+\infty} \left[\begin{array}{c} e_{i+1,j} + e_{i-1,j} + \dfrac{2\varepsilon_{i,j-1}}{\varepsilon_{i,j-1} + \varepsilon_{i,j}} e_{i,j-1} + \dfrac{2\varepsilon_{i,j+1}}{\varepsilon_{i,j+1} + \varepsilon_{i,j}} e_{i,j+1} \\[2ex] - \left\{ 4 + \dfrac{\varepsilon_{i,j} - \varepsilon_{i,j-1}}{\varepsilon_{i,j} + \varepsilon_{i,j-1}} + \dfrac{\varepsilon_{i,j} - \varepsilon_{i,j+1}}{\varepsilon_{i,j} + \varepsilon_{i,j+1}} - k_0^2 (\Delta y)^2 \varepsilon_{i,j} \right\} e_{i,j} \end{array} \right] e_{i,j}\, dx\, dy}{k_0^2 (\Delta y)^2 \displaystyle\int_{-\infty}^{+\infty}\int_{-\infty}^{+\infty} e_{i,j}^2\, dx\, dy} \tag{1.13}$$

In case of y-polarized mode, equations (1.12) and (1.13) are exactly followed with the indices i, j interchanged.

1.7. LEAD CHALCOGENIDE DEFECT CAVITY PBG STRUCTURE

The photonic semiconductor structure consists of PbSe-$Pb_{0.98}Sr_{0.02}$Se multiple quantum well structure, grown on BaF_2 substrate, which has arrays of dielectric air columns periodically distributed in hexagonal fashion on the surface (as shown in Figure 1.4(a)). Figure 1.4(b) demonstrates the symmetry points of the first Brillouin zone in the periodic hexagonal crystal. The structure has the following constituent material parameters: refractive indices are $n_{PbSe} = 5.0$ (bulk PbSe), $n_{PbSrSe} = 4.6$ ($Pb_{0.98}Sr_{0.02}$Se) and $n_a = 1$ (air). The active layer is considered to have a $\lambda/2$ thickness, where λ is the emission wavelength. In order to choose an optimized design along two-dimensional crystal structure, a scanning has been done to calculate modal band structure and the corresponding bandgaps in the crystal lattice for varying radius of air-hole.

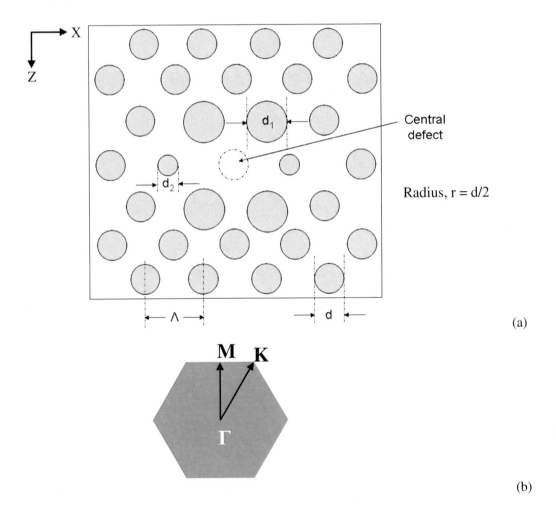

(a)

(b)

Figure 1.4. (a) Top view of triangular photonic crystal lattice structure, (b) Schematic of first Brillouin zone for the hexagonal lattice pattern.

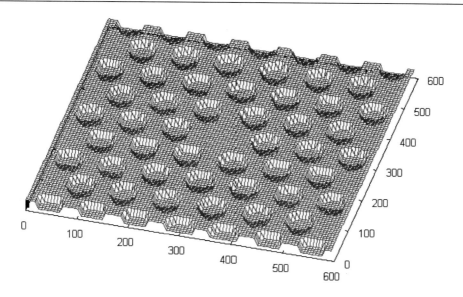

Figure 1.5.Top View of refractive index variation over the cross-section of hexagonal crystal lattice.

The cross section of the designed hexagonal crystal lattice structure constructed during FDM simulation is illustrated in Figure 1.5. The diagram illustrates the formation of a single defect inside the PC structure by omitting the central air-hole within a computational window of a 600 × 600 mesh points.

1.7.1. Mid Infrared Photonic Bandgap Formation

The bandgap structure is illustrated in Figure 1.6 where all the frequency eigenvalues from the dispersion curves are combined along the single vertical line for each specific scanned value of radius. The crystal is optimally designed to have a diameter of dielectric air-hole (d) of 0.63 μm with a periodic lattice constant (∧) of 0.96 μm. The air-fraction (d/∧) in the photonic crystal plays an important role in creating bandgap.

The band frequency tends to rise with the increase of air-fraction for a fixed lattice constant (∧) in the crystal as can be seen in Figure 1.6(a).

In Figure 1.7, both transverse electric (TE) as well as transverse magnetic (TM) band structures are plotted with the help of PWE method. However, it is mention-worthy that the resonating optical modes in such a periodically patterned crystal are not distinctively TE or TM but they can be thought as TE-like and TM-like [26]. The band diagrams in Figure 1.7(b) demonstrate one narrow and another quite broad TM-like photonic bandgaps exist in the structure, but no TE-like bandgaps. The bandgap values are populated in Table-1.1. Moreover, the broad TM bandgap covers the mid-infrared spectral region where room temperature photoluminescence from lead salts takes place. Therefore we focus to engineer a defect cavity mode in this frequency bandgap region.

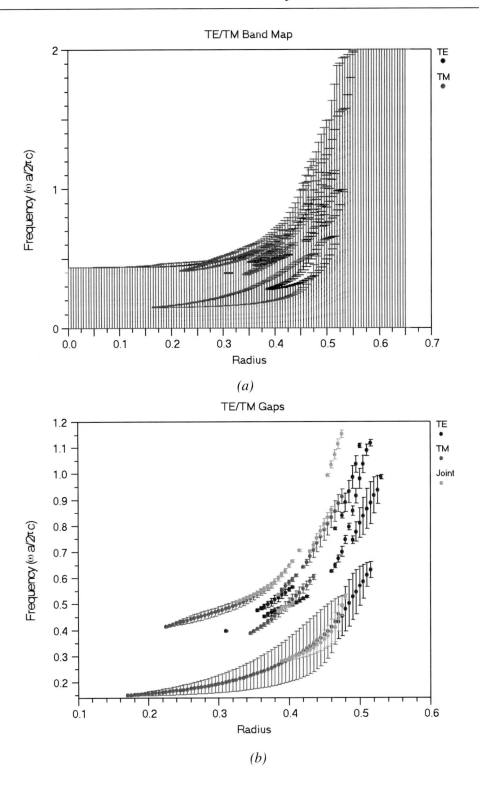

Figure 1.6. (a) Reduced band map for TE and TM mode, (b) band map for TE, TM, and joint mode.

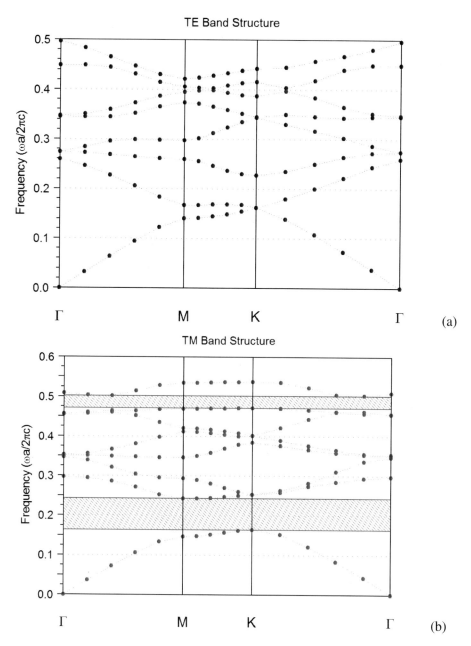

Figure 1.7. Band structure of the photonic crystal for (a) TE and (b) TM polarized light.

The defect cavity is designed (as illustrated in Figure 1.4(a)) by omitting the central air column, four vertical (two on top and two at bottom) lattice atoms along X-direction surrounding the central defect are made to have a diameter of 0.8 μm, two adjacent horizontal atoms along Z-direction surrounding the central defect are made to have a diameter of 0.46 μm.

Table 1.1. TM bandgaps in the hexagonal periodic photonic lattice structure on IV-VI lead salt semiconductor

Serial Number	Bandgap
TM1	1.916 μm – 2.038 μm
TM2	3.959 μm – 5.872 μm

1.7.2. Modal Analysis by FDM Scheme

To appreciate the confinement of mode to the central region, mode field distributions are calculated using the FDM technique. Different values of emission wavelength and crystal lattice parameters have been selected to optimize the variation in mode field distribution on changing these parameters. The final goal is to achieve a single modal laser emission from the lead chalcogenide defect cavity PC structure. It is noticed that the field distribution appears to be largely Gaussian at single modal condition. However, when the emission wavelength becomes multi-modal, the field assumes a distribution different from the usual gaussian shape. This can be seen in Figure 1.8, which depicts the distribution of electric field for the fundamental and multi mode at emission wavelengths of 4.17 μm [27].

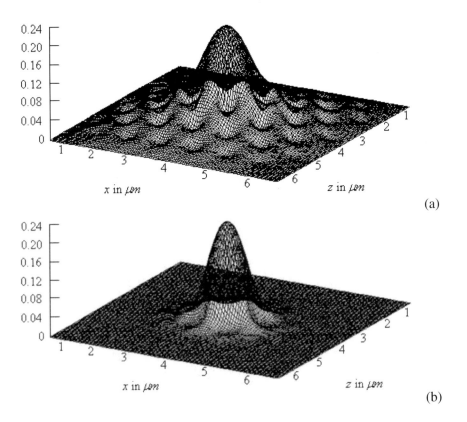

(a)

(b)

Figure 1.8. Resonating electric field distribution at λ = 4.17 μm (a) when multi-modal, (b) when single modal.

The effective modal refractive index at the resonating condition is formulated to be ~ 4.04. FDM perturbation technique determines the field distribution in the crystal at a fixed wavelength and therefore it is quite challenging while determining the resonating wavelength for a specific optical cavity. This necessitates the deployment of FDTD approach which produces a wideband response for the field distribution in the cavity by exciting the crystal with a gaussian pulse. Moreover, by applying FDTD approach we could verify the results of FDM approach.

1.7.3. Modal Analysis by FDTD Scheme

The uniaxial perfectly matched layer (UPML) absorbing boundary condition [28, 29, 30], which is a very efficient method in dealing with scattering by particles in vacuum, is implemented as the boundary layers outside the FDTD computational window. In order to excite the photonic crystal structure, a Gaussian modulated continuous wave (GMCW) point source is considered which can be expressed as:

$$E_{inc} = A.\exp\left[-0.5.(t-t_0)^2 / T^2\right]\sin(\omega t)$$

(1.14)

where A is the input wave amplitude, T is the half width, ω is the angular frequency, t_0 is the time offset. The rigorous numerical analysis produces precise and efficient time domain response for the field distribution in the defect cavity.

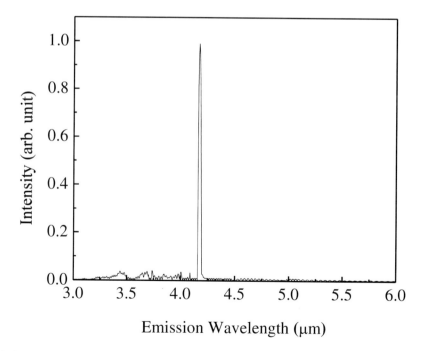

Figure 1.9. Spectral response from DFT calculations after FDTD analysis of the resonating defect cavity.

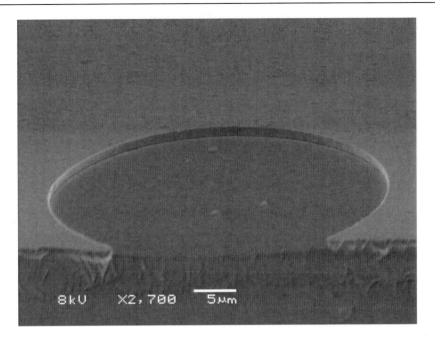

Figure 1.10. Realization of air-hole by photolithography and wet etching method.

To generate the spectral response corresponding to the time domain behavior, discrete Fourier transform (DFT) is applied. This calculates spectral response for a specific wavelength as shown by:

$$F(\omega) = \int_0^T F(t).\exp(-j\omega t)dt = \sum_{n=0}^N F(n).\exp(-j\omega n\Delta t).\Delta t \qquad (1.15)$$

where F(n) is the time domain response, N is the number of time steps. The spectral field obtained from the defect cavity (as in Figure 1.4(a)) in the photonic crystal is seen to resonate at 4.17 µm and is plotted in Figure 1.9. The result seems to be in exact correlation with the numerical results obtained from the FD perturbation correction analysis. The optimized quality factor for the resonating mode is calculated to be 5200.

1.8. EXPERIMENTAL STEPS FOR AIR HOLE FORMATION

In order to realize PC lasing cavity one has to perform lithography and etching in a clean-room environment. One of the most important parameters of PC air-hole is the inclination of its side wall with the base substrate. During the theoretical simulation, it is always considered that the side wall is perfect in terms of the surface roughness and inclination. Even a minute alteration in these parameters affects the modal behavior of the defect cavity emission inside the crystal lattice structure. The surface roughness is not a problem for wet etching method. The only challenge is the vertical inclination. In order to verify our presumption, we have selected a photo-mask which would lead to an air-hole with a diameter of 50 µm as shown in Figure 1.10. It is to be noted that the dimension of this mask has nothing to do with our original theoretical PC laser dimension. We considered a bigger air-hole mask just to study the perfection of wet etching procedure.

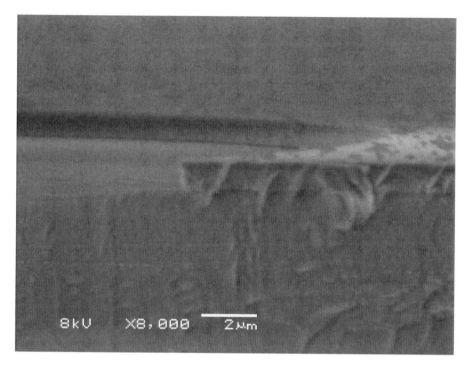

Figure 1.11. Air-hole side wall inclination to base not exactly vertical after wet etching method.

The air-hole is etched on a PbSe single layer grown on (111) BaF$_2$ substrate. From Figure 1.11, it can be seen that the air-hole is not perfectly vertical to the base substrate. Rather it poses an acute angle with the base. Therefore the wet etching method is proved to be unsuitable for PC laser structure fabrication. Thus dry etching or plasma etching seems to be the only feasible option remaining to fabricate PC lasing cavity on lead salt substrate. Regarding this we are doing a joint collaborative study with Penn State University and the preliminary results of their dry etching procedure on our MBE-grown lead salt material is illustrated in Figure 1.12.

CONCLUSION

In conclusion, a theoretical investigation of spontaneous mid-infrared emission from IV-VI semiconductor defect cavity in the hexagonal photonic crystal is elaborated in this chapter. The design is aimed to solve out challenges of the formation of resonating cavity for lead salt materials fabricated on Si(111) or BaF$_2$(111) substrates. The band structure calculations of the periodic crystal are performed using PWE method. Two approaches have been implemented to analyze and understand modal field distribution in the defect cavity. FD perturbation correction method and FDTD algorithms are very popular and well-established mathematical tools for optical waveguide analysis. It has been demonstrated that the FDTD results reasonably agree with that of FD perturbation method. A single TM-like mode working at

4.17 μm, having an optimized Q-factor of 5200, resonates in the designed defect cavity. The prospective practical applications of the single mid-infrared emission are mainly in industrial trace-gas sensing systems and emission monitoring.

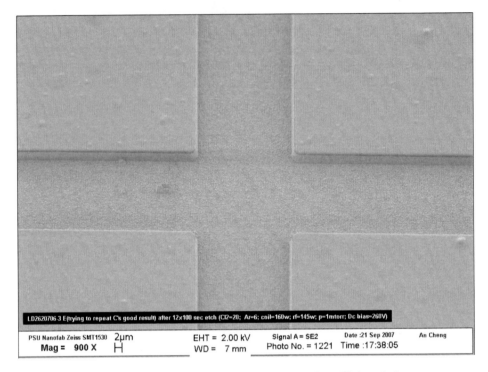

Figure 1.12. Plasma etching of lead salt material (Courtesy: Penn State University).

REFERENCES

[1]   Z. Shi, G. Xu, P. J. McCann, X. M. Fang, N. Dai, C. L. Felix, W. W. Bewley, I. Vurgaftman, and J. R. Meyer, "IV–VI compound midinfrared high-reflectivity mirrors and vertical-cavity surface-emitting lasers grown by molecular-beam epitaxy," *Appl. Phys. Lett.*, vol. 76, pp. 3688-3690, 2000.

[2]   C. L. Felix, W. W. Bewley, I. Vurgaftman, J. R. Lindle, J. R. Meyer, H. Z. Wu, G. Xu, S. Khosravani, and Z. Shi, "Low-threshold optically pumped λ = 4.4 μm vertical-cavity surface-emitting laser with a PbSe quantum-well active region," *Appl. Phys. Lett.*, vol. 78, pp. 3770-3772, 2001.

[3]   F. Zhao, H. Wu, L. Jayasinghe, and Z. Shi, "Above-room-temperature optically pumped 4.12 μm midinfrared vertical-cavity surface-emitting lasers," *Appl. Phys. Lett.*, vol. 80, pp. 1129-1131, 2002.

[4]   H. Xu, F. Zhao, A. Majumda, L. Jayasinghe, and Z. Shi, ""High power midinfrared optically pumped PbSe–PbSrSe multiple-quantum-well vertical-cavity surface-emitting laser operation at 325 K," *Electron. Lett.*, vol. 39, pp. 659-661, 2003.

[5]   J. D. Joannopoulos, R. D. Meade, and J. N. Winn, "Photonic Crystals, Modeling the Flow of Light," *Princeton University Press*, Princeton, N.J., 1995.

[6]   J. D. Joannopoulos, P. R. Villeneuve, S. Fan, "Photonic crystals: putting a new twist on light," *Nat.*, vol. 386, pp. 143-149, 1997.

[7]   T. Krauss, Y. P. Song, S. Thoms, C. D. W. Wilkinson, and R. M. DelaRue, "Fabrication of 2-D photonic bandgap structures in GaAs/AlGaAs," *Electron. Lett.*, vol. 30, pp. 1444-1446, 1994.

[8]   T. F. Krauss, R. M. D. L. Rue, and S. Brand, "Two-dimensional photonic-bandgap structures operating at near-infrared wavelengths," *Nat.*, vol. 383, pp. 699-702, 1996.

[9]   J. O'Brien, O. Painter, C. C. Cheng, R. Lee, A. Scherer, and A. Yariv, "Lasers incorporating 2D photonic bandgap mirrors," *Electron. Lett.*, vol. 32, pp. 2243-2244, 1996.

[10] T. Baba and T. Matsuzaki, "Fabrication and photoluminescence of GaInAsP/InP 2D photonic crystals," *Jpn. J. Appl. Phys.*, vol. 35, pp. 1348-1352, 1996.

[11] T. Hamano, H. Hirayama, and Y. Aoyyagi, "Optical characterization of GaAs 2D photonic bandgap crystal fabricated by selective MOVPE," in *Conference on Lasers and Electro-Optics*, vol. 11 of 1997 OSA Technical Digest Series (Optical Society of America, Washington, D.C.), pp. 528-529, 1997.

[12] E. Yablonovitch, "Inhibited spontaneous emission in solid state physics and electronics," *Phys. Rev. Lett.*, vol. 58, pp. 2059-2062, 1987.

[13] M. M. Sigalas, R. Biswas, Q. Li, D. Crouch, W. Leung, R. Jacobs-Woodbury, B. Laogh, S. Nielsen, S. McCalmont, G. Tuttle, K. M. Ho, "Dipole antennas on photonic band -gap crystals– experiment and simulation," *Microwave Opt. Tech. Lett.*, vol. 15, pp. 153-158, 1997.

[14] G. P. Gauthier, A. Courtay, and G. M. Rebeiz, "Microstrip antennas on synthesized low dielectric-constant substrates," *IEEE Trans. Antennas Propag.*, vol. 45, pp. 1310-1314, 1997.

[15] J. G. Maloney, M. P. Kesler, B. L. Shirley, and G. S. Smith, "A simple description for waveguiding in photonic bandgap materials," *Microwave Opt. Tech. Lett.*, vol. 14, pp. 261-265, 1997.

[16] Opti-FDTD user manual, Optiwave.

[17] K. S. Yee, "Numerical solution of initial boundary value problems involving maxwell's equations in isotropic media," *IEEE Trans. Antennas. Prop.*, vol. AP-14, pp. 302-307, 1966.

[18] A. Taflove and M. E. Brodwin, "Numerical solution of steady-state electromagnetic scattering problems using the time-dependent Maxwell's equations," *IEEE Trans. Microwave Theory Tech.*, vol. MTT-23, pp. 623-630, 1975.

[19] J. D. Joannopoulos, R. D. Meade, and J. N. Winn, "Photonic crystals, Molding the flow of light," *Princeton University Press*, 1995.

[20] S. G. Johnson, J. D. Joannopoulos, "Block-iterative frequency-domain methods for Maxwell's equations in a planewave basis," *Opt. Exp.*, vol. 8, pp.173-190, 2000.

[21] S. Guo, S. Albin, "Simple plane wave implementation for photonic crystal calculations," *Opt. Exp.*, vol. 11, pp.167-175, 2003.

[22] P. R. Chaudhuri, V. Paulose, C. Zhao, C. Lu, "Near-elliptic core polarization-maintaining photonic crystal fiber: modeling birefringence characteristics and realization," *IEEE Photon. Technol. Lett.*, vol. 16, pp. 1301-1303, 2004.

[23] P. R. Chaudhuri, A. K. Ghatak, B. P. Pal, C. Lu, "Fast convergence and higher-order mode calculation of optical waveguides: perturbation method with finite difference algorithm," *Opt. Las. Technol.*, vol. 37, pp. 61-67, 2005.

[24] A. W. Snyder, J. D. Love, "Optical waveguide theory," London:Chapman & Hall, pp. 487-513, 1983.

[25] K. Kawano, T. Kitoh, "Introduction to optical waveguide analysis," Chapter 4, New York:Wiley, 2001.

[26] O. Painter, J. Vuckovic, and A. Scherer, "Defect modes of a two-dimensional photonic crystal in an optically thin dielectric slab," *J. Opt. Soc. Am. B*, vol. 16, pp. 275-285, 1999.

[27] Shaibal Mukherjee, Gang Bi, Jyoti P. Kar, and Zhisheng Shi, "Two dimensional numerical analysis on mid-infrared emission from IV-VI lead salt photonic crystal microcavity," in Press, *Optica Applicata*, 2009.

[28] J. P. Berenger, "A perfectly matched layer for the absorption of electromagnetic waves," *J. Comput. Phys.*, vol. 114, pp. 185-200, 1994.

[29] D. S. Katz, E. T. Thiele, and A. Taflove, "Validation and extension to three dimensions of the Berenger PML absorbing boundary condition for FD-TD meshes," *IEEE Microwave Guid. Wave Lett.*, vol. 4, pp. 268-270, 1994.

[30] J. P. Berenger, "Three-dimensional perfectly matched layer for the absorption of electromagnetic waves," *J. Comput. Phys.*, vol. 127, pp. 363-379, 1996.

In: Light-Emitting Diodes and Optoelectronics: New Research   ISBN: 978-1-62100-541-4
Editors: J. T. Hall and A. O. Koshkinen   © 2012 Nova Science Publishers, Inc.

Chapter 9

MODELLING OF WIDEBANDGAP LIGHT EMITTING DIODES: FROM HETEROSTRUCTURE TO LED LAMP

K.A. Bulashevich[1], *I.Yu. Evstratov*[1], *S.Yu. Karpov*[1], *O.V. Khohlev*[1]
and A. I. Zhmakin[1,2*]
[1] STR-Softimpact Ltd., St. Petersburg, Russia
[2] A.F.Ioffe Physical Technical Institute, St. Petersburg, Russia

Abstract

Light-emitting diodes (LEDs) made from widebandgap materials, first of all, from group III nitrides, are among the most perspective optoelectronic devices serving both as sources of green, blue and ultraviolet light and as the basis of white LEDs for solid state lighting. Nowadays, growth technology of group-III nitride semiconductors and engineering of III-nitride advanced heterostructure devices has leaved far behind a detailed understanding of their operation. This is largely because of non-ordinary physical properties of III-nitrides like spontaneous electric polarization, strong piezoeffect, extremely low acceptor activation efficiency, and a high dislocation density inherent in epitaxial materials. All said, in combination with a complex multi-layer heterostructure and carrier degeneration typical for high power LEDs, hampers an intuitive designing of device based on analogies with conventional III-V compounds. Numerical simulation provides a deeper insight into physical mechanisms underlining the device operation and allows reduction in both time and cost of LED development and optimization. Difficulties of numerical simulation of widebandgap LEDs stem from the high anisotropy of the die, stiffness of the governing equations, extremely low concentrations of intrinsic carriers, a presence of strong build-in electrical fields, and large band offsets at the heterointerfaces.

The chapter describes approaches to III-nitride and II-oxide LED modelling including hybrid (1D/3D) ones and coupling of electrical and thermal processes and light generation/propagation. Results of modelling at different levels — a heterostructure, a die, a die with encapsulant, a LED lamp — are presented and used to analyze factors affecting internal quantum efficiency, high-current efficiency droop, current crowding, efficiency of light conversion in white LEDs, light extraction efficiency.

*E-mail address: ai@zhmakin.ru

PACS 77.65.Ly, 78.20-e, 78.66.Fd, 85.6b, 85.30.-z

Keywords: Current crowding, group-III nitrides, light extraction, light emitting diodes, modelling, quantum efficiency, white light

1. Introduction

The direct bandgaps of the group III-nitrides (AlN, GaN, InN) and their ternary and quaternary alloys span the range 0.7–6.2 eV - the whole visible spectrum and a good piece of ultraviolet (UV) one. The large part of this range could not be covered by conventional III-V semiconductor compounds. An early attempt was made to use another widebandgap semiconductor, indirect SiC, for the blue light emitting diode (LED) (with extremely low brightness) [1]. Thus, the value of nitrides for optoelectronic devices is evident.

The progress of light emitting diodes (LEDs), from being simply indicator lamps to efficient light sources, is impressive [2, 3, 4, 5]. The major breakthrough in LEDs operating in visible spectrum since demonstration by J.I. Pankove GaN LED in 1970 (see [6, 7]) was made in the early 90s by a development of high-brightness III-nitride blue and violet LEDs [8, 9] with external quantum efficiency (EQE) as high as a few per cent (the best SiC LEDs emitting blue light at 470 nm had efficiency about 0.03% [10]); a history of improving crystal quality (and thus electrical and luminescence properties) of InGaN and development of III-nitride LEDs can be found in the review paper by I. Akasaki and H. Amano [11].

A few years ago, similar to well-known Moore's Law of IC packaging, Haitz's law was formulated [12] stating that in the past thirty years the lumen flux per package has increased twenty fold while the cost has decreased ten fold each decade; now these trends are even surpassed [13] (the typical improvement of luminous efficacy of traditional lamps is about 1.1 - 1.2 times per decade [14]). A record luminous efficacy of 249 lm/W has been recently reported which is two times higher than a sodium-vapour lamp; using three phosphors (bluish-green, yellow and red) enhanced significantly luminescence in the blue-green and red regions producing the the best colour reproduction of all white light sources [14].

Still, especially for LEDs for general lighting [4, 15, 16, 17, 18, 19], considerable efforts are being devoted now to further increase of the efficiency of light-emitting diodes. Another concern is the reduction of the operating voltage [20] while keeping the same current densities and, hence, the same luminance that is mainly achieved by reducing the charge injection barriers, effective sheet resistance of contact layers and the LED thickness. At present, probably, the two most evident roadblocks in the LED development are the so called "green gap" (also "yellow-green gap") and the low efficiency of UV LEDs.

The green gap refers to the interval of the wavelength approximately from 550 nm to 590 nm: there is a lack of sufficiently efficent LEDs for this range [21, 22, 23, 24] while "green droop" is sometimes used to describe the corresponding efficiency drop with current [25]. Another problem related to LEDs emitting in this wavelength range is a strong temperature and current dependence of the emission wavelength [26, 27].

III-nitride LEDs cover UV subranges UV-A (315–400 nm, "black light"), UV-B (280–315 nm, "erythermal") and a large part of UV-C (180–280 nm, "germicidal") [28, 29], with the shortest reported wavelength being 210 nm [30]. However, the achieved EQE is about 2 to 6 % for subranges UV-A and UV-B and around 1% for UV-C [30].

Nowadays, growth technology of group-III nitride semiconductors and engineering of III-nitride LEDs has leaved far behind a detailed understanding of their operation. This is largely because of non-ordinary physical properties of III-nitrides like spontaneous electric polarization, strong piezoeffect, extremely low acceptor activation efficiency, and a high dislocation density inherent in epitaxial materials. All said, in combination with a complex multi-layer heterostructure, carrier degeneration typical for high power LEDs and influence of the fabrication process on the material parameters [7], hampers an intuitive designing of device based on analogies with conventional III-V compounds. Numerical simulation provides a deeper insight into physical mechanisms underlining the device operation and determines quantitative links between the device structure and performance allowing to reduce both time and cost of LED development and optimization.

An early attempt of 1D modelling using the drift-diffusion (DD) mode of the band diagram, performance and emission spectra of a blue AlGaN/InGaN LED (fabricated by S. Nakamura [8]) was reported in 1996 [31]. Since such a principle feature of nitride semiconductors as strong spontaneous and piezo polarization had not been recognized at that time, it was not accounted for by P. Shah et al. [31]. Note also that the value of the bandgap of InN (three times smaller than was assumed) was reliably established only in 2002 [32]. Last years a number of 2D & 3D simulations of LEDs and laser diodes using DD model were performed exploiting APSYS and LASTIP packages by Crosslight Software [33] (e.g., [34, 35, 36, 37]). Since 2004 the software package Simuled by STR Group Ltd. [38] is used widely both to study the fundamental processes in III-nitride LEDs such as dominant recombination mechanisms and reasons for the efficiency droop at high operation current densities [39, 40, 41] and to optimize LED and LED array design, thermal management and light extraction (see, eg., [42, 43, 44, 45, 46, 47, 48, 49, 50, 51, 52, 53]). Operation of LEDs based on other widebandgap material, ZnO[54], also was simulated [55] as well as processes in hybrid ZnO/III-nitride LEDs [56, 57, 58].

To develop a high performance LED, a number of issues should be considered such as an optimization of the LED die for the maximum internal quantum efficiency (IQE), control of the current spreading to provide uniformity of the light emission, effective light extraction, thermal management at both die and lamp levels.

This chapter deals with the numerical simulation of the stationary operation of LED, from die to lamp, using Simuled. Neither the transient processes [59, 60, 61] nor the degradation issues [62, 63] are considered.

2. Carrier Injection and Light Emission in LED Heterostructures

Characterisation of light sources   The intensity of the light source is assessed using either *radiometry* (the measurement of electromagnetic radiation in the very wide wavelength range; the radiant flux Φ_e is defined as the radiant energy per unit time in Watt; radiometric units are also referred to as "energetic" ones) or *photometry* (the measurement of visible light as perceived by human in the wavelength range of 360-830 nm; the luminous flux Φ_v is defined as the total light emission per unit time detected by the human eye in lumen units that is equivalent to the radiant flux of 1/683 W at the most sensitive wavelength for human photopic vision of 555 nm) [64]. The luminous flux Φ_v is related to the radiant flux Φ_e by the spectral luminous efficacy K_λ as $K_\lambda = \Phi_v(\lambda)/\Phi_e(\lambda)$. Evidently, K_λ reaches maximum $K_{\max} = 683 lm/W$ at a wavelength λ of 555 nm. At other wavelength, $K_\lambda = K_{\max}V_\lambda$ is expressed via the relative sensitivity of the human eye called the luminous function or spectral luminous efficiency V_λ [65].

LED efficiencies   Light emitting diodes (LEDs) considered below are frequently referred to as 'LED devices"[1], i.e. the packaged light-emitting chips or dice including the mounting substrate, encapsulant, phosphor if applicable, and electrical connections. Packaging main purpose is to protect LED chips from electrostatic discharge, high temperature (that is considered, in particularity, as the cause of generation of defects in the active region [67], especially in the devices subjected to high injection currents [68]), moisture that can induce delamination [69], chemical oxidation [16]. The standard LED package is the 5 mm epoxy dome [5]. In addition, packaging could enhance light extraction [13]: in particularity, encapsulant shape and its material are optimized for this goal.

The external quantum efficiency (EQE) η_{ext} of light emission from a light emitting diode is defined as [70] $\eta_{\text{ext}} = N_{\text{emit}}^{\text{free}}/N_{\text{inj}}$, where $N_{\text{emit}}^{\text{free}}$ is the number of photons emitted into free space per second, N_{inj} is the number of electrons injected into LED per second.

The wall plug efficiency (WPE) η is obtained by multiplying η_{ext} by the electrical efficiency η_{el} which is the ratio of the energy of the emitted quantum E_{photon} to the applied forward voltage V (q is the electron charge) $\eta_{\text{el}} = \hbar\omega/qV$.

The external quantum efficiency could be written as a product of the injection efficiency η_{inj} (the ratio of the number of electrons injected into the active region to the number of electrons injected into the device), the internal quantum efficiency (IQE) $\eta_{\text{int}} = N_{\text{emit}}/N_{\text{inj}}$ where N_{emit} is the total number of photons emitted from active layer per second, and the light extraction efficiency (LEE) $\eta_{\text{extr}} = N_{\text{emit}}^{\text{free}}/N_{\text{emit}}$ Since the internal quantum efficiency η_{int} is the ratio of the radiative recombination rate k_{rad} to the total recombination rate $k_{\text{rad}} + k_{\text{non}}$, the equation for η_{ext} could be re-written as

$$\eta_{\text{ext}} = \eta_{\text{extr}} \times \eta_{\text{inj}} \times \eta_{\text{int}} = \eta_{\text{extr}} \times \eta_{\text{inj}} \times \frac{k_{\text{rad}}}{k_{\text{rad}} + k_{\text{non}}}.$$

The measurements of the total light emission from the LED are relatively easy since they are essentially the same as for other light sources; the recently developed method of confocal microscopy based on the ability to analyze the object properties at a specified depth provide a unique possibility to study light emission from the LED in detail, including different elements of the emitter and the near-field light intensity distribution [71] as well as angular beam divergences [72]. On the problem of the experimental separation of different contribution to the total efficiency see, e.g., Ref. [73].

LED efficiency enhancement   There are different ways to increase the light output such as to decrease k_{non}, to increase k_{rad}, to increase η_{extr}.

---

[1]There also other names used depending on the presence of additional elements such as LED Module, LED Lamp, LED Light Engine and LED Luminaire, defined in Ref. [66].

Evidently, the main approach aiming at maximizing η_{int} is the LED design involving the choice of active layer structure (the sequence of layers, their thickness, composition, doping, the number of QWs, the nature of the electron blocking layer (EBL) etc.) [74, 36, 45, 49, 75, 76, 77, 78, 79], advancements in design of contacts with better optical (higher transparency) and electrical properties (smaller specific contact resistance) [80, 81, 82, 83, 84, 85, 86, 87]; barriers could be compositionally graded [88, 89, 90]. To reduce the polarization effects - spatial separation of the electron and hole wavefunctions that suppresses the radiative recombination and current-dependent red-shift due to the quantum confined Stark effect (QCSE) - nitride layers are grown along non-polar or semipolar wurzite directions [91, 92, 93, 94, 95] or polarization-matched quaternary layers are used [96]. In order to improve carrier confinement, additional elements of the chip structure are implemented as, e.g., a SiO_2 current blocking layer beneath the p-pad electrode [97], short-period superlattice cladding layers [98], an n-doped InGaN electron reservoir layer inserted to influence the electron capture process [99] (an increase of the efficiency of electron capture into QWs could also be achieved via so called "phonon engineering" that exploits high polar optical energy in AlInGaN materials for confining electrons in the QWs [100]), tunnel junctions to increase electron and hole injection [101], electron tunneling barrier beneath the MQW [102]; the aim of staggered QWs [103, 104, 105] or "stair case electron injector" [106, 107] is to assist the thermallization of injected hot electrons; stair-shaped QW are exhibit stronger localization effect [108].

Decreasing k_{non} is achieved by growing higher quality layers using advanced buffer layers [109] (such as double MgN/AlN [110] and GaN/SiN [111] buffer layers on the low-defect foreign substrates or using native GaN ones [112][2] and bulk AlN substrates.

There is still no agreement on the role of the In composition fluctuations observed in the active layers made from the ternary solid solution InGaN and in the corresponding QWs [115, 116, 117] (these fluctuations also, to a lesser extent, affect emission of laser diodes [118]). Radiative recombination in InGaN QW is about 50 times more efficient than in GaN [119]. In composition fluctuations (also referred to as *indium-rich nano-clusters* - see, e.g. [120]) resulting from the poor In incorporation in the epitaxial layer during the growth or from the phase separation due to the large miscibility gap [121, 122, 123] are considered to be responsible for the unusual properties of the nitride-based light emitters —their ability to emit light in spite of the extremely high (compared to the devices based on the conventional III-V materials) level of the dislocation density[3]—via the localized states related to the composition fluctuations [130, 131] or the quantum-dot-like structures [132, 133]. This carrier localization - *indium-rich nano-clusters* are potential minima for the electrons in the conduction band nad holes in the valence band - suppresses the in-

---

[2]While GaN substrates provide significant improvement over sapphire ones [113], they are inferior in comparison to AlN substrates for the UV LEDs due to light absorption into the GaN substrate [114].

[3]Threading dislocations with vacancies at Ga sites [124] acting as non-radiative recombination centers [125] present more serious problem for GaN/AlGaN system than for GaN/InGaN one [11]. Some experiments, however, indicate that threading dislocation density does not directly affect the photoluminescence decay time and the observation results could be explained by the reduction of the net volume of the light-emitting region [126]. V-defects and the associated dislocations are responsible for the leakage current observed in LEDs [127]. Linear defects are also considered to be the source of the failure of nitride-based LEDs, that was demonstrated for GaN RCLEDs by experiments of B. Roycroft et al. [128]. H. Amano et al. showed that in AlGaN not only screw or mixed dislocation could act as nonradiative recombination centers, but edge ones as well [129].

plane carrier diffusion reducing the probability of the non-radiative recombination [134]. The average size of InN cluster depends on the growth method, being somewhat larger for MOCVD growth compared to MBE [135]. The density of the localized states depends on the In composition, e.g. in green LED it is two order of magnitude higher than in UV LED [136], thus introducing a small amount of In into the quaternary active layer of UV LED increases emission intensity [137]. Note, however, that large-scale (micron-sized) clusters formed in InGaN QWs with high In composition (green LEDs) could themselves contain non-radiative recombination centers [138].

It should be stressed that evidence of In composition fluctuation in InGaN layers is partly circumstantial [139]. The presence of such features is registered as spatial inhomogeneities of both photoluminescence intensity and spectra with scale of the order of 200 nm or less by confocal scanning laser microscopy [140], by confocal microphotoluminescence [141] revealing submicron-scale inhomogeneities of both PL intensity and spectrum and by near-field optical microscopy [142][4] as well as by transmission electron microscopy [146]. The presence of In-rich separated phases in cubic InGaN epitaxial layers is evident from resonant Raman scattering and X-ray diffractoscopy experiments [147]. The existence of the distribution of InGaN localized states is also supported by the analysis of the temperature and current dependence of the optical intensity and energy shift in InGaN-based LEDs via comparison between electroluminescence and cathodoluminescence [148]. Van Daele et al. [149] reported the existence of metallic In disk-shaped platelets in a GaN matrix and large clusters of metallic In near the treading dislocations detected by high-resolution TEM, but did not suggest an explanation how the metallization process could be triggered.

Quantum dots could be introduced into InGaN active layer intentionally exploiting the strain and the affinity difference between the InGaN and SiCN layer [150]; self-assembled InGaN quantum dots as strongly localized recombination sites are exploited in UV LEDs [151]. Compositional fluctuation of In could also be created by anti-sufracant effect using SiN nano-islands [152].

There is, however, an opinion that there is no need to invoke the concept of the localized states to explain, using the spontaneous and piezo polarization effects [153, 154, 155], all the effects observed in InGaN quantum well (QW) such as the emission energy shifts towards the lower energy even below the bulk bandgap and increase of the lifetime of emission lines with increasing the well width [156, 157]. This point of view is supported by the experimental evidence of similar behaviour of the emission from the GaN/AlGaN QW, where neither composition fluctuations nor quantum-dot structures occur. Still, H. Hirayama et al. [158] observed carrier confinement in the In segregation region from cathodoluminescence measurements and suggested an indirect evidence supporting the existence of localized states — a drastic increase of room temperature ultraviolet emission by introducing a small concentration of In into AlGaN active region. However, note that the localized states could also be related to the variation of the thickness of the QW [159, 160] that can be more important then the composition fluctuations [139]. As was proved recently in ref.

---

[4]Wang et al. [143, 144] suggested to exploit this phenomenon - blue and yellow emission from In-rich clusters and low-indium regions, respectively, with wavelengths about 440 and 570 nm - as a basis of a *single emission layer* white LED. One more exotic option to develop a single-chip multicolour LED was proposed Azuhata et al. [145] - to apply a pulse current containing two pulses of different amplitude in a cycle to get, via the current-induced blueshift, emission at two wavelengths.

[161], the nature of localization is of secondary importance. A model has been proposed based on the exponential decay of the density of states in the bandgap that also explanes strange IQE dependencies on the current density at low temperatures. Direct evidence of nanocluster-induced luminescence reported in [162]. H. Sch'omig et al. [163] were able to probe individual localization centers in an InGaN/GaN QW using PL spectroscopy with subwavelength lateral resolution. Experiments at low (up to a few tens of K)temperatures showed a blueshift of PL emission with temperature due to a temperature dependent screening of the electric field but no blueshift of emission form a single localization center with increasing excitation intensity. The authors speculate that increasing excitation power leads to activation of additional localization centers that emit at shorter wavelength resulting in the macroscopic behaviour of blueshift with increasing power.

2.1. LED die structure

An LED structure considered is a stack of uniform or graded-composition epitaxial layers pseudomorhically grown on an underlying template layer (usually a buffer layer) in the direction z corresponding to the hexagonal (c) axis of the wurzite crystal. Thus all layers have the in-plane lattice constant equal to that of the template layer a_s. The thickness of the latter is assumed to be much greater than that of the LED heterostructure, hence, the bending of the structure can be neglected.

2.1.1. Strain and piezoeffect

The misfit strain components in a planar biaxially stressed [0001] layer are $\epsilon_1 = \epsilon_2 = (a - a_s)/a_s, \epsilon_3 = -2\epsilon_1 C_{13}/C_{33}$, where C_{ij} are the components of the elastic stiffness tensor in Voigt notation [164]. The unstrained lattice constant a of $Al_x Ga_y In_{1-x-y}N$ alloy is defined by Vegard rule as

$$a = x a_{\text{AlN}} + y a_{\text{GaN}} + (1 - x - y) a_{\text{InN}}$$

The strain affects the LED operation. Firstly, the tension (compression) decrease (increase) the energy bandgap. The variation could be significant: e.g., the difference of the bandgap for $Ga_{0.8}In_{0.2}N$ under tensile and compressive strain equal (in magnitude) to 0.02 is about 0.5 eV. The dependence is more steep under tension. While the energy of heavy holes only weakly depends on the strain, a large shift is observed in light hole band under tensile strain [165]. Strain influences the optical properties of nitride QW via valence band mixing [166, 167].

The compressive strain relaxation with increasing thickness of the n-GaN layer results in generation of numerous stacking fault just under the active layer that could act as a current bypass reducing the LED efficiency [168].

The strain also results in the strong built-in electric field due to the piezoeffect. Only z-component of the piezoelectric polarization is sufficient to consider in the LED heterostructure [2]

$$P^{\text{pz}} = 2\epsilon_1 \left(E_{13} - E_{33} \frac{C_{13}}{C_{33}} \right)$$

here E_{ij} are the components of the piezoelectric tensor.

The other source of the polarization field is spontaneous polarization present in polar semiconductors with wurzite or lower symmetry crystal structure, having a fixed direction along c-axis:

$$P^{\mathrm{sp}} = x P_{\mathrm{AlN}}^{\mathrm{sp}} + y P_{\mathrm{GaN}}^{\mathrm{sp}} + (1 - x - y) P_{\mathrm{InN}}^{\mathrm{sp}}$$

The total polarization field P^{tot} is the sum of these two contributions, the first being usually greater (except in the lattice-matched alloy). The sign of piezoelectric field is defined by the kind of strain (tensile or compressive) while that of spontaneous polarization by the layer polarity (Ga-polarity or N-polarity).

2.1.2. Electron, hole and impurity statistics

The electron n and hole p concentrations obey the Fermi-Dirac statistics

$$n = N_{\mathrm{c}} \mathcal{F}_{1/2} \left(\frac{\varphi_n - E_{\mathrm{c}} + q\varphi}{kT} \right), \qquad p = N_{\mathrm{v}} \mathcal{F}_{1/2} \left(\frac{E_{\mathrm{v}} - \varphi_p - q\varphi}{kT} \right) \qquad (1)$$

where E_c and E_v are the conduction band bottom and valence band top, respectively, $\mathcal{F}_{1/2}$ is the Fermi integral

$$\mathcal{F}_{1/2}(\xi) = \frac{2}{\sqrt{\pi}} \int_0^\infty \frac{x^{1/2}}{1 + \exp(x - \xi)} dx$$

k is the Boltzmann constant, T is temperature, q is the electron charge, and the effective densities of states in conduction, N_{c}, and valence, N_{v}, bands are

$$N_{\mathrm{c}} = 2 \left(\frac{m_n^{\mathrm{av}} kT}{2\pi\hbar^2} \right)^{3/2}, \qquad N_{\mathrm{v}} = 2 \left(\frac{m_p^{\mathrm{av}} kT}{2\pi\hbar^2} \right)^{3/2}$$

where $\hbar$ is the Plank constant, m_n^{av} and m_h^{av} are the averaged electron and hole masses. With account of anisotropy of the conduction band, the averaged electron effective mass is $m_n^{\mathrm{av}} = (m_n^{\perp}(m_n^{\parallel})^{1/2})^{2/3}$, where $m_n^{\perp}$ and $m_n^{\parallel}$ are the in-plane and normal (along the z-axis) effective masses. For holes, one has to consider the complex valence band structure and splitting of the valence bands in the centre of the Brilluene zone:

$$N_{\mathrm{v}} = N_{\mathrm{hh}} + N_{\mathrm{lh}} + N_{\mathrm{so}}, \qquad N_s = 2 m_{\mathrm{p,s}}^{\perp} (m_{\mathrm{p,s}}^{\parallel})^{1/2} \left(\frac{kT}{2\pi\hbar^2} \right)^{3/2}$$

where s = hh, lh, so refers to heavy, light, and split-off (also called crystal-field split [169]) holes, respectively.

The concentrations of ionized donors N_{D}^+ and acceptors N_{A}^- are given by

$$N_{\mathrm{D}}^+ = \frac{N_{\mathrm{D}}}{1 + g_{\mathrm{D}} \exp\left(\frac{\varphi_n - E_c + E_D + q\varphi}{kT} \right)}, \qquad N_{\mathrm{A}}^- = \frac{N_{\mathrm{A}}}{1 + g_{\mathrm{A}} \exp\left(\frac{E_v + E_A - q\varphi_p - \varphi}{kT} \right)}.$$

Here g_D and g_A are the degeneracy factors[5] and E_D and E_A are the activation energies of electron and holes, respectively.

---

[5]Generally g_A varies from 4 to 6 in the nitride semiconductors due to small valence band splitting.

2.2.   Carriers transport

There are two main approaches to simulation of carrier transport in semiconductor devices [170]. The first is a particle–based one such as the full band Ensemble Monte Carlo method [171]. Its undeniable advantage is the correct description of the carrier dynamics in the first Brilloiuin zone of momentum space. The price for the accuracy is the need of huge computational resources, somewhat alleviated by both advances in algorithms [172] and development of the low-cost hardware for parallel computations.

The second approach relies on continuum (macroscopic) models. The complexity of the carrier transport description reduces from kinetic models (the Boltzmann equation in the semiclassical case or its quantum mechanical equivalent involving the Wigner function) through numerous quasi-hydrodynamical models down to the DD one [173, 174]. The family of continuum models is further extended by the development of hybrid models that couple different level of description in different domains of the device [175]. Continuum models, when adequate, provide well over two orders of magnitude reduction of CPU time compared to the particle-based approaches [176].

DD model [177] could be derived formally from the Boltzmann equation (see, e.g., [178]). Introduced over a half century ago in Ref. [179], it has been based on the older ionic transport model of Nernst and Plank. Still, up to now it represents a reasonable compromise between the computational efficiency and the accuracy of the description of the underlying physical phenomena for a great variety of semiconductor devices [180].

DD model for simulation of the carrier transport in nitride-based LEDs should account for the above mentioned specific features of these materials. It represents a coupled nonlinear system consisting of the Poisson equation for the electrostatic potential φ and continuity equations for the concentrations of electrons n and holes p:

$$\nabla \cdot (\mathbf{P}^{\text{tot}} - \epsilon_0 \epsilon \, \nabla \varphi) = q \, (N_{\text{d}}^+ - N_{\text{a}}^- + p - n) \tag{2}$$

$$\nabla \cdot \left(\frac{\mu_n \, n}{q} \nabla(\varphi_n) \right) = R \tag{3}$$

$$\nabla \cdot \left(\frac{\mu_p \, p}{q} \nabla(\varphi_p) \right) = -R \tag{4}$$

Here ϵ_0 is dielectric permittivity of vacuum, ϵ is the static dielectric permittivity tensor.

Electron and hole current densities are, respectively, $\mathbf{j}_n = q\mathbf{J}_n$, $\mathbf{j}_p = q\mathbf{J}_p$, where carrier fluxes are defined as

$$\mathbf{J}_n = -\frac{\mu_n \, n}{q} \nabla(\varphi_n), \qquad \mathbf{J}_p = \frac{\mu_p \, p}{q} \nabla(\varphi_p)$$

The DD model Eqs. (2) - (4), being the singular pertubed problem, can exhibit both boundary and interior layers with the rapid variation of the carrier concentrations and/or electric field [181, 182]. The stiffness of the DD equations nonlinearly increases with the value of the bandgap. Additional difficulty specific for nitride devices is the strong build–in electric field due to polarization. Although DD model have been used for decades (see, e.g., [183]), a rigorous analysis of the underlying system has been carried out under quite restrictive assumptions only (decoupled problems for potential and carrier concentrations, no generation-recombination, Boltzmann statistics etc.) and mostly deals with existence rather than uniqueness of the solution [184, 185, 186]. Moreover, it is established that in certain cases the solutions of one dimensional "voltage driven" problem are not unique in general [187].

2.2.1. Recombination of non-equilibrium charge carriers

The recombination rate accounts for both non-radiative and radiative channels $R = R^{\text{nr}} + R^{\text{rad}}$. One of the non-radiative recombination channels is the recombination through the threading dislocation cores [188]

$$R^{\text{nr}} = \frac{np}{\tau_p(n + n_{\text{D}}) + \tau_n(p + p_{\text{D}})} \cdot \left[1 - \exp\left(-\frac{\varphi_n - \varphi_p}{kT} \right) \right]$$

where

$$n_{\text{D}} = n \cdot \exp\left(\frac{E_{\text{D}} - \varphi_n}{kT} \right), p_{\text{D}} = p \cdot \exp\left(\frac{\varphi_p - E_{\text{D}}}{kT} \right)$$

and carrier lifetimes are

$$\tau_{n(p)} = \frac{1}{4\pi D_{n(p)} N_{\text{D}}} \left[\ln\left(\frac{1}{\pi a^2 N_{\text{D}}} \right) - \frac{3}{2} + \frac{2D_{n(p)}}{aV_{n(p)}S} \right]$$

Here $D_{n(p)}$ is the diffusion coefficient of electrons (holes), $V_{n(p)} = (3kT/m_{n(p)}^{\text{av}})^{1/2}$ is the carrier thermal velocity, N_{D} is the dislocation density, S is the fraction of electrically active sites on the surface of a dislocation core, and E_{D} is the energy level associated with the dislocations. For more detail, see [188].

The importance of the Auger recombination

$$R^{\text{Auger}} = (C_n n + C_p p)np \left[1 - \exp\left(-\frac{\varphi_n - \varphi_p}{kT} \right) \right]$$

was realized only some years ago [189]. Now it is recognized as one of the major reason for the efficiency droop at large current densities [39], along with the carrier leakage [190]. The early underestimate of the effect of the Auger recombination was based on theoretical coefficients that seem to be too small. However, computations do not account for the Auger recombination involving excited levels and phonons. The role of the Auger recombination is confirmed by careful experiments on saturation of electro- and photoluminesce in the nitride MQWs [191].

The bimolecular radiative recombination rate is defined by

$$R^{\text{rad}} = Bnp \cdot \left[1 - \exp\left(-\frac{\varphi_n - \varphi_p}{kT} \right) \right]$$

The boundary conditions for the Poisson equation (2) are stated as follows:

$$\varphi_{\text{l}} = 0, \quad \varphi_{\text{r}} = \triangle\varphi.$$

Here, $\triangle\varphi$ is the sum of the applied bias and the contact potential, the subscripts 'l' and 'r' refer the position of the left and right edges of the heterostructure, respectively. The boundary conditions for the continuity equation for the majority carriers are derived from the electric neutrality $N_d^+ - N_a^- + p - n = 0$ at the edges of the heterostructure:

$$\begin{cases} \varphi_n = E_c + kT\,\mathcal{F}_{1/2}^{-1}\left(\dfrac{N_d^+ - N_a^- + p}{N_c}\right) - q\varphi, & N_d > N_a \\[3mm] \varphi_p = E_v - kT\,\mathcal{F}_{1/2}^{-1}\left(\dfrac{N_a^- - N_d^+ + n}{N_v}\right) - q\varphi, & N_a > N_d \end{cases}$$

while Neumann boundary conditions are set for the minority carrier quasi-Fermi levels.

Probably, the most important effect that this model does not account for is the tunneling current. There is an evidence that carrier tunneling enhanchanced by the deep-level states associated with the dislocations [192] is important at low to medium injection levels only, while at the typical LED operation current densities diffusion-recombination processes dominate [193, 194]. Recent experiments [195] also show that even at the intermediate forward bias the tunneling current is suppressed in the high quality structures due to the defect reduction. However, the existense of strong, up to 3 MV/cm, built-in electric field originated from polarization charges appearing at the structure interfaces could result in local potential barriers near the active region hindering the carrier injection due to the formation of highly resistant regions. It is anticipated that in such cases the contribution of the tunneling current could be significant.

2.3. Light emission spectra

Computation of the emission spectra is decoupled from the carrier transport. Only the principle channel - radiative recombination between the electron and hole states confined in the QW active region - is considered. The Schrödinger equations for electron and holes are solved with the potential energy determined from self-consistent solution of the Poisson and DD transport equations (2 - 4).

The complex structure of the valence band of group-III nitride materials is taken into account within the 6x6 Bir-Pikus Hamiltonian [196]. The splitting of the heavy, light and split-off branches of the valence bands in the centre of the Brilluene zone is assumed to be independent of build-in electric field. Thus, the profiles of the valence subbands are equal to that obtained from the solution DD equations.

The wave functions of electrons and holes are chosen in the form

$$u_n\Psi(z)\exp(\mathbf{k_n}\cdot\mathbf{r}), \qquad u_{p,s}\Psi_s(z)\exp(\mathbf{k_{p_n}}\cdot\mathbf{r}),$$

where u_n and $u_{p,s}$ are the Bloch amplitudes of electrons and holes corresponding to the centre of the Brilluene zone, $\mathbf{k}$ and $\mathbf{k_p}$ are their in-plane quasi-moment vectors, Ψ and Ψ_s are the envelope functions. The Bloch amplitudes can be found from the basis wave functions derived in [197]. The Schrödinger equation for the electron envelope function can be written in the form

$$-\frac{\hbar^2}{2m_n^{\parallel}}\frac{d^2\Psi}{dz^2} + (E_c - q\varphi)\Psi = E\Psi$$

where $m_n^{\parallel}$ is the electron effective mass in z direction averaged over the QW with weights proportional to the quasi-classical electron concentration (1).

For heavy, light, and split-off holes similar equations are written

$$\frac{\hbar^2}{2m_{p,s}^{\parallel}} \frac{d^2\Psi_s}{dz^2} + (E_{v,s} + q\varphi)\Psi_s = E\Psi_s$$

where $m_{p,s}^{\parallel}$ is the hole effective mass.

The problem is solved in the domain that includes QW and parts of barriers on its sides large enough to set the homogeneous Dirichlet boundary conditions for the equations (2.3.,2.3.). The semiclassical wavefunction with the energy equal to the Fermi level in QW is used to safely estimate the necessary size of the domain. The fourth order finite difference (FD) scheme has been used for the solution of the eigenvalue problem for the Shrödinnger equation.

Assuming the vertical band-to-band transitions, the emission rate of photons with the frequency ω is

$$w(\omega) = \frac{\pi}{\hbar}\left(\frac{q}{m_0 c}\right)^2 \sum_{\mathbf{k},s,\lambda} \sum_{s_e,s_h} |\langle \Psi_{\mathbf{k}}^{s_e}|\Psi_{\mathbf{k},s}^{s_h}\rangle|^2 |\langle u_n|\mathbf{A}_\lambda \cdot \hat{\mathbf{p}}|u_{p,s}\rangle|^2 \tag{5}$$

$$\times f_{\mathbf{k}}^e f_{\mathbf{k}}^{h,s} \delta\left(\hbar\omega - E_g - \frac{\hbar^2 k_{\parallel}^2}{2\mu_s^{\parallel}} - \frac{\hbar^2 k_{\perp}^2}{2\mu_s^{\perp}}\right)$$

where m_0 is the electron mass in vacuum, c is the light velocity in vacuum, $\mathbf{A}_\lambda$ is the vector potential of the electromagnetic wave with polarization denoted by the subscript λ, $\hat{\mathbf{p}}$ is the momentum operator, $f_{\mathbf{k}}^e$ and $f_{\mathbf{k}}^{h,s}$ are the distribution functions of electrons and holes corresponding to the quasi-momentum vector $\mathbf{k}$, E_g is the bandgap, and $\mu_s^{\perp} = m_n^{\perp} m_{p,s}^{\perp}/(m_n^{\perp}+m_{p,s}^{\perp})$ and $\mu_s^{\parallel} = m_n^{\parallel} m_{p,s}^{\parallel}/(m_n^{\parallel}+m_{p,s}^{\parallel})$ are the in-plane and normal reduced effective masses of the carriers, $\delta(x)$ is the delta-function. In Eq.(5) summation is made over all k-states, valence subbands, two orthogonal light polarizations, and spin states of electron and holes, s_e and s_h .

2.4. Examples

For two reasons validation of the one-dimensional numerical model of the LED heterostructure is not easy. Firstly, some assumptions to account for the in-plane variation of the light emission should be made. Two kinds of the emission nonuniformity have been reported. The first is the macroscale one resulting from the nonuniform current spreading (see, e.g., [198]), being defined mainly by the LED layout and the geometry of the contacts. The emission nonuniformity due to the composition fluctuations has a much smaller scale (about a few tens of nanometers), but the intensity variation could be huge: for example, fluctuations of the internal quantum efficiency from 10% to 50% has been reported [199]. Secondly, the published descriptions of the experiments rarely are complete, first of all in respect to the active layer doping. This issue is extremly important if co-doping is used. e.g., simultaneous use of Si and Zn with the difference in the ionization energy about 0.35 eV could result in the large uncertainty of the emission peak position if the details of doping are not known. The next crucial issue is the shape and the composition of QW s that should be known with sufficient accuracy.

Material properties Material properties of group III-nitrides are not so well established as those of III-V compounds, to say nothing about silicon. There are considerable differences between theoretical estimates via Monte Carlo method (see, e.g., Refs. [200, 201], published in the same journal issue). Experimentally measured carrier mobilities strongly depend on the growth conditions. The material properties used for computations of transport and light emission in the LED structures are described below.

In the present paper all material parameteres of nitride alloys are determined by Vegard rule

$$f_{Al_xGa_yIn_{1-x-y}N} = x f_{AlN} + y f_{GaN} + (1 - x - y) f_{InN}$$

the only exception being the bandgap for which a quadratic approximation is used:

$$(E_g)_{Al_xGa_yIn_{1-x-y}N} = x(E_g)_{AlN} + y(E_g)_{GaN} + (1 - x - y) E_{g_{InN}} \qquad (6)$$
$$- xy b_{Al-Ga} - x(1 - x - y) b_{Al-In} - y(1 - x - y) b_{Ga-In}$$

where the values for the bowing parameters b_{Al-Ga}, b_{Al-In}, b_{Al-In}, based on the analysis of Refs. [202, 203, 204, 205, 206, 207], are 1.0 eV, 4.5 eV,1.2 eV for AlGaN, AlInN, GaInN, respectively. Parameters of binary nitrides used in computations are shown in Table 1. These values has been chosen as a result of the analysis of the published data: Refs.

Table 1. Properties of binary nitrides

Parameter	Symbol	Unit	AlN	GaN	InN
Lattice constant	a	nm	0.3112	0.3188	0.3540
Static dielectric constant	ϵ_{33}		8.5	8.9	15.3
Spontaneous polarization	P^{sp}	C/m^2	-0.081	-0.029	-0.032
Piezoelectric constant	E_{13}	C/m^2	-0.58	-0.33	-0.22
Piezoelectric constant	E_{33}	C/m^2	1.55	0.65	0.43
Elastic constant	C_{13}	GPa	115	105	95
Elastic constant	C_{33}	GPa	385	395	200
bandgap	E_g	eV	6.2	3.4	0.7

[202, 208, 209] on the lattice constant, Refs. [202, 208, 210, 211] on the elastic constants, Refs. [212, 213] on the dielectric constant. bandgap values are taken from Ref.[214] while parameters related to polarization - from Ref. [215]. Effective electron and hole masses in binary nitrides at 300 K are taken from Ref. [216] for AlN, InN and from Ref. [197] for GaN.

The global parameters used for the whole structure are: the acceptor and donor ionization energy $E_d = 13$meV, $E_a = 170$meV, the acceptor and donor g-factors $g_d = 2, g_a = 4$, the dislocation density, the radiative recombination coefficient $B = 2.4 \times 10^{-11}$cm^3/s [217], the temperature, the substrate lattice constant, the crystal polarity, and spectrum broadening $\Gamma = 20$meV. Reducing nominal threading dislocation density at least qualitatively accounts for the effect of In composition fluctuations responsible for the carrier capture by the localized energy states before they nonradiatively recombine on the dislocations core [218].

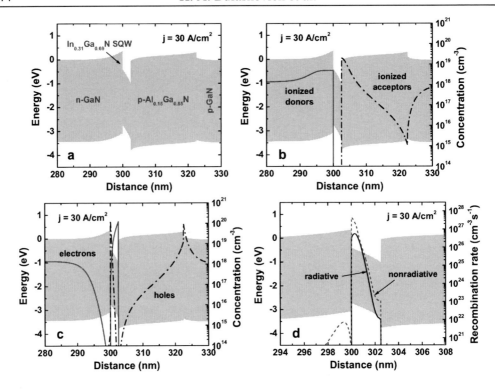

Figure 1. Room-temperature band diagram (a) and distributions of ionized donor/acceptor concentrations (b), electron and hole concentrations (c), and radiative and non-radiative recombination rates (d) of 2.5 nm $In_{0.31}Ga_{0.69}N$ SQW LED structure with 20 nm $p - Al_{0.15}Ga_{0.85}N$ EBL sandwiched between thick $n - GaN$ and $p - GaN$ contact layers.

Blue single-QW (SQW) LED   As a first example results of computations for the SQW LED structure having Ga-polarity with compositionally uniform layers are presented in Fig. 1.

An example of the InGaN SQW with graded emitters at two values of operating current is presented in Fig. 2. Bandstructure (a,b), carrier concentrations (c,d) and rate of both radiative and nonradiative recombination are plotted.

Comparison of spectral properties of the considered SQW LED structures with layers of different nature is presented in Fig. 3.

The next example is a multiple-QW (MQW) LED with low or high barrier doping (Fig. 4).

Effect of the dislocation density in nitride layers on the LED efficiency is illustrated by Fig. 5.

To asses the advantages of non-polar nitride structures a comparison of polar and non-polar SQW LEDs have been performed (Fig. 6).

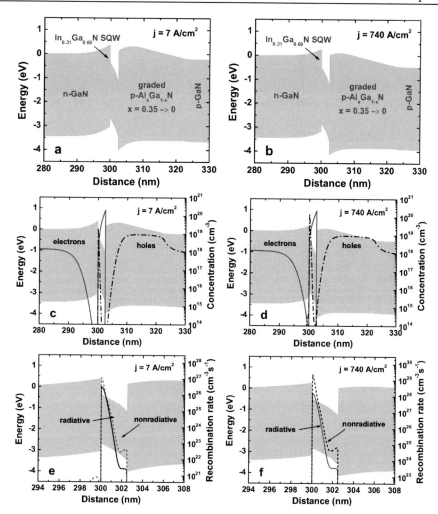

Figure 2. Room-temperature band diagrams (a,b) and distributions of electron and hole concentrations (c,d) and radiative and non-radiative recombination rates (e,f) of 2.5 nm $In_{0.31}Ga_20.69N$ SQW LED structure with 20 nm graded-composition $p - Al_xGa_{1-x}N$ EBL at the current density of 7 A/cm^2 (a,c,e) and 740 A/cm^2 (b,d,f).

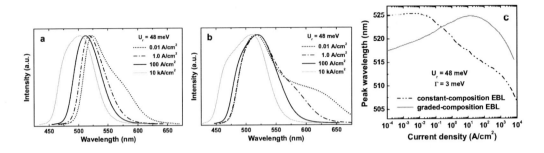

Figure 3. Emission spectra from the 2.5 nm In − 0.31Ga − 0.69N SQW LED structures at various current densities computed for constant-composition (a) and graded-composition (b) EBLs specified in Figs.1 and 2. Spectral shifts of the peak emission wavelengths as a function of current density for both structures (c).

3.   Current Spreading in LED Die and Light Extraction

Current crowding (also *current crowding effect*) is a nonhomogenous in-plane distribution of current density, especially at the vicinity of the edge of the metal electrodes. It is one of the limiting factors of efficiency of light emitting diodes, but it is also of concern in other semiconductor devices, e.g., bipolar transistors and Schottky diodes [219]. Materials with low mobility of charge carriers and devices on the insulating substrates that enploy mesa structures with lateral geometry of the anode and cathode electrodes (with both contacts placed in the same plane) are especially prone to current crowding effect - current crowding is essentially suppressed in vertical LEDs [220, 221]. Current crowding can lead to localized overheating and formation of thermal hot spots, lowering IQE and affecting the series resistance of the diode as well as can result in a premature device failure [48]. The two evident ways to cope with this problem are to reduce the sheet resistivity of the top (usually p-type - due to low p-type doping capability of GaN [222]) contact layer [223] (e.g., by inserting a tunnel junction into the the upper cladding layers allowing to use n-type GaN instead of p-type GaN as the top contact [224]) or to reduce the length of the current path by modification of the geometry of the contacts (e.g., using the so called *interdigitated* mesa geometry wherein each of p-type contacts is surrounded by two n-type fingers [225, 226, 227, 228], employing multiple thin ohmic metal patches ("electrode tails") running along the edges of the chip [229] or using an interconnected array of micro LEDs (μLEDs [230, 231]).

3.1.   Hybrid approach to LED die modeling

Earlier studies of the current crowding were primarily based on analytical quasi-2D models. The roles of the *n*-contact layer [232] and a semitransparent *p*-electrode [233], as well as the electrode configurations producing a more uniform lateral distribution of the current density [234, 225, 233, 235, 236] were the main focuses of the studies. Such models rely on simplification assumptions (e.g., an assumption that the p-type contact has the same electrostatic potential at every point [234, 225]) or represent the lateral resistance of the

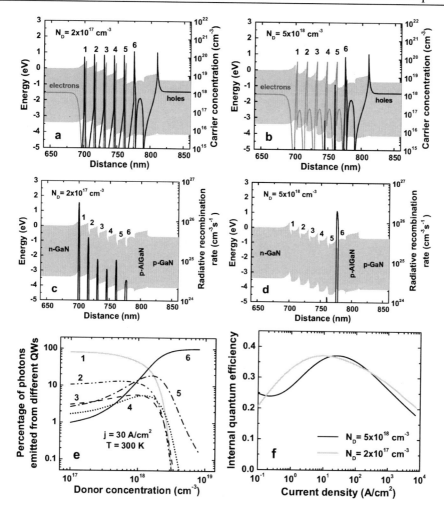

Figure 4. Band diagrams and distributions of non-equilibrium carrier concentrations (a,b) and radiative recombination rates (c,d) in $3nmIn_{0.18}Ga_{0.82}N/12nmn - GaN$ MQW LED structures with low (a,c) and high (b,d) donor concentrations in the MQW GaN barriers (various wells are numbered in the plots). Fractions of light emitted from different quantum wells as a function of n-GaN barrier doping (e). The spacer separating the top QW and p-AlGaN EBL are assumed to be doped with Mg acceptors up to the concentration of 1×10^{19} cm^{-3} Comparison of IQE computed for the MQW LED structures with different barrier doping (f).

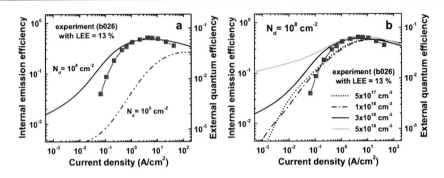

Figure 5. IQE of the MQW LED structure specified in Fig.4 as a function of current density computed for different densities of threading dislocations (a) and different doping levels of the n-GaN MQW barriers (b). Circles are experimental data on the external quantum efficiency for such structures borrowed from [193].

transparent electrode via an equivalent electrical circuit [226, 237, 238, 229].

While these studies were useful, introducing a notion of the critical transparent electrode thickness [236] and suggesting perspective designs of the contact geometry, a much deeper insight into the current-spreading phenomena was provided by first 3D simulations using APSYS of a 340 nm UV LED with a simple electrode geometry, demonstrating, in particular, a dramatic current localization near the electrode edges and its strong effect on the carrier leakage from the active region [239].

State-of-the-art III-nitride LEDs normally utilize chips with rather complex electrode configurations. A brute-force approach to the solution of a conjugated electro-thermal problem of current spreading and heat transfer in the practically used LED dice provided by general-purpose software is frequently inefficient. First, to get a sufficient accuracy, extremely large non-uniform grids are needed (evidently, grid steps for a QW and contact layers should differ greatly, as these layers actually have very different dimensions). Second, Cartesian grids could not practically resolve small local geometry features; any grid refinement is immediately extended all over the computational domain to the boundary, thus greatly increasing the number of grid cells. Unstructured grid for a such anisotropic domain (aspect ratio is ∼100-1000) will contain numerous highly distorted tetrahedra that could result in a violation of the discrete maximum principle for elliptic problems.

In order to reduce the computing time, we have proposed an approximate hybrid approach [47, 48] that enables full modeling of the LED dice with an arbitrary complex design. On the one hand, this approach is close to that successfully used for analysis of mid-infrared InAs LEDs [240]. On the other hand, it may be considered as a particular case of a general approach to multi-scale multi-physics problems [241], utilizing a coupled model that consists of a few constituent submodels in corresponding subdomains that could overlap. There are two advantages of the hybrid approach: (i) the computational grid may be generated and optimized in each subdomain (and, respectively, for each submodel) independently and (ii) the physical models may be chosen and properly simplified in every subdomain with account of its specific properties. As a result, the conjugated solution of the whole problem would require much less computer resources and much shorter com-

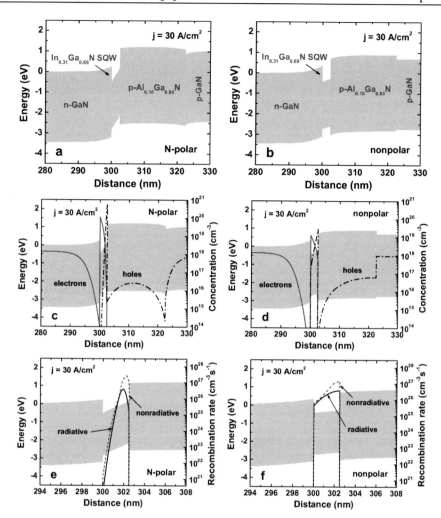

Figure 6. Room-temperature band diagrams (a,b) and distributions of electron and hole concentrations (c,d), and radiative and non-radiative recombination rates (e,f) of 2.5 nm $In_{0.31}Ga_{0.69}N$ SQW LED structure with 20 nm p $-$ $Al_{0.15}Ga_{0.85}N$ EBL grown on N-polar (a,c,e) and non-polar (b,d,f) facets.

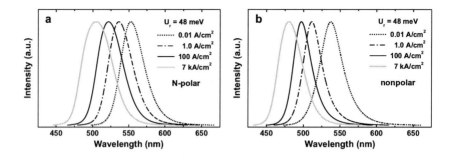

Figure 7. Emission spectra from the N-polar (a) and nonpolar (b) 2.5 nm $In_{0.31}Ga_{0.69}N$ SQW LED structures at various cirrent densities.

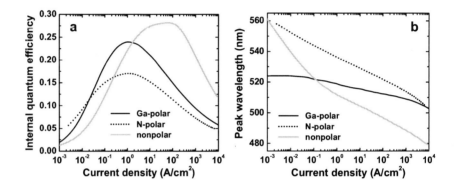

Figure 8. IQE (a) and spectral shifts of the peak emission wavelengths (b) of the 2.5 nm $In_{0.31}Ga_{0.69}N$ SQW LED structures as a function of current density for the structures of various polarities.

puting time, making feasible coupled analysis of electrical, thermal, and optical processes underlying the LED operation.

3.2. Current crowding in LED die

To simulate the current spreading in an LED die, we identify first in the LED heterostructure electrically neutral n- and p-regions (see Fig. 9) where the carrier drift dominates over diffusion and the carrier recombination do not affects the transport of the majority carriers. The electron current density $\mathbf{j}_n$ obeys in the n-region the continuity equation

$$\nabla \cdot \mathbf{j}_n = 0 \quad , \quad \mathbf{j}_n = \mu_n n \nabla F_n \quad , \tag{7}$$

while a similar equation is valid for the hole current density $\mathbf{j}_p$ in the p-region

$$\nabla \cdot \mathbf{j}_p = 0 \quad , \quad \mathbf{j}_p = \mu_p p \nabla F_p \quad . \tag{8}$$

Here, n and p are the concentrations of electrons and holes in the neutral n- and p-region, respectively, μ_n and μ_p are the mobilities of the carries. The expressions for $\mathbf{j}_n$ and

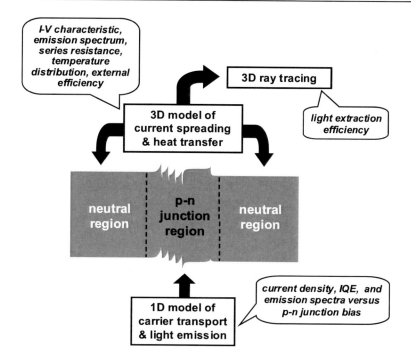

Figure 9. Schematic of hybrid approach to LED modeling. Gray region indicate the band diagram of a multiple-quantum-well LED structure where electrically neutral n- and p-contact regions with flat bands and a *p-n* junction region are selected.

$\mathbf{j}_p$ in Eqs.(7) and (8) actually represent the Ohm law extended to the case of isotype n-n^+ or p-p^+ junctions with negligible non-linearity in their current-voltage characteristics. In a uniform neutral region, the gradient of a quasi-Fermi level becomes equal the gradient of electric potential that governs the majority carrier transport. Being linear, Eqs.(7) and (8) can be easily solved on a 3D-grid, using advanced numerical algorithms.

The neutral n- and p-regions are separated from each other by a thin p-n junction zone that contains a single-quantum-well (SQW) or multiple-quantum-well (MQW) active layer with adjacent space-charge regions. The processes in the p-n junction zone are considered within a 1D drift-diffusion approach that accounts for the non-equilibrium carrier injection, their non-radiative and radiative recombination, and light emission [49]. Only a part of the information coming from the 1D modeling is required for simulating current spreading in the LED die. First of all, that is the dependence of the vertical current density j_z on the p-n junction bias $U_{p-n} = (F_n - F_p)/q$ and temperature T, where q is the electron charge, and F_n and F_p are the electron and hole quasi-Fermi levels taken at the n- and p-sides of the p-n junction zone, respectively. Another important characteristic is the dependence of the internal quantum efficiency (IQE) η_{int} of the LED structure on j_z and T. The IQE is defined as

$$\eta_{\text{int}} = \frac{q}{j_z} \int_{p-n \text{ zone}} dz \cdot R_{\text{rad}}(z) \quad , \qquad (9)$$

where R_{rad} is the radiative recombination rate integrated in (9) over the p-n junction zone.

Neglecting the in-plane carrier transport in the thin p-n junction zone, one can parameterize results of the 1D modeling and replace this subdomain by an effective interface that couples n- and p- neutral-region domains via the following boundary conditions

$$\mathbf{n} \cdot \mathbf{j}(\pm 0) = j_z(T, U_{p-n}) \quad , \quad qU_{p-n} = F_n - F_p \tag{10}$$

Here, $\mathbf{n}$ is the normal vector to the interface plane directed from the p- to n-neutral region, and the notations '-0' and '$+0$' correspond to the values of variables at the interface in the n- and p-neutral region, respectively. The boundary conditions at free semiconductor surfaces imply the electron or hole current density to equal zero.

Thin metallic electrodes semitransparent for emitted light are frequently deposited on the top of the p-contact layer, if the light is extracted though the top surface of the LED die. To predict correctly the LED series resistance in this case, it is important to account for the current spreading in the metallic electrode too. This is made by solving 2D equation with respect to the electron Fermi level F_m in the metal:

$$\nabla(\sigma_m \nabla F_m) = -q j_{\mathrm{ex}}/h \quad , \quad j_{\mathrm{ex}} = (F_m - F_p)/q\rho_p \quad , \tag{11}$$

where j_{ex} is the density of exchange current outgoing from the metallic electrode to the semiconductor contact layer. The exchange current density is determined within the model of Ohmic contact having the specific resistance ρ_p and the thickness of the electrode h. In this case, the boundary condition for Eq.(8) set at the interface between the p-contact layer and p-electrode reads

$$\mathbf{n} \cdot \mathbf{j_p} = j_{\mathrm{ex}} \quad . \tag{12}$$

The normal vector $\mathbf{n}$ is here directed from the metal to p-semiconductor. A similar boundary condition is also set at the interface between the n-contact layer and n-electrode.

The n- and p-electrode pads are considered as equipotential areas inside the n- and p-electrode domains where $F_m = 0$ and $F_m = qU_f$, respectively, U_f is the forward voltage (bias) applied to the LED.

Thus, the current spreading problems is reduced to self-consistent solution of Eqs.(7), (8), and (11) in the corresponding domains with the boundary conditions (10) and (12) accounting for the current exchange among the n- and p-neutral regions and n- and p-metallic electrodes. The exchange currents j_z and j_{ex} connect the current densities in the neutral n- and p-regions with the differences of the Fermi levels on both sides of the respective interfaces.

To verify the hybrid approach a model 2D problem was considered. Current spreading in a 100×100 μm^2 vertical LED die (Fig. 10 a) was computed using a full 2D model and the described hybrid approach that combines 1D analysis of carrier injection in the active region of LED structure with unipolar electron and hole transport in the n-GaN and p-GaN contact layers, respectively. The full operating current predicted by simulations was 15.3 mA in the case of direct modeling and 15.0 mA in the case of hybride approach at the forward voltage of 3.4 V.

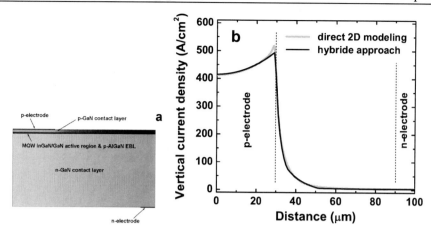

Figure 10. Schematic design of a $100 \times 100 \ \mu m^2$ vertical LED die (a) and comparison of vertical current density distributions obtained by direct 2D modeling and hybride approach (b).

3.3.   Heat transfer in LED die

Heat generation and dissipation is one of the factors limiting LED performance (reduction of efficiency with temperature, see, e.g. [242, 136]) and lifetime [243] (such as failure due to arising thermal stresses [244] leading, e.g., to delamination of the interface layers [245, 246]). LEDs are more sensitive to temperature than standard silicon chips [247], thus thermal management - from chip level to system level - is extremely important [248, 249, 250, 251]. Usually LED systems are cooled passively, but increase of heat dissipation in power LED arrays/lamps makes thermal management the development bottleneck of LED system [16, 252, 253] and forces one to consider active cooling systems, e.g., closed microjet system [254] or a cooling system with a fan [255].

Thermal design of LED lamps has been considered in a number of papers, e.g. [256, 244, 257], with special attention being paid to the design of the cooling systems such as finned ones [258]. Experimental data on the thermal regimes in LEDs and LED lamps are scarce and frequently are limited to integral characteristics such as the total thermal resistance of the multi-chip package [259], with a few exceptions, e.g. study by Luo et al. [256] of an 80 W LED street lamp where the temperature was measured at 16 positions of the lamp.

At low current densities a junction temperature rises linearly with the applied forward current [260]. There are different direct and indirect methods to measure it (see, e.g. [261, 262]). Increased temperature of the active region (it could exceed 100 °C [260, 263]) lowers IQE that depends almost exponentially on the temperature, shifts spectra [264] and enhances diffusion of impurities as well as migration of dislocations [265, 266]. Experiments show that output power of InGaN/GaN MQW LEDs decreases more rapidly for devices with higher In composoition [267]. Heat generation due to the colour conversion in phosphor particles also results in significant light output reduction [268]. Heating is one of the reasons for the efficiency droop at high operation currents as a comparison between dc and low duty pulsed operation proves [269]. Sometimes secondary elements of the LED could be important for the thermal design, e.g., heat drain by bond wires could be significant [266] while a die attach layer could give a dominant contribution to the internal resistance of LED [270].

The heat transfer is simulated within a conventional 3D approach that considers the non-uniform current density distribution in the LED die. Both Ohmic heating of the materials bulk and the heating caused by the non-equilibrium carrier thermalization in the *p-n* junction region are included in the model as the principal heat sources. At the external chip surfaces, either temperature or a heat transfer coefficient and an environment temperature are prescribed. The thermal analysis is coupled with the current-spreading simulation via temperature-dependent electrical conductivity of the neutral regions and the dependence $j_z(T, U_{p-n})$.

Importance of thermal effects increases with the forward voltage and significant even for the low-power LEDs. Such LED based on the the MQW die structure shown in Fig.4 with the chip size 190×250 μm^2 is shown in Fig. 11 (top), the mesa depth was $0.7 \mu m$. Computations were performed with and without account for the thermal effects (Fig. 11). The current crowding near the electron edge is clearly seen producing significant temperature non-uniformity in the die that, in turn, affects the conductivity of the contact layers.

From the shown I-V curves it is evident that ignoring thermal effects results in considerable overestimation of the LED series resistance, mainly via the temperature dependence of the n-contact layer conductivity (in this LED design p-contact layer is of minor importance due to its small thickness).

3.4.   Light extraction from LED die

3.4.1.   "Light escape" problem

The large amount of emitted light being trapped inside the semiconductor structure is the consequence of the large value of the refractive index. This value for nitrides of elements of the group III (about 2.5 at blue wavelengths [271]) is smaller than for the conventional III-V materials (between 3.0 and 3.5 for AlGaInP system [272]) but it is still large enough for the total internal reflection (TIR) to be the major factor responsible for the small light extraction efficiency (other important contributions to the losses are the internal absorption and blocking of the light by contacts). The problem is even more sever for organic LEDs (OLEDs): there are two TIRs at the active region/substrate and at substrate/air interfaces [273], thus sometimes one distinguish waveguide and glass (substrate) modes [274].

Light incident on a planar semiconductor/air interface is totally internally reflected if the angle of incidence exceeds the critical value $\theta_c = \arcsin(n_1/n_2)$ determined by Snell's law that describes the relationship between the angles of incidence θ_i and refraction θ_t when light is passing the boundary between media with refractive indices n_1 and n_2 [275]
$$\sin \theta_i / \sin \theta_t = n_2/n_1$$
The fraction of the incident power that is reflected from the boundary is given by the reflectance R and the fraction that is refracted is given by the transmittance T that depend on the light polarization. Evidently, $R + T = 1$. If the media are non-magnetic, the reflectance

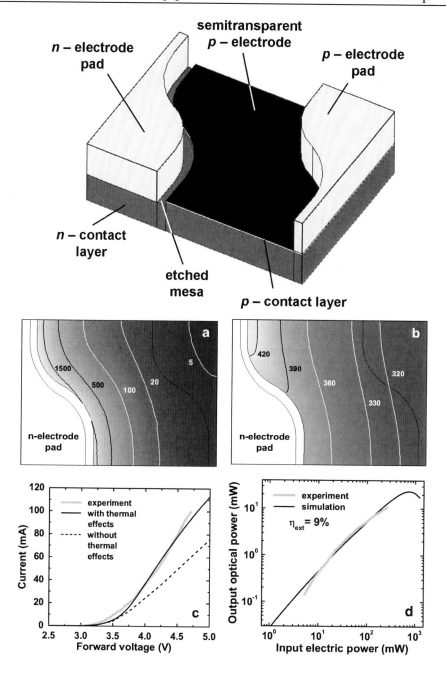

Figure 11. Low-power LED chip design (top). Distrubutions of current density (a) and temperature (b) in the active region of the LED chip at the operating current of 80 mA. Current-voltage characteristics of the LED, experimental and simulated with and without account of thermal effects (c). Output optical power *versus* input electric power: comparison of the theoretical predictions with experiment (d).

is determined by the Fresnel equations

$$R_s = \left[\frac{n_1 \cos\theta_i - n_2\sqrt{1 - \left(\frac{n_1}{n_2}\sin\theta_i\right)^2}}{n_1 \cos\theta_i + n_2\sqrt{1 - \left(\frac{n_1}{n_2}\sin\theta_i\right)^2}}\right]^2,$$

$$R_p = \left[\frac{n_1\sqrt{1 - \left(\frac{n_1}{n_2}\sin\theta_i\right)^2} - n_2\cos\theta_i}{n_1\sqrt{1 - \left(\frac{n_1}{n_2}\sin\theta_i\right)^2} + n_2\cos\theta_i}\right]^2$$

where subscripts "s" and "p" refer to s- and p-polarized light.

These relations are valid as the short wavelength limit for a plane wave incident on an infinite flat interface; the wavelength-dependent corrections should be introduced for the incident beam having curved wavefronts or/and boundaries with finite curvature or sharp corners [276].

For angles of incidence greater than θ_c the value of θ_t becomes complex. The total internal reflection generate an evanescent wave in the low index material. This wave propagates along the interface and decreases exponentially with distance perpendicular to the interface; this wave does not transfer energy along the normal to the interface and thus all the power is reflected.

The critical angle corresponding to TIR defines the so called *escape cone* (sometimes named *collection* cone [5]). The solid angle of the escape cone is $\Omega_c = 2\pi(1 - \cos\theta_c)$. Photons emitted outside the escape cone get trapped in the structure.

The typical LED structure comprising a number of layers most of which have high refractive index could be considered as a multilayer waveguide [277] that could support a rather large number (up to over fifty [278]) of the trapped *guided modes*.

The extraction efficiency is determined by the fraction of the photons emitted into the escape cone and for the semiconductor source with the refractive index n emitting into air ($n_{air} = 1$) is

$$\eta_{\text{extr}} = \frac{\int_0^{\arcsin(1/n)} p(\theta)2\pi\sin(\theta)d\theta}{\int_0^\pi 2\pi\sin(\theta)d\theta}$$

where $p(\theta)$ is the light power distribution; for an isotropic emission (which is characteristic for the conventional double heterostructure structures) extraction efficiency is about $1/(4n^2)$ If losses due to the Fresnel reflections are considerable, this value should be multiplied by the Fresnel transmission factor [279].

For an anysotropic emission (e.g., in the case of QWs —the preferable emission is in the plane of the QW [280]) the dependence on the refractive index is conserved while the numerical coefficient is different [281].

For the case of GaN bounded by air critical angles θ_c are approximately 21°, 24° and 25° at the wavelength of 365, 450 and 520 nm, respectively [5]. The situation with emission from Al-rich AlGaN alloys (UV emission) differs considerably from the case of blue and green InGaN LEDs. The reason is in the unique optical properties of AlN: the recombination between conduction band electrons and the holes in the top valence band produced photons that are polarized along the direction of $\vec{E}\|\vec{c}$ in contrast to GaN where polarization is along $\vec{E}\perp\vec{c}$; thus UV photon with polarization parallel to the c axis could no be easily extracted through the escape cone [282].

The trapped optical modes —*whispering gallery* modes (the authorship of this term usually is attributed to Lord Raylegh who studied the sound propagation close to the curved walls without being audible in its centre in the circular hall in St. Paul cathedral in London [283]) —that are confined by multiple total internal reflections could be observed in the LED chip layers as well as propagate along the circumference of the encapsulant in white light-emitting diode lamps with remote phosphor [284]. Various means introducing stochasticity into the light propagation (a diffuse reflector cap, a textured encapsulant dome, *diffuser* added to the encapsulant) are used for suppression of these modes [285].

Y. Xi et al. [286] estimated the probability of extraction of a guided mode from a waveguide with a flat top surface and a bottom diffuse omnidirectional reflector after a reflection event as

$$p = R\frac{P_{\text{diff}}}{P_{\text{diff}} + P_{\text{spec}}}\frac{\int_0^{\theta_c} I_{\text{diff}}\cos(\theta)\sin(\theta)2\pi d\theta}{\int_0^{\pi/2} I_{\text{diff}}\cos(\theta)\sin(\theta)2\pi d\theta},$$

where R is the mirror reflectivity, P_{diff} and P_{spec} are the powers of diffusive and specular reflections, I_{diff} is the intensity of the diffusive reflection along the normal to the interface. For $R = 1$ and value of θ_c given by Snell's law this expression is simplified to [286]

$$p = \frac{P_{\text{diff}}}{P_{\text{diff}} + P_{\text{spec}}}\frac{n_e^2}{n_s^2},$$

where subscripts "e" and "s" refer to the environment and semiconductor, respectively.

Each reflection reduces the power of light trapped in the waveguide $(1 - p)$ times, thus the number of reflections needed to reduce it e times is given as

$$N = -\left[\ln\left(1 - \frac{P_{\text{diff}}}{P_{\text{diff}} + P_{\text{spec}}}\frac{n_e^2}{n_s^2}\right)\right].$$

Evidently, a diffusive component of the reflection should be maximized to increase the extraction.

The angle of reflectance is equal to the angle of incidence for a specular reflector surface while for the diffuse reflector the intensity of the reflected light could be written as [286]

$$I(\theta, \phi) = I_{\text{diff}}\cos(\theta) + I_{\text{spec}}\frac{1}{\sigma^2 2\pi}\exp\left[-\frac{1}{2}\left(\frac{\theta - \theta_i}{\sigma}\right)\right]\exp\left[-\frac{1}{2}\left(\frac{\phi - \phi_i}{\sigma}\right)\right]$$

where θ and ϕ are the polar and azimuthal angles of reflection and θ_i and ϕ_i are the polar and azimuthal angles of incidence, I_{diff} is the maximum intensity for diffuse reflection, $I_{\text{spec}}\cos(\theta_i)/(\sigma^2 2\pi)$ is the maximum intensity for specular reflection. The diffuse reflection intensity follows the Lambertian distribution; the specular reflection intensity is assumed to be broadened as a Gaussian function.

3.4.2. Approaches to LEE enhancement

There are a number of approaches aimed to the increasing of light extraction efficiency from the LED chip (references could be found in a review [287]):

- chip-shaped LED design (chip shaping);

- placing a back-surface mirror —either a distributed Bragg reflector (DBR) or omnidirectional reflector (ODR) —between the lower cladding layer and the substrate; such LEDs are sometimes are referred to as to Reflective Submount LEDs (RSLEDs) [288];

- surface treatment:

  - one-dimensional or two-dimensional periodic corrugation acting via Bragg scattering; these surface corrugations could either cover all the chip area or be etched around the periphery of the light emitting structure (e.g., circular Bragg gratings [289]), avoiding thus a possible drop of the internal quantum efficiency due to the introduction of new recombination centers during fabrication [290], e.g. via plasma etching [277];

  - random surface texturing or roughening [291, 292, 293] that reduces internal light reflection and scatters the light outward; roughening could have a characteristic scale in the nano range [294, 295] and be applied to the different surfaces involved in the light propagation, e.g., to one or both sides of the sapphire substrate (as a by-product an improvement of the quality of the grown nitride layers on the patterned sapphire was observed [296, 297, 298]); to the bottom of 6H-SiC substrate; to the tin oxide electrode; to the undoped-GaN, p-type GaN or n-type GaN surface; to encapsulant; to sidewalls; to both substrate sides and p-GaN layer ("triple light-scattering layer") [299].

  - placing arrays of the hexagonally packed silica microspheres [300] or other optical diffractive elements (ODE), or simply *diffractives* such as plastic micropyramids [301], ellipsoidal Ag nano-particles [302], ZnO nanowire array [303], concave microstructures [304] or microstructured films [305] on the top emitting surface [306];

  - integrating LEDs with microlenses made from either sapphire [307], diamond [308], GaN [309], Si_2/polysterene [310] or glass (plastic) lens-shaped elements [311];

  - placing a close-packed monolayer of monodispersive CdSe/ZnS core/shell nanocrystals on the top emitting surface [312] (these nanocrystals are pumped by Förster-like non-radiative energy transfer based on Coulomb interactions rather than a wavefunctions overlap and activate an additional relaxation channel for the QW excitations, also acting as the colour-conversion mechanism);

  - placing either a patterned high-index layer [277] that provides the deeper penetration of guided mode profile into the upper layer of the structure or a layer with graded refractive index (GRIN layer) that eliminates Fresnel reflection [313] onto the top emitting surface;

- exploiting photonic crystals (PhC) in LEDs sometimes referred by an acronym PXLED [314]) with the best results (a close to unity LEE) achieved via effective coupling of guided modes with PhC by embedding PhC within the LED structure [315];

- placing the light-emitting region inside an optical cavity where enhancement of spontaneous emission occurs [316] if mirror separation is of the order of the optical wavelength (the microcavity enclosing the light emitting region could be formed "unintentionally" by the multilayer structure designed from other considerations [317][6]);

- the so-called 'thin-film' (TF) LED design, including ThinGaN^{TM} [319] LEDs based on the reducing internal absorption and the use of buried micro-reflectors (BMRs); the disadvantage of the TF-LED structure is the small ratio of the current-injection area to the chip area that could be less than 10 %[320]; moreover, if the microreflectors penetrate the active region, the nonradiative recombination could be increased. Nevertheless, this approach proved to be highly efficient for both AlGaInP [321] and InGaN [322] LEDs;

- exploiting the surface plasmon resonance (it should be noted that the larger IQE the smaller the enhancement of EQE - for InGaN LEDs the efficiency enhancement is more significant in the green-red range [323];

- placing self-assembly of poly(sterene-b methyacrylate) block copolymers as nanoporous pattern on the GaN surface [324].

These approaches could be grouped into two sets depending on whether their application results in the change in the spontaneous emission (either the spontaneous emission rate or the angular distribution, or both):

- modification of the emitted light (die shaping, photonic crystals, surface texturing/roughening, reflective submount, buried micro-reflectors, microlenses & diffusive optical elements);

- modification of the spontaneous emission itself (microcavity effect, photonic crystals, surface plasmons).

This classification is, certainly, not perfect since, e.g., photonic crystals could be used to either just mold the light emitted by the unaffected source acting as a diffraction grating to extract the guided modes or to modify both the emission by preventing emission into certain modes and modify the far-field pattern [325]; the microcavity effects both increase the spontaneous recombination rate and considerably change the directionality of the emitted light; the regular surface corrugations could act as photonic crystals if their characteristic period length is apopriately related to the light wavelength. Note that while surface texturing applied to outer layers such as, e.g. tin oxide one [326], does not alter electrical properties

---

[6]For example, Fabri-Perot effects in InGaN/GaN heterostructures on Si-substrate results from the high refractive index contrasts at the GaN/Si and air/InGaN interfaces [318], no bottom or top mirrors are included. A simple three-layer Fabry-Perot resonator model consisting of Si/(GaN + InGaN)/air stack satisfactorily describes measurements of the modulatin of light intensity.

of LED (exhibit unchanged I-V characteristics), modification of layers close to the active region or the active region itself could deteriorate electrical properties and reduce IQE due to the introduction of new recombination centers during fabrication [290], e.g. via plasma etching [277].

It is hardly possible to indicate the single best approach, except for applications where the high directionality of the light is of primary importance (as, for example, for short-haul polymer optical fiber (POF) [327] data transmission systems [328, 279] and low-power high-speed photonic circuits [329] —in this case RCLED is an evident winner). Fortunately, most of the LEE enhancement approaches are not mutually exclusive and thus can be combined (examples could be found in ref. [287]).

Two additional issues could be considered: (1) possible damage to the active layer or/and degradation of the electrical properties of the LED[7] and (2) the fabrication cost unless the additional cost is small compared to the cost of the conventional LED die, as, for example, is the case of 2D photonic crystal patterning by imprint lithography which is estimated to cost less than $0.01 per square millimeter [333].

There are two approaches to simulation of light propagation in optoelectronic devices. The first one is the high frequency asymptotic approximation such as ray tracing and geometrical optics. It is valid if the geometric features are large compared to the wavelength (more detailed analysis of the corresponding conditions, including necessary and sufficient ones formulated via the so called "Fresnel volume" could be found in [334]). Nevertheless, it is widely used beyond its region of applicability because of the simplicity, producing sometimes reasonable results. The strict approach to the problem is based on the solution of Maxwell equations.

Ray and Beam Tracing   The best known method of this class is the geometric optics used in applications most frequently in the framework of ray tracing method [335] (sometimes called *photon* tracing [336]). Its main weak point is the danger of the "undersampling" errors [337]. Ray tracing method is also known to be unstable in case of geometry displaying numerous small features [334]. This method frequently used to study light extraction on both chip [338, 339, 340] and lamp [336, 341, 342, 343] levels. Application of the ray tracing in this context is straightforward except two elements encountered in LED structures that require a special treatment. The first one is the textured surfaces that can not be strictly modelled if the the feature size is small compared to the wavelength, the second is the semitransparent ohmic metal layer that cannot be simply classified as metallic or dielectric [338].

Beam ray tracing, or simply beam tracing, method [344, 345] is free from undersampling artifacts of ray tracing at the price of loosing simplicity: geometric operations needed to trays polyhedral beam (e.g., intersection, clipping) are rather complex.

---

[7]Dry etched-induced damage could be sometimes recovered by thermal annealing [330]; electrical properties could also be improved due to the pattern fabrication, for example, the corrugated photonic crystal structure fabricated by the etching of an indium-tin-oxide anode layer in OLED is reported to lower the operating voltage (by 30% at constant current compared to the conventional structure) due to the partial reduction of the thickness of the organic layer [331] as well similar results for InGaN-based LED due to the increased the effective area of contact between the n electrode and the n-GAN layer [332]; however, B.S. Cheng et al. reported a slight increase of the forward voltage of the InGaN-based grebe LED due to the nano-imprinted photonic crystal pattern [325].

Geometric Optics Extensions The geometric theory of diffraction [346] as well as its modification — uniform theory of diffraction [347]— supplements the standard geometric optics with the ability to model diffraction by introducing additional (diffracted) rays when the original ray strikes an edge or a vertex. These methods incorporate closed-form analytical solutions for a number of diffraction cases [348]. Diffraction is a local phenomenon at high frequencies. Thus, the behaviour of the diffracted wave at edges, corners etc. could be approximately determined from the exact solutions for simple canonical problems. For example, the diffraction around an edge is based on the asymptotic form of the solution for an infinite wedge.

An efficient algorithm for the determination of the propagation path of multiple diffracted rays was proposed recently by P. Bagnerini et al. [349]. B.V. Budaev and D.B. Bogy re-formulate the problem as the mathematical expectation of some functional in the space of Brownian trajectories with the Wiener probabilistic measure [350].

One more extension of the geometric optics should be mentioned - "complex geometric optics" [351] that is based on consideration of complex values of the wave vector as means to describe nonuniform waves.

Computational Electromagnetics Computational Electromagnetics (CEM) problems include both low-frequency and high-frequency ones that mathematically differ by the nature of the kernel in the integral equation (a fundumental solution or Green function) that could be either smooth or oscillatory [352]. In the former case the higher-order derivatives become increasingly smaller as the distance from the source increases while in the case of an oscillatory kernel all derivatives decay as the inverse of the distance. Thus, the spatial bandwidth of an oscillatory kernel does not decrease away from the source enhacing the complexity of the numerical solution. The field of Computational Electromagnetics could be subdivided into three subfields [353, 354, 355]: *Frequency-domain eigensolvers* (find the bandstructure $\omega(\vec{k})$), *Frequency-domain simulations* (find the amplitudes $\vec{E}_0(\vec{r})$ and $\vec{H}_0(\vec{r})$ of the fields $\vec{E}(\vec{r}, t) = \vec{E}_0 \exp^{-i\omega t}$ and $\vec{H}(\vec{r}, t) = \vec{H}_0 \exp^{-i\omega t}$ for a given current distribution $\vec{J} \exp^{-i\omega t}$ at a fixed frequency ω) and *Time-domain simulations* (simulate the time evolution of the fields $\vec{E}(\vec{r}, t)$ and $\vec{H}(\vec{r}, t)$, usually for a given $\vec{J}(\vec{r}, t)$).

A so called *complex frequency hopping method* based on the low-order multi-point Padé approximation is in some sense intermediate since it is able to produce simultaneously both the frequency- and time-domain results [356].

Maxwell equations for the frequency-domain methods are, as a rule, re-written in terms of the second-order equation for one field ($\vec{E}$ or $\vec{H}$) (see, however, [357]). In contrast, in time-domain methods initial (first-order) Maxwell equations are usually discretized. A single simulation by the time domain method can be used to study the material response over a wide bandwidth. To simulate harmonic (sinusoidal) steady-state, one can either directly prescribe a single-frequency incident plane wave or perform a Fourier transform step on the pulse waveform response [358]. In the latter case, a too short pulse suffers from accumulating errors due to overshoot and ringing during propagation requiring filtration of the numerical noise before Fourier transformation. Anyway, integration time should be not lesser than several wave periods until the sinusoidal steady-state is achieved in every element.

To reduce the continuum problem to the discrete one, a number of approaches are used

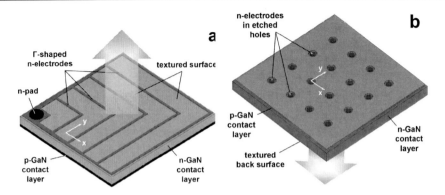

Figure 12. Scheme of vertical 815×875 μm^2 ThinGaN^{TM} (a) and planar $1\times$ mm^2 TFFC (b) LED dice.

such as finite differences (finite difference time domain - FDTD), finite elements, spectral methods, boundary elements giving several closely related methods (see, e.g., Appendix to [287]). FDTD is the major approach to the study of the LEE enhacement via photonic crystals [359] and the most accurate tool for the simulation of the effect of the subwavelength features on the light propagation [360].

3.5. Current crowding effect on light extraction efficiency

As soon as the coupled current-spreading and heat-transfer problems are solved, the local power density P_{rad} of light emitted in the *p-n* junction region can be determined as

$$P_{\text{rad}} = (\hbar\omega/q)\eta_{\text{int}} j_z \quad , \tag{13}$$

where $\hbar\omega$ is the energy of emitted photons. 3D ray tracing accounting for the non-uniform current density distribution is then applied to estimate the light extraction efficiency from the LED die and the radiation pattern. The number of rays emitted from an elementary part of the active layer is chosen to be proportional to the emission intensity. Ray propagation through the chip is monitored until the ray leaves the computational domain or decays considerably because of the light absorption inside the die.

Generally, there are two alternative approaches to treat the ray reflection and refraction at the interface separating media with different refraction indexes. It is possible either to trace both reflected and refracted rays outgoing from every interface or to imply a special random process into the procedure to decide which of these two rays should be further traced in accordance with the value of the transmission and reflection coefficients. The latter approach is used in the present work.

Current crowding effect on light extraction efficiency will be illustrated using two advanced LED dice - a vertical ThinGaN^{TM} and (a) and a planar thin-film flip-chip (TFFC) LED dice (Fig. 12). Distribution of current density for these two dice with the same average current density is presented in Fig. 13 and Local probabilities of light extraction from the active regions of the LED - in Fig. 14.

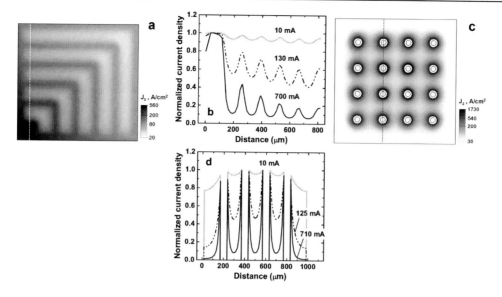

Figure 13. Current density distributions in the active regions of ThinGaN^{TM} (a) and TFFC (c) LED dice at the operation currents 700 and 960 mA, respectively (in both cases the mean current density is avout 96 A/cm^2). Normalized current density distributions in the cross-sections marked by dotted lines in (a,c) computed at various operation currents for ThinGaN^{TM} (b) and TFFC (d) LED dice.

In a vertical LED the current localization near or under the low reflective electrodes and the pad becomes stronger with current leading to the decrease of LEE (Fig. 15. The dependencies for an improved design (making the electrodes narrower and inserting an insulation layer under the n-pad) suggested in ref. [361] are also plotted.

Effect of the thickness of the indium-tin oxide (ITO) spreading layer on the p-contact on the performance of blue LEDs is illustrated in Fig. 16 where the current density distributions in the active region of the square-shaped $300 \times 300 \ \mu\text{m}^2$ LED die si presented (the operating current is 20 mA).

Dependence of LEE on the ITO film thickness and of EQE on the current have been computed for the square-shaped LED die (Fig. 17).

A comparison of Interdigitated multipixel array (IMPA) [228] was made with conventional square chip design. Both LEDs utilized the same heterostructure grown on a basal-plane sapphire substrate. The structure contained a 2 μm Si-doped n-GaN contact layer followed by the active region, which consisted of a five 5 nm $\text{In}_{0.08}\text{Ga}_{0.92}\text{N}$ QWs separated by 32 nm Si-doped GaN barriers. A 15 nm Mg-doped $\text{Al}_{0.2}\text{Ga}_{0.8}\text{N}$ layer was used as an electron blocking layer. The structure was capped with a 200 nm Mg-doped p-GaN contact layer. The donor concentration in the n-GaN contact layer ([Si] = 3×10^{19} cm^{-3}) was estimated from the sheet resistance reported in [228]. We also assume the donor concentration in the barriers to be of 5×10^{18} cm^{-3} and the acceptor concentration in the p-AlGaN and p-GaN layers to be of 1.5×10^{19} cm^{-3} and 2×10^{19} cm^{-3}, respectively.

The IMPA die consisted of a hundred of $30 \times 30 \ \mu\text{m}^2$ pixel-LEDs integrated on the same substrate (for a detailed description of the IMPA design and its fabrication procedure

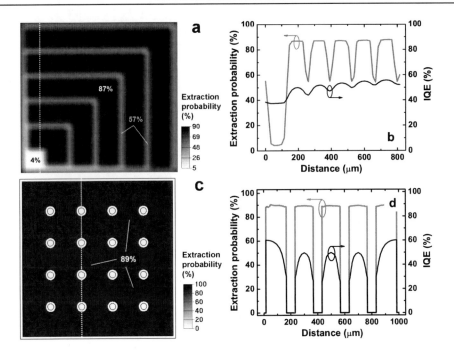

Figure 14. Local probabilities of light extraction from the active regions of ThinGaN[TM] (a) and TFFC (c) LED dice at the operation currents 700 and 960 mA, respectively. Probability of light extraction and IQE in the cross-sections marked by dotted lines in (a,c)simulated for ThinGaN[TM] (b) and TFFC (d) LED dice.

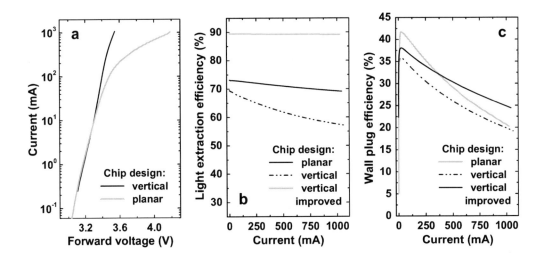

Figure 15. Comparison of characteristics of the ThinGaN[TM], ThinGaN[TM] of improved design, and TFFC LED dice: current-voltage characteristics (a), light extraction efficiencies (b), and wall-plug efficiencies (c).

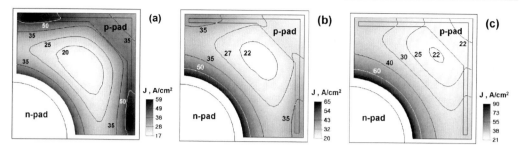

Figure 16. Current density distributions for different thicknesses of current-spreading ITO film deposited on top of the p-GaN contact layer: (a) 20 nm, (b) 50 nm, and (c) 100 nm. Digits near the isolines indicate the corresponding current density in A/cm².

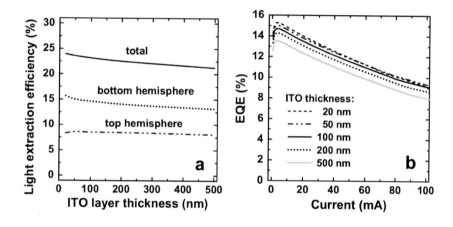

Figure 17. Dependence of light extraction efficiency on the ITO film thickness (a) and external quantum efficiency on operating current at various thicknesses (b).

see ref. [228]. Both dice had equal light emitting areas but the total IMPA die area was approximately ten times larger than that of the square die. The *n*- and *p*-contact resistances were $5 \times 10^{-5} \, \Omega \cdot cm^2$ and $1 \times 10^{-3} \, \Omega \cdot cm^2$, respectively, as reported in [228]. Computation show that output power dependence on the current is much weaker for the IMPA (Fig. 18 a) while polar far-field radiation pattern in the cross-section coinciding with the chip plane essentially the same (Fig. 18 b). IMPA also has a much smaller serial resistance.

4.   Modeling of LED Lamps

An important issue for white LEDs is the colour control (*colour-rendition*) [362, 363] that is characterized by the correlated colour temperature (CCT) and the colour rendering index (CRI) [13], the most stringent requirements to the colour properties being that of a retail store [364]. Thus sometimes to increase CRI up to four different phosphors are used simultaneously [365].

To assess the efficiency of white LED, one has to consider both the primary light inten-

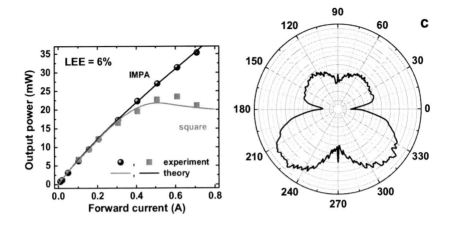

Figure 18. Output optical power as a function of forward current simulated for conventional square-shaped LED die and multipixel LED array (IMPA) having the same light-emitting area (a) and polar far-field radiation pattern (b).

sity from the LED chip and the intensity of the light converted (emitted) by the phosphor [13]. Some energy is inevitably lossed due to the conversion of the UV or blue photons to the photons of longer wavelength — the so called Stokes shift $\eta_{stokes} = \lambda_{chip}/\lambda_{phosphor}$, e.g., for typical InGaN emission of 455-465 nm and YAG:Ce phosphor [366, 367] with the wavelength of the converted light 550–560 nm, η_{stokes} is lower than 85 %. Evidently, UV LEDs as the basis of phosphor-converted white light emitting diodes are less efficient than blue or green one assuming comparable EQE, e.g., WPE of UV LED should be at least twice that of green LED to provide the same luminous efficacy. There are a number of parameters that determine the colour conversion by phosphor: geometry and size of the matrix loaded with the phosphor particles, the particle diameter, the phosphor concentration and the refractive indices of the matrix material and phosphor. Ray tracing method is a natural approach to simulation of the conversion process [368, 369, 370, 371].

Note that an alternative to phosphor conversion is the use of integration of CdSe/ZnS core/shell nanocrystal quantum dots with blue nitride LED (Warm-white LED - WLED - with CRI close to 90) [372] or of polychromatic LEDs [373] designed either as multiple emitters mounted in one package [374] possibly with feedback control schemes using thermal, electrical or optical sensors to maintain the desired colour properties within acceptable tolerances [364] or as stacked QW structures (called *monolithic* [375], *dual-wavelength* [376] or *cascade* [377] LEDs) grown sequentially onto the same substrate that emit basic colours [378, 379] with electrical pumping of active layers with the widest bandgap (usually blue) and exploiting re-emission of light absorbed by other active regions (photon recycling) [380].

The study of the phosphorescence efficiency for different proximate and remote phosphor configurations for the cases of flat, convex or semispherical shapes of the encapsulant top surface showed that the remote phosphor [381, 382] and the semispherical encapsulant dome provide the highest phosphorescence extraction efficiency [285]. The ray tracing simulation had been used neglecting the scattering of light by the phosphor particles that

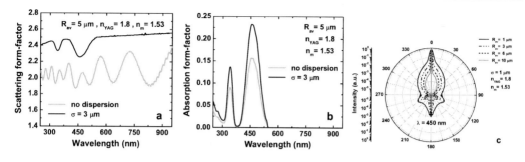

Figure 19. Spectral dependence of the scattering (a) and absorption (b) form-factors and scattering patterns (c) for the YAG : Ce^{3+} phosphor particles with the mean radius of 8 μm.

is justified if either the size of the particle is small compared to the light wavelength or the refractive indices of phosphor and the encapsulant are approximately equal. If in the computations of the radiation pattern the computational domain is bounded by the exterior phosphor surface, the latter is usually assumed to be the homogeneous [383] or inhomogeneous [384] Lambertian light source. The blue passing/yellow reflecting filter made from quarter-wave films of alternate high and low refractive index dielectric layers (TiO_2/SiO_2) placed between the phosphor and a blue LED provides 1. 64 and 1.95 fold increase in efficiency and luminous efficacy of the forward white emission by recycling of yellow light emitted by phosphor downwards toward the InGaN chip [385].

An crucial stage of the LAMP modelling is the correct description of the light scattering by absorbing phosphor particles. The well-known Mie theory is a most appropriate approach accounting for wave nature of light. Its main assumption – the spherical shape of the particles – seems to be acceptable for the analysis of the light conversion by the phosphor particles.

Fig. 19 show a spectral dependence of the scattering (a) and absorption (b) form-factors for the most common yellow phosphor based on yttrium-aluminum garnet doped with cerium with the mean particle radius of 8 μm computed by Mie theory with and without account of particle size dispersion. Note the absence of interferential oscillations of the scattering form-factor in the case of gaussian radius dispersion with $\sigma = 3\mu m$. Scattering patterns of the dispersed ($\sigma = 1\mu m$) phosphor particles with different mean radia at the wavelength of 450 nm (c). Note the dominating forward scattering in all the cases considered.

The heating of the matrix containing phosphor particles due to the Stokes shift and phosphor quantum efficiency are shown in Fig. 20.

There are two approaches to the development of LED lamps for high voltage AC operation: using AC-DC transforming scheme or a capacitance coupling of integrated on-chip arrays of μ LEDs [231]. The advantages of the latter one are the elimination of multiple soldering points and doubling of the light on/off frequency that reduce flickering.

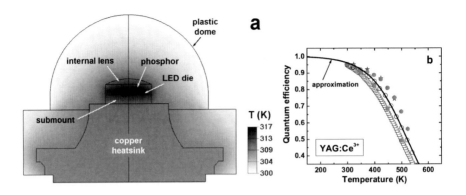

Figure 20. Temperature distribution in a white-light LED lamp having desing similar to that of Luxion K2 produced by Philips Lumileds (a). Temperature dependence of the YAG : Ce^{3+} phosphor quantum efficiency (b): symbols are data compiled from various literature sources, line is the approximation used in simulations.

5. Conclusion

Numerical simulation of widebandgap LEDs is now an efficient tool helping both to elucidate remaining fundamental issues and to assist researches and engineers in designing and/or optimization of the optoelectronic devices. Still, there is *plenty of room* left for the models and methods development. Two general comments are due on modelling. The first note on the model completeness is inspired by the well known passage from a letter by Blaise Pascal (14 Dec 1656): Je n'ai fait celle-ci plus longue que parce que je n'ai pas eu le loisir de la faire plus courte[8]. Model is simpler than reality by definition and its development should be considered completed not when all the relevant physical phenomena are accounted for, but when you could not simplify the model without corrupting it. This final model could be referred to as a "minimal model".

The second one concerns the *reliability* of a computer prediction. One should be aware that this "ultimum kriterion" characterizes not the model itself but simulation (i.e. model application) being dependent on the purpose of the study. The same computations could be consider as successful if the task was to unveil some trend and as a failure if one, e.g., planned to find an optimal thickness and doping of some layer in the multilayer structure.

Acknowledgments

The authors are gratefull to M.V. Bogdanov, V.F. Mymrin and M.S. Ramm for usefull discussions.

---

[8]I have made this letter so long only because I did not have the leisure to make it shorter.

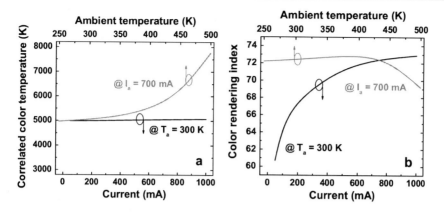

Figure 21. Variation of correlated color temperature (a) and color rendering index (b) of the white-light LED lamp with the ambient temperature and operation current of a blue LED pumping the yellow YAG : Ce^{3+} phosphor.

References

[1] J. Orton and C. Foxon, Rep. Prog. Phys. **61**, 1 (1998).

[2] B. Monemar and G. Pozina, Prog. Quantum Electron. **24**, 239 (2000).

[3] E. F. Schubert, *Light-Emitting Diodes (2nd ed.)*, Cambridge University Press, 2006.

[4] A. Zukauskas, M. S. Shur, and R. Gaska, *Introduction to Solid State Lighting*, Wiley-Interscience, New York, 2002.

[5] H. Morkoc, *Handbook of Nitride Semiconductors and Devices: GaN-Based Optical and Electronic Devices*, volume 3, Wiley-VCH Verlag GmbH & Co. KGaA, 2009.

[6] J. I. Pankove, GaN for LED applications, NASA CR-132263, 1973.

[7] J. Piprek, Introduction, in *Nitride Semiconductor Devices: Principles and Simulation*, edited by J. Piprek, pages 3–11, Weinheim, 2007, WILEY-VCH Verlag GmbH & Co. KGaA.

[8] S. Nakamura, T. Mukai, and M. Senoh, Appl. Phys. Lett. **64**, 1687 (1994).

[9] S. Nakamura, M. Senoh, N. Iwasa, and S. I. Nagahama, Appl. Phys. Lett. **67**, 1868 (1995).

[10] D. M. Brown et al., IEEE Trans. El. Dev. **40**, 325 (1993).

[11] I. Akasaki and H. Amano, Jap. J. Appl. Phys. **454**, 9001 (2006).

[12] F. Wall, P. S. Martin, and G. Harber, Proc. of SPIE **5187**, 85 (2004).

[13] Z. Liu, S. Liu, K. Wang, and X. Luo, Front. Optoelectron. China **2**, 119 (2009).

[14] Y. Narukawa, M. Ichikawa, D. sanga, and M. Sano, J. Phys. D.: Appl. Phys. **43**, 354002 (2010).

[15] S. Muthu, F. J. P. Schuurmans, and M. D. Pashley, IEEE J. Selected Topics Quantum Electron. **8**, 333 (2002).

[16] N. Narendran, Y. Gu, J. P. Freyssnier, H. Yu, and L. Deng, J. Crystal Growth **268**, 449 (2004).

[17] N. Narendran, Proc. of SPIE **5941**, 594108 (2005).

[18] M. G. Craford, Proc. of SPIE **5941**, 594101 (2005).

[19] E. F. Schubert and J. K. Kim, Science **308**, 1274 (2005).

[20] E. H. Park, S. K. Jeon, C. T. Kim, D. H. Kim, and J. S. Park, Proc. SPIE **6669**, 66690J (2007).

[21] A. Walsh and S. H. Wei, phys. stat. sol. (c) **6**, 2326 (2008).

[22] Y. Xia et al., IEEE Trans. Electron. Dev. **57**, 2639 (2010).

[23] M. Razeghi et al., J. Light Emitting Diodes **2**, 1 (2010).

[24] I. K. Park and S. J. Park, Appl. Phys. Express **4**, 042102 (2011).

[25] J. R. Oh et al., Optics Express **19**, 4188 (2011).

[26] M. Funato et al., Jpn. J. Appl. Phys. **45**, L659 (2006).

[27] K. Fujiwara, H. Jimi, and K. Kaneda, phys. stat. sol. (c) **6**, S814 (2009).

[28] R. Gaska and J. Zhang, Proc. of SPIE **6037**, 603706 (2005).

[29] M. S. Shur and R. Gaska, Proc. of SPIE **6894**, 689419 (2008).

[30] M. Kneissl et al., Semicond. Sci. Technol. **26**, 014036 (2011).

[31] P. Shah, V. Mitin, M. Grupen, H. Song, and K. Hess, J. Appl. Phys. **79**, 2755 (1996).

[32] V. Y. Davydov et al., Phys. stat. sol. (b) **234**, 787 (2002).

[33] http://www.crosslight.ca.

[34] M. Hansen et al., Appl. Phys. Lett. **81**, 4275 (2002).

[35] J. Piprek, *Semiconductor Optoelectronic Devices: Introduction to Physics and Simulation*, Academic Press, San Diego, 2003.

[36] Y. K. Kuo, S. H. Yen, and J. R. Chen, Proc. of SPIE **6368**, 636812 (2006).

[37] Y. K. Kuo, S. H. Yen, and Y. W. Wang, Proc. SPIE **6669**, 66691J (2007).

[38] http://www.str-soft.com/products/SimuLED/.

[39] K. A. Bulashevich and S. Y. Karpov, phys. stat. solidi (c) **5**, 2066 (2008).

[40] S. Y. Karpov, Phys. Stat. Sol. RRL **4**, 320 (2010).

[41] S. Y. karpov, Proc. of SPIE **7939**, 79391C (2011).

[42] S. Y. Karpov et al., Phys. Stat. Sol. (a) **241**, 2668 (2004).

[43] V. F. Mymrin et al., Phys. Stat. Sol. (c) **15**, 2928 (2005).

[44] V. F. Mymrin, K. A. Bulashevich, N. I. Podolskaya, and S. Y. Karpov, J. Cryst. Growth **281**, 115 (2005).

[45] K. A. Bulashevich, V. F. Mymrin, S. Y. Karpov, I. A. Zhmakin, and A. I. Zhmakin, J. Comput. Phys. **213**, 214 (2006).

[46] I. Y. Evstratov, V. F. Mymrin, S. Y. Karpov, and Y. N. Makarov, Phys. stat. sol.(c) **3**, 1645 (2006).

[47] I. Y. Evstratov, V. F. Mymrin, S. Y. Karpov, and Y. N. Makarov, Phys. Stat. Solidi (c) **3**, 1645 (2006).

[48] K. A. Bulashevich, I. Y. Evstratov, V. F. Mymrin, and S. Y. Karpov, Phys. Stat Solidi (c) **4**, 45 (2007).

[49] S. Y. Karpov, Visible light-emitting diodes, in *Nitride Semiconductor Devices: Principles and Simulation*, edited by J. Piprek, pages 303–325, Weinheim, 2007, WILEY-VCH Verlag GmbH & Co. KGaA.

[50] M. V. Bogdanov, K. A. Bulashevich, I. Y. Evstratov, A. I. Zhmakin, and S. Y. Karpov, Semicond. Sci. Technol **23**, 125023 (2008).

[51] M. V. Bogdanov, K. A. Bulashevich, I. Y. Evstratov, and S. Y. Karpov, phys. stat. solidi (c) **5**, 2070 (2008).

[52] M. V. Bogdanov et al., phys. stat. sol. (c) **7**, 2127 (2010).

[53] K. A. Bulashevich, M. S. Ramm, and S. Y. Karpov, phys. stat. solidi (c) **6**, S804 (2009).

[54] U. ʿOzgur et al., J. Appl. Phys. **98**, 041301 (2005).

[55] A. Osinsky and S. Karpov, ZnO-based light emitters, in *Zink Oxide. Bulk, Thin Films and Nanostructures*, edited by C. Jagadish and S. J. Pearton, pages 525–554, Elsevier, 2006.

[56] K. A. Bulashevich, I. Y. Evstratov, V. N. Nabokov, and S. Y. Karpov, Appl. Phys. Lett **87**, 243502 (2005).

[57] K. A. Bulashevich, I. Y. Evstratov, and S. Y. Karpov, phys. stat. solidi (a) **204**, 241 (2007).

[58] J. W. Mares et al., J. Appl. Phys. **104**, 093107 (2008).

[59] A. B. Brailovsky and V. V. Mitin, Solid-State Electron. **44**, 713 (2000).

[60] C. D. McGuinness, K. Sagoo, D. Mc.Loskey, and D. J. S. Birch, Meas. Sci. Technol. **15**, L19 (2004).

[61] J. W. Shi, J. K. Sheu, C. H. Chen, G. R. Lin, and W. C. Lai, IEEE El. Dev. Lett. **29**, 158 (2008).

[62] N. Narendran, Y. Gu, J. P. Freyssinier, H. Yu, and L. Deng, J. Cryst. Growth. **268**, 449 (2004).

[63] G. Meneghesso, M. Meneghini, and E. Zanoni, J. Phys. D.: Appl. Phys. **43**, 354007 (2010).

[64] Y. Ohno, Proc. of SPIE **6046**, 604625 (2006).

[65] http://cvision.ucsd.edu/.

[66] U. S. Department of Energy, Transformations in lighting, SSL R &D Workshop, Atlanta, 2008.

[67] A. Uddin, A. C. Wei, and T. G. Andersson, Thin Solid Films **483**, 378 (2005).

[68] X. A. Cao, P. M. Sandvik, S. F. LeBoeuf, and S. D. Arthur, Microelectronics Reliability **43**, 1987 (2003).

[69] J. Hu, L. Yang, and M. W. Shin, Proc. of SPIE **6355**, 635516 (2006).

[70] E. F. Schubert, T. Gessmann, and J. K. Kim, Inorganic semiconductors for light-emitting diodes, in *Organic Light Emitting Devices. Synthesis, Properties and Applications*, edited by K. Müllen and U. Scherf, pages 1–33, Weinheim, 2006, Wiley VCH.

[71] L. Kuna et al., Appl. Phys. B **91**, 571 (2008).

[72] C. Griffin et al., Appl. Phys. Lett. **86**, 041111 (2005).

[73] S. L. Rumyantsev, C. Wetzel, and M. S. Shur, Proc. of SPIE **6600**, 660001 (2007).

[74] C. Huh and S. J. Park, Electrochem. Solid-State Lett. **7**, G266 (2004).

[75] J. Piprek, editor, *Optoelectronic Devices: Advanced Simulation and Analysis*, Springer, 2004.

[76] Y. A. Chang et al., Semicond. Sci. Techn. **21**, 598 (2006).

[77] Y. K. Kuo, M. C. Tsai, and S. H. Yen, Optics Comm. **282**, 4252 (2009).

[78] Y. K. Kuo, M. C. Tsai, S. H. Yen, T. C. Hsu, and Y. J. Shen, IEEE J. Quant. Electron. **46**, 1214 (2010).

[79] S. H. Yen, M. L. Tsai, S. C. Tsai, S. J. Chang, and Y. K. Kuo, IEEE Photonics Technol. Lett. **22**, 1787 (2010).

[80] Y. C. Lin et al., Solid-State Electron. **47**, 849 (2003).

[81] C. H. Kuo et al., Mater. Sci. Engineer. B **106**, 69 (2004).

[82] K. M. Chang, J. Y. Chu, and C. C. Cheng, Solid-State Electron. **49**, 1381 (2005).

[83] C. Huh, W. J. Schaff, L. F. Eastman, and S. J. Park, Electrochem. Solid-State Lett. **6**, G79 (2003).

[84] C. H. Kuo et al., Mater. Sci. Engineer. B **106**, 69 (2004).

[85] H. Kim, D. J. Kim, S. J. Park, and H. Hwang, J. Appl. Phys. **89**, 1506 (2001).

[86] S. P. Jung et al., Appl. Phys. Lett. **987**, 181107 (2005).

[87] K. M. Chang, J. Y. Chu, and C. C. Cheng, Solid-State Electron. **49**, 1381 (2005).

[88] T. Onuma et al., Jpn. J. appl. Phys. **42**, L1369 (2003).

[89] Y. J. Lee, C. H. Chen, and C. J. Lee, IEEE Photon. Technol. Lett. **22**, 1506 (2010).

[90] T. H. Wang, J. Y. Chang, M. C. Tsai, and Y. K. Kuo, Proc. SPIE **7954**, 79541F (2011).

[91] U. T. Schwarz and M. Kneissl, phys. stat. sol.(RRL) **1**, A44 (2007).

[92] P. Waltereit, (Al,Ga,In)N heterostructures grown along polar and non-polar directions by plasma-assisted molecular beam epitaxy, Dr. rer. nat. thesis, Paul-Drude-Institut für Festkörperelektronik, Berlin, 2001.

[93] Y. J. Sun, Growth and characterization of M-plane GaN and (In,Ga)N/GaN multiple quantum wells, Ph. D. thesis, Humboldt-Universität, Berlin, 2004.

[94] M. Stutzmann et al., phys. stat. sol. (b) **228**, 505 (2001).

[95] T. Koida et al., Appl. Phys. Lett. **84**, 3768 (2004).

[96] Y. K. Kuo, M. C. Tsai, and S. H. Yen, Optics Commun. **282**, 4252 (2009).

[97] C. Huh, J. M. Lee, D. J. Kim, and S. J. Park, J. Appl. Phys. **92**, 2248 (2002).

[98] S. P. Lepkowski and S. Krukowski, J. Appl. Phys. **100**, 016103 (2006).

[99] Y. Takahashi et al., Physica E **21**, 876 (2004).

[100] M. S. Shur and R. Gaska, Proc. of SPIE **6894**, 689419 (2008).

[101] T. E. Nee et al., Jpn. J. Appl. Phys. **47**, 7148 (2008).

[102] T. G. Kim et al., J. Crystal Growth **272**, 264 (2004).

[103] R. A. Arif, Y. K. Ee, and N. Tansu, phys. stat. sol. (a) , 1 (2007).

[104] H. Zhao and R. A. Arif, IEEE J. Select. Topics Quant. Electr. **15**, 1104 (2009).

[105] B. T. Liou, M. C. Tsai, C. T. Liao, S. H. Yewn, and Y. K. Kuo, Proc. SPIE **7211**, 72111D (2009).

[106] X. Ni et al., J. Appl. Phys. **108**, 033112 (2010).

[107] X. Ni et al., Appl. Phys. Lett. **97**, 031110 (2010).

[108] J. H. Chen et al., J. Cryst. Growth **287**, 354 (2006).

[109] H. Amano, N. Sawaki, I. Akasaki, and Y. Toyoda, Appl. Phys. Lett. **48**, 353 (1986).

[110] C. W. Kuo et al., J. Crystal Growth **311**, 249 (2009).

[111] C. H. Liu et al., Mater. Sci. Engineer. **B 111**, 214 (2004).

[112] T. Paskova, D. A. Hanser, and K. R. Evans, Proc. IEEE **98**, 1324 (2010).

[113] X. A. Cao et al., Appl. Phys. Lett **84**, 4313 (2004).

[114] T. Nishida, N. Kobayashi, and T. Ban, Appl. Phys. Lett. **82**, 1 (2003).

[115] J. Christen et al., phys. stat. sol. (c) **0**, 1795 (2003).

[116] J. B. Limb, Design, fabrication and characterization of III-nitride pn junction devices, PhD thesis, Georgia Institute of Technology, 2007.

[117] T. Mukai, S. Nagahama, N. Iwasa, M. Senoh, and T. Yamada, J. Phys.: Condens. Matter **13**, 7089 (2001).

[118] M. Godlewski et al., Acta Physica Polonica A **108**, 675 (2005).

[119] W. D. Herzog, R. Singh, T. D. Moustakas, B. B. Goldberg, and M. S. Ünlü, Appl. Phys. Lett. **70**, 1333 (1997).

[120] N. A. Shapiro, *Radiation transition in InGaN quantum-well structures*, PhD thesis, Univ. of California, Berkley, 2002.

[121] I. hsiu Ho and G. B. Stringfellow, Appl. Phys. Lett. **69**, 2701 (1996).

[122] S. Y. Karpov, MRS Internet J. Nitride Semicond. Res. **3**, 16 (1998).

[123] S. Korcak et al., Surf. Sci. **601**, 3892 (2007).

[124] L. B. Hovakimian, Appl. Phys. A **96**, 255 (2009).

[125] Y. S. Choi et al., Mater. Lett. **58**, 2614 (2004).

[126] S. F. Chichibu et al., Appl. Phys. Lett. **74**, 1460 (1999).

[127] X. A. Cao, J. A. Teetsov, F. Shahedipour-Sandvik, and S. D. Arthur, J. Crystal Growth **264**, 172 (2004).

[128] B. Roycroft et al., Opt. Express **12**, 736 (2004).

[129] H. Amano, S. Takanami, M. Iwaya, S. Kamiyama, and I. Akasaki, phys. stat. sol. (a) **195**, 491 (2003).

[130] S. Chichibu, T. Azuhata, T. Sota, and S. Nakamura, Appl. Phys. Lett. **69**, 4188 (1996).

[131] I. A. Pope et al., Appl. Phys. Lett **82**, 2755 (2003).

[132] Y. Narukawa et al., Appl. Phys. Lett. **70**, 981 (1997).

[133] C. S. Xia et al., Opt. Quantum Electron. **38**, 1077 (2006).

[134] Y. H. Cho et al., Phys. Rev. B **61**, 7571 (2000).

[135] P. Ruterana, S. Kret, A. Vivet, G. Maciejewski, and P. Dluzewski, J. Appl. Phys. **91**, 1 (2002).

[136] X. A. Cao, S. F. LeBoeuf, L. R. Rowland, C. H. Yan, and H. Liu, Appl. Phys. Lett. **82**, 3614 (2003).

[137] A. Kinoshita et al., Mat. Res. Soc. Symp. **639**, G12.6 (2001).

[138] Y. H. Cho et al., Appl. Phys. Lett. **83**, 2578 (2003).

[139] J. Narayan et al., Appl. Phys. Lett. **81**, 841 (2002).

[140] K. Okamoto et al., Jap. J. Appl. Phys. **43**, 839 (2004).

[141] K. Okamoto et al., J. Appl. Phys. **98**, 064503 (2005).

[142] A. Kaneta et al., Jap. J. Appl. Phys. **40**, 110 (2001).

[143] X. H. Wang et al., Appl. Phys. Lett **91**, 161912 (2007).

[144] X. H. Wang et al., Appl. Phys. Lett **94**, 111913 (2009).

[145] T. Azuhata, T. Homma, Y. Ishikawa, and S. F. Chichibu, Jpn. J. Appl. Phys. **42**, L497 (2003).

[146] Y. D. Qi et al., Appl. Phys. Lett. **86**, 101903 (2005).

[147] J. R. Leite, Microelectronics J. **33**, 323 (2002).

[148] F. Rossi et al., Semicond. Sci. Technol. **21**, 638 (2006).

[149] B. V. Daele, G. V. Tendeloo, K. Jacobs, I. Moerman, and M. R. Leys, Appl. Phys. Lett. **85**, 4379 (2004).

[150] E. H. Park et al., IEEE Photonics Technol. Lett. **19**, 24 (2007).

[151] I. K. Park, M. K. Kwon, S. B. Seo, and J. Y. Kim, Appl. Phys. Lett. **90**, 111116 (2007).

[152] UK Department of Trade and Industry, LED lighting technology: lessons from the Japan, Global watch mission report, 2005.

[153] O. Mayrock, H. J. Wünsche, and F. Henneberger, Phys. Rev. B **62**, 16870 (2000).

[154] S. Heikman et al., J. Appl. Phys. **93**, 10114 (2003).

[155] E. T. Yu, X. Z. Dang, P. M. Asbeck, S. S. Lau, and G. J. Sullivan, J. Vac. Sci. Technol. **B 17**, 1742 (1999).

[156] J. S. Im et al., Phys. Rev. B **57**, R9435 (1998).

[157] J. S. Im, Spontaneous recombination in group-III nitride quantum wells, Dr. rer. nat. thesis, TU Braunschweg, 2000.

[158] H. Hirayama, A. Kinoshita, and Y. Aoyagi, RIKEN Rev. , 28 (2001).

[159] X. Zhou et al., Appl. Phys. Lett. **85**, 407 (2004).

[160] X. Zhou et al., J. Vac. Sci. Technol. **B 23**, 1808 (2005).

[161] S. Y. Karpov, Phys. Stat. Sol. RRL **11**, 320 (2010).

[162] H. J. Chang et al., Appl. Phys. Lett. **86**, 021911 (2005).

[163] H. Sch′omig et al., Phys. Rev. Lett. **92**, 106802 (2004).

[164] J. F. Nye, *Physical Properties of Crystals. The Representation by Tensors and Martices*, (Oxford at the Clarendon Press, 1964.

[165] K. Xiao, D. J. Pommerenke, and J. L. Drewniak, IEEE Trans. Antennas Propag. **55**, 1981 (2007).

[166] L. H. Peng, Y. C. Hsu, and C. W. Chuang, IEEE J. Select. Topics Quantum Electron. **5**, 756 (1999).

[167] J. Bhattachryya, S. Ghosh, and H. T. Grahn, Appl. Phys. Lett. **93**, 051913 (2008).

[168] C. S. Kim, H. G. Kim, C. H. Hong, and H. K. Cho, Appl. Phys. Lett. **87**, 013502 (2005).

[169] J. Piprek and S. Li, Gan-based light-emitting diodes, in *Optoelectronic Devices - Advanced Simulation and Analysis*, edited by J. Piprek, pages 293–312, Springer, 2007.

[170] D. Vasileska and S. M. Goodnick, Mater. Sci. Engin. **R 38**, 181 (2002).

[171] M. V. Fischetti and S. E. Laux, Phys. Rev. **38**, 9721 (1988).

[172] M. Saraniti, J. Tang, S. Goodnick, and S. J. Wigger, Math. Comput. in Simulation **62**, 501 (2003).

[173] P. Degond and C. Schmeiser, J. Math. Phys. **39**, 4634 (1998).

[174] A. Jüngel, *Quasi-Hydrodynamic Semiconductor Equations*, Birkhauser Verlag, 2000.

[175] A. el Ayyadi and A. Jüngel, Semiconductor simulations using a coupled quantum drift-diffusion schrödinger-poisson model, der Johannes Gutenberg-Universität, Mainz, Preprint 09/04, 2004.

[176] O. M. J.A. Carrillo, I.M. Gamba and C.-W. Shu, Comparison of monte carlo and deterministic simulations of a silicon diode, Università degli Studi di Catania, Preprint, 2000.

[177] S. M. Sze, *Physics of Semiconductor Devices*, Wiley, New York, 1981.

[178] N. B. Abdallah and M. L. Tayer, Discrete and Cont. Dynam. Systems - Series B **4**, 1129 (2004).

[179] W. V. Roosbroeck, Bell System Tech. J. **29**, 560 (1950).

[180] V. Palankovski, R. Quay, and S. Selberherr, IEEE J. Solid-State Circuits **36**, 1365 (2001).

[181] E. P. Doolan, J. J. H. Miller, and W. H. A. Shielders, *Uniform Numerical Methods for Problems with Initial and Boundary Layers*, Boole, Dublin, 1980.

[182] J. J. H. Miller, W. H. A. Shielders, and S. Wang, Rep. Prog. Phys. **62**, 277 (1999).

[183] D. R. R.E. Bank and W. Fichtner, IEEE Trans. Electron. Dev. **ED-30**, 1031 (1983).

[184] M. S. Mock, Comm. Pure Appl. Math. **25**, 781 (1972).

[185] P. Markowich, SIAM J. Appl. Math. **5**, 896 (1984).

[186] J. W. Jerome and T. Kerkhoven, SIAM J. Numer. Anal. **28**, 403 (1991).

[187] F. Alabau, Trans. Am. Math. Soc. **348**, 823 (1996).

[188] S. Y. Karpov and Y. N. Makarov, Appl. Phys. Lett. **81**, 4721 (2002).

[189] Y. C. Shen, G. O. M. ans S. Watanabe, N. F. Gardner, A. Munkholm, and M. R. Krames, Appl. Phys. Lett. **91**, 141101 (2007).

[190] J. Piprek, phys. stat. sol. A , 1 (2010).

[191] A. laubsch et al., Phys. Sta. Sol. C **6**, 5913 (2009).

[192] X. A.Cao, J. A.Teetsov, F. Shahedipour-Sandvik, and S. D. Arthur, J. Crystal Growth **264**, 172 (2004).

[193] K. G. Zolina, V. E. Kudryashov, A. N. Turkin, A. E. Yunovich, and S. Nakamura, MRS Internet J. Nitride Semicond. Res. **1** (1996).

[194] X. A. Cao et al., (2002).

[195] X. A. Cao, J. M. Teetsov, M. P. D'Evelyn, and D. W. Merfeld, Appl. Phys. Lett. **85**, 7 (2004).

[196] S. Chuang and C. Chang, Phys. Rev. B **54**, 2491 (1996).

[197] C. H. Liu et al., Mater. Sci. Eng. B **112**, 10 (2004).

[198] H. Kim, S. J. Park, and H. Hwang, IEEE Trans. El. Dev. **48**, 1065 (2001).

[199] A. Kaneta et al., Jpn. J. Appl. Phys. **40**, 110 (2001).

[200] M. Farahmand et al., IEEE Trans. El. Dev. **48**, 535 (2001).

[201] A. F. M. Anwar, S. Wu, and R. T. Webster, IEEE Trans. El. Dev. **48**, 567 (2001).

[202] O. Ambacher, J. Phys. D: Appl. Phys. **31**, 2653 (1998).

[203] S. Stepanov et al., MRS J. Nitride Semicond. Res. **6**, 6 (2001).

[204] H. Angerer et al., Appl. Phys. Lett. **71**, 1504 (1997).

[205] O. Katz, B. Meyler, U. Tisch, and J. Salzman, Phys. Stat. Solidi (a) **188**, 789 (2001).

[206] F. Yun et al., 2002 **92**, 4837.

[207] K. Kim, A. Saxler, P. Kung, M. Razeghi, and K. Y. Lim, Appl. Phys. Lett. **71**, 800 (1997).

[208] I. Akasaki and H. Amano, Jpn.J.Appl.Phys. **36**, 5393 (1997).

[209] M. Leszcynski et al., J.Appl.Phys. **76**, 4909 (1994).

[210] M. van Schilfgaarde, A. Sher, and A. B. Chen, J.Cryst.Growth **178**, 8 (1997).

[211] S. Mohammad and H. Morkoc, Prog. Quant. Electron. **20**, 361 (1996).

[212] D. Cherns, J. Barnard, and F. A. Ponce, Solid State Commun. **111**, 281 (1999).

[213] V. W. L. Chin, T. L. Tansley, and T. Osotchan, J.Appl.Phys. **75**, 7365 (1994).

[214] M. E. Levinshtein, S. L. Rumyantsev, and M. S. Shur, editors, *Properties of advanced semiconductor materials. GaN, AlN, InN, BN, SiC, SiGe*, Wiley-Interscience, NY, 2001.

[215] www.iiiv.cornell.edu/www/foutz/nitrid.html.

[216] D. J. Dugdale, S. Brandt, and R. Abram, Phys.Rev. B **61**, 12933 (2000).

[217] J. S. Im et al., Appl. Phys. Lett. **70**, 631 (1997).

[218] S. Nakamura, Semicond. Sci. Technol. **14**, R27 (1999).

[219] J. R. Chen et al., Chin. Phys. Lett. **24**, 2112 (2007).

[220] X. A. Cao and S. D. Arthur, Appl. Phys. Lett. **85**, 3971 (2004).

[221] K. Kawasaki, C. Koike, Y. Aoyagi, and M. Takeuchi, Appl. Phys. Lett. **89**, 261114 (2006).

[222] W. Nakwaski, Opto-Electronics Rev. **6**, 93 (1998).

[223] X. A. Cao et al., Solid State Electron. **46**, 1235 (2002).

[224] S. R. Jeon et al., Appl. Phys. Lett. **78**, 3265 (2001).

[225] X. Guo, Y. L. Li, and E. F. Schubert, Appl. Phys. Lett. **79**, 1936 (2001).

[226] H. Kim, S. J. Park, and H. Hwang, Appl. Phys. Lett. **81**, 1326 (2002).

[227] Y. S. Ting, C. C. Chen, J. K. Sheu, G. C. Chi, and J. T. Hsu, J. Electron. Mater. **32**, 312 (2003).

[228] A. Chakraborty, L. Shen, H. Masui, S. P. DenBaars, and U. Mishra, Appl. Phys. Lett. **88**, 181120 (2006).

[229] S. J. Lee, Proc. SPIE **5530**, 338 (2004).

[230] H. W. Choi and M. D. Dawson, Appl. Phys. Lett. **86**, 053504 (2005).

[231] Z. Y. Fan, J. Y. Lin, and H. X. Jiang, J. Phys. D: Appl. Phys. **41**, 094001 (2008).

[232] I. Eliashevich, A. Osinsky, C. A. Tran, M. G. Brown, and K. A. Karlicek, Proc. SPIE **3621**, 28 (1999).

[233] H. Kim et al., Appl. Phys. Lett. **77**, 1903 (2000).

[234] X. Guo and E. F. Schubert, Appl. Phys. Lett. **78**, 3337 (2001).

[235] H. Kim, S. Park, and H. Hwang, IEEE Trans. Electron. Dev. **48**, 1065 (2001).

[236] H. Kim, S. Park, and H. Hwang, IEEE Trans. Electron. Dev. **49**, 1715 (2002).

[237] A. Ebong et al., GE Global Research Tech. Rep. 2002GRC173, 2002.

[238] A. Ebong et al., Solid State Electron. **47**, 1817 (2003).

[239] J. Piprek, T. Katona, and S. P. DenBaars, Proc. SPIE **5365**, 127 (2004).

[240] A. Monakhov, A. Krier, and V. V.Sherstnev, Semicond. Sci. Technol. **19**, 480 (2004).

[241] J. W. Larson, Some organising principles for coupling in multiphysics and multiscale models, Preprint ANL/MCS-P1414-0207, 2007.

[242] T. Stephan et al., phys. stat. sol. (a) **194**, 568 (2002).

[243] N. Narendran, Y. Gu, J. P. Freyssinier, H. Yu, and L. Deng, J. Crystal Growth **36**, 449 (2004).

[244] M. K. Yeh and Y. L. Huang, J. Chin. Inst. Engineer. **31**, 271 (2008).

[245] J. Hu, L. Yang, W. J. Hwang, and M. W. Shin, J. Crystal Growth **288**, 157 (2006).

[246] J. Hu, L. Yang, and M. W. Shin, Microelectron. J. **38**, 157 (2007).

[247] A. Arik, J. Petroski, and S. Weaver, Thermal challenges in the future generation solid state lighting applications: Light emitting diodes, GE Global Research Technical Report 2002GRC078, 2002.

[248] A. Zukauskas, F. Ivanauskas, F. Vaicekauskas, M. S. Shur, and R. Gaska, Poc. SPIE **4445**, 148 (2001).

[249] H. Chen, Y. Lu, Y. Gao, H. Zhang, and Z. Chen, Thermochimica Acta **488**, 33 (2009).

[250] C. J. Weng, Comm. Heat Mass Transfer **36**, 245 (2009).

[251] M. Shatalov et al., Appl. Phys. Lett. **86**, 201109 (2005).

[252] B. Yu and Y. Wang, Chin. J. Luminescence **26**, 761 (2005).

[253] W. Dai, J. Wang, and Y. Li, Semicond. Optoelectron. **29**, 324 (2008).

[254] X. Luo and S. Liu, IEEE Trans. Adv. Packaging **30**, 475 (2007).

[255] R. Wang and J. Li, J. Mod. Phys. **1**, 196 (2010).

[256] X. Luo, T. Cheng, W. Gan, and S. Liu, IET Optoelectronics **1**, 191 (2007).

[257] T. Xiaogai, C. wei, and Z. Jiyong, J. Semicond. **32**, 014009 (2011).

[258] X. Xuliang, Appl. Mech. Mater. **55–57**, 2135 (2011).

[259] L. Kim, W. J. Hwang, and M. W. Shin, IEEE Electronics Components and Technology Conf.

[260] C. Winnewisser, J. Schreider, M. Börsch, and H. W. Rotter, J. Appl. Phys. **89**, 3091 (2001).

[261] Y. Xi and E. F. Schubert, Appl. Phys. Lett. **85**, 2163 (2004).

[262] W. J. Hwang, T. H. Lee, L. Kim, and M. W. Shin, phys. stat. sol. (c) **1**, 2429 (2004).

[263] N. Narendran, Y. Gu, and R. Hosseinzadeh, Proc. SPIE **5366**, 158 (2004).

[264] H. S. Lee, H. W. Shin, and S. B. Jung, Mater. Sci. Forum **654–656**, 2811 (2010).

[265] L. Siguira, J. Appl. Phys. **81**, 1633 (1997).

[266] V. Schwegler et al., MRS Internet J. Nitride Semicond. Res. **5S1**, W11.18 (2000).

[267] C. Huh, W. J. Schaff, L. F. Eastman, and S. J. Park, Proc. SPIE **5187**, 330 (2004).

[268] M. Arik, S. Weaver, C. Becker, M. Hsing, and A. Srivastava, GE Global Research Tech. Rep. 2003GRC187, 2003.

[269] C. Pan, C. M. Lee, J. W. Liu, G. T. Chen, and J. I. Chyi, Appl. Phys. Lett. **84**, 5249 (2004).

[270] Y. Bumai, A. Vaskou, and V. Kononenko, Metrology and Measur. Systems **17**, 39 (2010).

[271] X. A. Cao, III-nitride light-emitting diodes on novel substrates, in *Wide Bandgap Light Emitting Materials and Devices*, edited by G. F. Neumark, I. L. Kuslovsky, and H. Jiang, pages 3–48, Weinheim, 2007, WILEY-VCH Verlag GmbH & Co. KGaA.

[272] K. Streubel, N. Linder, R. Wirth, and A. Jaeger, IEEE J. Select. Topics Quantum. Electron. **8**, 321 (2002).

[273] S. Mladenovski, K. Neyts, D. Pavicic, A. Werner, and C. Rothe, Optics Express **17**, 7562 (2009).

[274] S. Jeon et al., Appl. Phys. Lett. **92**, 223307 (2008).

[275] M. Born and E. Wolf, *Principles of Optics: electromagnetic theory of propagation, interference and diffraction of light, 7 ed.*, Cambridge University Press, Cambridge, 2005.

[276] J. U. Nöckel and R. K. Chang, 2-d microcavities: theory and experiments, arXiv:physics/0406134, 2004.

[277] S. K. Kim et al., Appl. Phys. Lett. **92**, 241118 (2008).

[278] M. D. B. Charlton, M. E. Zoorob, and T. Lee, Proc. of SPIE **6486**, 64860R (2007).

[279] D. Ochoa, R. Houndre, M. Ilegems, C. Hanke, and B. Borchert, C. R. Physique **3**, 3 (2002).

[280] M. F. Schubert, S. Chajed, and J. M. Redwing, Appl. Phys. Lett. **91**, 051117 (2007).

[281] H. Y. Chen et al., Appl. Phys. Lett. **81**, 574 (2002).

[282] K. B. Nam, J. Li, M. L. Nakami, J. Y. Lin, and H. X. Jiang, Appl. Phys. Lett. **84**, 5264 (2004).

[283] J. S. Anderson and M. Bratos-Anderson, J. Sond Vibr. **236**, 209 (2000).

[284] H. Luo et al., Appl. Phys. Lett. **89**, 0411125 (2006).

[285] E. F. Schubert, J. K. Kim, H. Luo, and J. Q. Xi, Rep. Prog. Phys. **69**, 3069 (2006).

[286] Y. Xi et al., J. Vac. Sci. Techn. A **24**, 1627 (2006).

[287] A. I. Zhmakin, Physics Reports **498**, 189 (2011).

[288] T. Gessmann, E. F. Schubert, J. W. Graff, and K. Streubel, SPIE Proc. **4996**, 26 (2003).

[289] M. Y. Su and R. P. Mirin, Appl. Phys. Lett. **89**, 033105 (2006).

[290] M. Y. Su and R. P. Mirin, Appl. Phys. Lett. **89**, 033105 (2006).

[291] I. Schnitzer, E. Yablonovich, C. Caneau, T. J. Gmitter, and A. Scherer, Appl. Phys. Lett. **63**, 2174 (1993).

[292] C. Huh, K. S. Lee, E.-J. Kang, and S.-J. Park, J. Appl. Phys. **93**, 9383 (2003).

[293] Y. H. Cheng, J. L. Wu, C. H. Cheng, K. C. Syao, and M. C. M. Lee, Appl. Phys. Lett. **90**, 091102 (2007).

[294] C. C. Wang, H. C. Lu, F. L. Jeng, Y. H. Wang, and M. P. Houng, IEEE Photonics Technol. Lett. **20**, 428 (2008).

[295] S. Chajed, W. Lee, J. Cho, E. F. Schubert, and J. K. Kim, Appl. Phys. Lett. **98**, 071102 (2011).

[296] Y. J. Lee, H. C. Kuo, T. C. Lu, B. J. Su, and S. C. Wang, J. Electrochem. Soc. **153**, G1106 (2006).

[297] K. Tadatomo et al., SPIE Proc. **5187**, 243 (2004).

[298] Y. Li et al., Appl. Phys. Lett. **98**, 151102 (2011).

[299] C. E. Lee, Y. C. Lee, H. C. Kuo, T. C. Lu, and S. C. Wang, IEEE Photonics Technol. Lett. **20**, 659 (2008).

[300] T. Yamasaki, K. Sumioka, and T. Tsutsui, Appl. Phys. Lett. **76**, 1243 (2000).

[301] L. Lin, T. K. Shia, and C. J. Chiu, J. Micromech. Microeng. **10**, 395 (2000).

[302] B. Butun, J. Cesaro, S. Enoch, R. Quidant, and E. Ozbay, Photonics and Nanostructures — Fundamentals and Appl. **5**, 86 (2007).

[303] X. M. Zhang, M. Y. Lu, Y. Zhang, L. J. Chen, and Z. L. Wang, Adv. Mater. **21**, 2767 (2009).

[304] Y. K. Ee et al., IEEE J. Select. Topics Quantum Electron. **15**, 1218 (2009).

[305] H. Y. Lin et al., Opt. Commun. **275**, 464 (2007).

[306] T. Tuohioja, MS Thesis, Helsinki University of Technology, Espoo, 2006.

[307] H. W. Choi et al., Appl. Phys. Lett. **84**, 2253 (2004).

[308] E. Gu et al., Appl. Phys. Lett. **84**, 2754 (2004).

[309] D. Kim, H. Lee, N. Cho, Y. Sung, and G. Yeom, Jap. J. Appl. Phys. **44**, L18 (2005).

[310] Y. K. Ee, R. A. Arif, N. Tansu, P. Kumnorkaew, and J. F. Gilchrist, Appl. Phys. Lett. **91**, 221107 (2007).

[311] C. F. Madigan, M.-H. Lu, and J. C. Sturm, Appl. Phys. Lett. **76**, 1650 (2000).

[312] M. Achermann et al., Nature **429**, 642 (2004).

[313] D. S. Liu et al., Appl. Phys. Lett. **94**, 143502 (2009).

[314] J. J. Wierer et al., Appl. Phys. Lett. **84**, 3885 (2004).

[315] E. Matioli and C. Weisbuch, J. Phys. D.: Appl. **43**, 354005 (2010).

[316] F. D. martini, M. Marrocco, P. Mataloni, L. Crescentini, and R. Loudon, Phys. Rev. A **43**, 2480 (1991).

[317] J. Chan et al., Proc. of SPIE **5277**, 311 (2004).

[318] C. Hums et al., J. Appl. Phys. **101**, 033113 (2007).

[319] V. Haerle et al., phys. stat. sol. (a) **201**, 2736 (2004).

[320] T. Gessmann and E. F. Schubert, J. Appl. Phys. **95**, 2203 (2004).

[321] S. Illek et al., SPIE Proc. **4996**, 18 (2003).

[322] K. Bao et al., Appl. Phys. Lett. **92**, 141104 (2008).

[323] W. H. Chuang, J. Y. Wang, C. C. Wang, and Y. W. Kiang, IEEE Photonics Technol. Lett. **20**, 1339 (2008).

[324] Y. H. Cho et al., Korean J. Chem. Eng. **26**, 277 (2009).

[325] B. S. Cheng et al., Semicond. Sci. Techn. **23**, 055002 (2008).

[326] R. H. Horng, Y. L. Tsai, T. M. Wu, D. S. Wuu, and C. H. Chao, IEEE J. Select. Topics Quantum Electron. **15**, 1327 (2009).

[327] I. E. Bergman, Optics and Photonics news **9**, 29 (1998).

[328] M. Pessa et al., Semicond. Sci. Technol. **17**, R1 (2002).

[329] T. M. Benson et al., Micro-optical resonators for microlasers and integrated opto-electronics, in *Frontiers in Planar Lightwave Circuit Technology*, edited by S. Janz, J. Ctyroki, and S. Tanev, pages 39–70, Springer, 2006.

[330] H. G. Hong et al., phys. Stat. sol. (a) **204**, 881 (2007).

[331] M. Fujita et al., Appl. Phys. Lett. **85**, 5769 (2004).

[332] S. K. Kim et al., Appl. Phys. Lett. **92**, 241118 (2008).

[333] R. Hershey et al., 2D photonic crystal patterning for high volume LED manufacturing, SPIE Optics and Photonics manuscript, 2006.

[334] Y. A. Kravtsov and Y. I. Orlov, Sov. Phys. Uspekhi **23**, 750 (1980).

[335] A. S. Glassner, *An Introduction to Ray Tracing*, Academic, San Diego, 1989.

[336] F. Hu, K. Y. Qian, and Y. Luo, Appl. Opt. **44**, 2768 (2005).

[337] H. Lehnert, Appl. Acoust. **38**, 207 (1993).

[338] S. J. Lee, Optical Engineer. **45**, 014601 (2006).

[339] M. Yamada et al., Jap. J. Appl. Phys. **41**, L1431 (2002).

[340] X. H. Wang, P. T. Lai, and H. W. Choi, J. Appl. Phys. **108**, 023110 (2010).

[341] S. J. Lee, Appl. Opt. **40**, 1427 (2001).

[342] D. Z. Ting and T. C. McGill, Opt. Eng. **34**, 3345 (1995).

[343] D. Z. Y. Ting and T. C. McGill, VLSI Design **6**, 363 (1998).

[344] P. Heckbert and P. Hanrahan, ACM Computer Graphics **18**, 119 (1984).

[345] T. Funkhouser et al., J. Acoust. Soc. Am. **115**, 739 (2004).

[346] J. B. Keller, J. Opt. Soc. Am. **52**, 116 (1962).

[347] R. G. Kouyoumjian and P. H. Pathak, Proc. IEEE **62**, 1448 (1974).

[348] N. D. Taket and R. E. Burge, IEEE Trans. Antennas Propagat. **39**, 719 (1991).

[349] P. Bagnerini, A. Buffa, and A. Cangiani, IEEE Trans. Antennas Propagat. **55**, 1416 (2007).

[350] B. V. Budaev and D. B. Bogy, The Quart. J. Mech. Appl. Math. **55**, 209 (2002).

[351] A. V. Timofeev, Phys. Usp. **48**, 609 (2005).

[352] W. C. Chew, Phil. Trans. R. Soc. Lond. A **362**, 579 (2004).

[353] E. K. Miller, IEEE Trans. Antennas and Propagat. **36**, 1288 (1988).

[354] L. B. Felsen and L. Sevgi, Turk. J. Elec. Engin. **10**, 131 (2002).

[355] J. D. Joannopoulos, S. G. Johnson, J. N. Winn, and R. D. Meade, *Photonic Crystals: Molding the Flow of Light (2nd Ed.)*, Princeton Univ. Press, 2008.

[356] M. A. Kolbehdari, Progr. Electromagn. Res., PIER **19**, 93 (1998).

[357] J. Schefter, Discretization of the Maxwell equations on tetrahedral grids, WIAS, Berlin, Technical Report no. 6, 2003.

[358] A. Taflove, IEEE Trans. Electromagn. Compatib. **22**, 191 (1980).

[359] B. Wang, Y. Jin, and S. He, J. Appl. Phys. **106**, 014508 (2009).

[360] R. Zheng and T. Taguchi, Proc. of SPIE **4996**, 105 (2003).

[361] M. V. Bogdanov et al., phys. stat. sol. (c) **7**, 2124 (2010).

[362] Y. Ohno, Proc. of SPIE **5530**, 88 (2004).

[363] A. Žukauskas, R. Vaicekauskas, and M. S. Shur, J. Phys. D: Appl. Phys. **43**, 354006 (2010).

[364] S. Muthu, F. J. P. Schuurmans, and M. D. Pashley, IEEE J. Select. Topics Quatum. Electron. **8**, 333 (2002).

[365] K. Sakuma et al., IEICE Trans. Elecgtron. **E88-C**, 2057 (2005).

[366] J. H. Yum, S. Y. Seo, S. Lee, and Y. E. Sung, Proc. of SPIE **4445**, 60 (2001).

[367] K. Yamada, Y. Imai, and K. Ishii, J. Light Visual Environment **27**, 70 (2003).

[368] C. C. Chang et al., Jap. J. Appl. Phys. **44**, 6056 (2005).

[369] N. Kimura et al., Appl. Phys. Lett. **90**, 051109 (2007).

[370] N. T. Tran and F. G. Shi, J. Lightwave Technol. **26**, 3556 (2008).

[371] A. Borbély and S. G. Johnson, Optical Engineer. **44**, 111308 (2005).

[372] S. Nizamoglu, T. Erdem, X. W. Sun, and H. V. Demir, OPtics Lett. **35**, 3372 (2010).

[373] K. N. Hui, X. H. Wang, Z. L. Li, P. T. Lai, and H. W. Choi, Optics Express **17**, 9873 (2009).

[374] M. Yamada, Y. Narukawa, H. Tamaki, Y. Murazaki, and T. Mukai, IEICE Trans. Electron. **E88-C**, 1860 (2005).

[375] B. Damilano, N. Grandjean, C. Pernot, and J. Massies, Jpn. J. Appl. Phys. **40**, L918 (2001).

[376] I. Ozden, E. Makarona, A. V. Nurmikko, T. takeuchi, and M. Krames, Appl. Phys. Lett. **79**, 2532 (2001).

[377] C. H. Chen et al., IEEE Photonics Technol. Lett. **14**, 908 (2002).

[378] B. Damilano, N. Grandjean, C. Pernot, and J. Massies, Jap. J. Appl. Phys. **40**, L918 (2001).

[379] M. Yamada, Y. narukawa, and t. Mukai, Jap. J. Appl. Phys. **41**, L246 (2002).

[380] V. V. Nikolaev, M. E. Portnoy, and I. Eliashevich, Photon recycling white emitting diode based on InGaN multiple quantum well heterostructure, arXiv:cond-mat/0111386, 2001.

[381] N. Narendran, Proc. of SPIE **5941**, 45 (2005).

[382] M. Zachau et al., Proc. of SPIE **6910**, 691010 (2008).

[383] A. Boroley and S. G. Johnson, Opt. Eng. **44**, 111308 (2005).

[384] W. T. Chien, C. C. Sun, and I. Moreno, Opt. Express **15**, 7572 (2007).

[385] J. R. Oh, S. H. Cho, Y. H. Lee, and Y. R. Do, Optics Express **17**, 7450 (2009).

INDEX

A

absorption spectra, 125, 127, 132
access, 55, 77
accounting, 262, 267
acrylate, 28
activation energy, 71, 74
adaptation, 137
adjustment, 55
advancements, 235
AFM, 31, 32
age, 33
algorithm, 88, 89, 98, 112, 215, 217, 218, 219, 230, 261
alters, 89, 101, 119
aluminium, 18
ammonia, 161, 164
amplitude, 26, 52, 54, 56, 66, 225, 236
anisotropy, x, 213, 231, 238
annealing, ix, 61, 151, 156, 158, 159, 160, 161, 164, 165, 166, 260
annihilation, 8, 198
anti-reflection coating, 40
Asia, 114
assessment, 48, 49, 81, 219
atmosphere, 156, 157
atoms, 55, 176, 177, 182, 183, 185, 198, 212, 223
avoidance, 8
awareness, 89, 98
azimuthal angle, 20, 257

B

ban, 120, 122, 133
bandgap, vii, 196, 212, 221, 229, 233, 236, 237, 239, 242, 243, 266
bandwidth, viii, 70, 83, 90, 91, 105, 120, 261
banks, 107

barriers, 4, 11, 57, 89, 119, 120, 125, 132, 133, 137, 232, 235, 241, 242, 247, 248, 263
base, 19, 36, 98, 104, 115, 135, 157, 226, 227
bending, 119, 237
benzene, 160, 164
beryllium, 80
bias, 20, 43, 44, 48, 51, 53, 56, 65, 66, 69, 70, 71, 73, 74, 78, 129, 133, 134, 137, 138, 178, 186, 190, 191, 240, 241, 251, 252
birefringence, 229
blueshift, 236, 237
Boltzmann constant, 238
bonding, 44
bonds, 158, 200
boundary value problem, 229
breakdown, 180, 186, 187
browsing, 89, 98

C

calculus, 51, 54, 55, 56, 57, 77
calibration, 109, 111
candidates, ix, 113, 117, 118, 129
catalyst, 158, 160
cation, 260
CBP, 5, 6, 8
cerium, 267
challenges, 90, 105, 227, 280
charge density, 31, 119
chemical, 20, 45, 46, 66, 69, 158, 159, 182, 189, 234
chemical etching, 45, 189
chemical vapor deposition, 20, 45, 46
China, 1, 195, 208, 269
chlorobenzene, 160, 164
chloroform, 160, 164
circularly polarized light, 102
cladding, 16, 46, 53, 58, 235, 246, 258

cladding layer, 46, 53, 235, 246, 258
classes, 24
classification, 259
cleaning, 175
cleavage, 212
clusters, 12, 197, 200, 205, 206, 235, 236
coherence, 199, 206
color, iv, 11, 12, 13, 19, 21, 22, 32, 35, 84, 100, 108, 152, 153, 154, 269
commercial, 2, 24, 44, 45, 46, 77, 89, 98
communication, ix, 80, 117, 120, 121, 123, 124, 125, 126, 135, 137, 146, 152
communication systems, 80, 120, 121, 125, 152
compensation, ix, 173, 178, 186
competitive process, 121
complexity, 211, 239, 261
composition, 45, 51, 65, 71, 182, 235, 236, 237, 242, 243, 245, 246
compounds, x, 71, 78, 118, 125, 152, 231, 232, 233, 243
compression, 237
computational grid, 248
computer, 99, 104, 248, 268
computing, 135, 248
condensation, 158, 159, 160
conduction, 20, 52, 53, 55, 57, 66, 118, 119, 120, 121, 122, 124, 126, 127, 129, 130, 132, 133, 134, 138, 139, 200, 212, 235, 238, 256
conductivity, ix, 48, 173, 174, 182, 190, 212, 254
conductor, 45, 46
conductors, 44, 233
configuration, 78, 90, 100, 113, 120, 200, 201, 203, 204, 205, 206
confinement, vii, ix, 1, 32, 117, 118, 120, 122, 127, 130, 138, 212, 213, 224, 235, 236
constituent materials, 212
construction, 96
consumption, vii, 1, 84, 89, 98
convergence, 119, 218, 219, 230
cooling, 253
copolymers, 259
correlation, 45, 119, 226
cost, x, 10, 18, 27, 84, 87, 89, 93, 98, 152, 153, 174, 212, 231, 232, 233, 239, 260
Coulomb interaction, 258
covering, 12, 123, 124, 125, 126, 146, 218
CPU, 239
cracks, 43, 190
critical value, 126, 206, 254
crystal growth, 43, 212
crystal quality, 232
crystal structure, 118, 153, 212, 220, 225, 238, 260

crystalline, 159
crystals, 27, 35, 36, 40, 80, 196, 212, 229, 259
CVD, 45, 46
cycles, 43, 45, 69, 110

D

danger, 260
data transfer, viii, ix, 83, 91, 105, 117
decay, 235, 237, 261
decomposition, 159
defects, 20, 43, 44, 50, 61, 63, 64, 65, 66, 68, 69, 70, 71, 77, 80, 81, 93, 94, 113, 234, 235
degradation, viii, 20, 43, 44, 45, 59, 60, 61, 67, 70, 71, 72, 75, 77, 78, 79, 80, 81, 151, 156, 159, 160, 164, 166, 260
degradation mechanism, viii, 43, 61, 77, 79
Department of Energy, 272
deposition, 19, 20, 37, 45, 46, 179, 180, 182, 183, 184, 185, 186, 187, 188, 189
depth, viii, 18, 84, 90, 108, 176, 177, 178, 180, 182, 184, 185, 189, 234, 254
derivatives, 213, 215, 261
designers, 84
detection, 104, 131
deviation, 43
DFT, 225, 226
dichloroethane, 160, 164, 166
dielectric constant, 119, 121, 212, 213, 214, 243
dielectric permittivity, 215, 217, 239
differential equations, 119
diffraction, 26, 28, 213, 259, 261, 281
diffusion, ix, 43, 57, 61, 62, 63, 65, 66, 68, 71, 77, 80, 100, 130, 134, 173, 174, 176, 178, 180, 181, 182, 183, 184, 185, 187, 189, 190, 193, 233, 236, 240, 241, 250, 251, 253, 277
diffusion process, 181
diffusion time, 184, 185, 187
dimerization, 200, 201, 205
diodes, iv, vii, ix, x, 1, 19, 33, 34, 35, 36, 37, 38, 39, 40, 60, 79, 80, 81, 91, 146, 154, 173, 174, 196, 211, 231, 232, 233, 234, 235, 246, 271, 272, 276, 280, 281
dipoles, 199
directionality, 259, 260
discontinuity, 118
discretization, 218, 219
discrimination, 215
dislocation, x, 71, 211, 231, 233, 235, 240, 243, 244
disorder, 4, 80, 196, 197
dispersion, 45, 213, 214, 217, 221, 267
displacement, 198, 200

distilled water, 160
distribution, 5, 8, 10, 16, 24, 44, 62, 70, 72, 74, 75, 76, 77, 78, 111, 112, 119, 129, 130, 159, 197, 199, 200, 202, 203, 205, 206, 207, 208, 212, 215, 217, 218, 219, 224, 225, 227, 234, 236, 242, 246, 254, 256, 257, 259, 261, 262, 268
distribution function, 242
diversity, vii, 1
DMF, 160, 161, 164
donors, 119, 121, 174, 238, 244
dopants, 8, 65
doping, 3, 7, 31, 41, 58, 61, 63, 64, 65, 80, 119, 128, 132, 133, 138, 139, 140, 141, 142, 143, 144, 145, 174, 178, 180, 186, 235, 242, 244, 246, 247, 248, 268
double bonds, 200

E

editors, iv, 278
education, 89, 98
electric field, 25, 57, 125, 126, 127, 132, 134, 136, 146, 180, 199, 202, 207, 217, 224, 237, 239, 241
electrical characterization, 134
electrical conductivity, 48, 182, 254
electrical fields, x, 231
electrical properties, 20, 29, 79, 235, 259, 260
electricity, 12
electrodes, 15, 94, 113, 188, 246, 252, 263
electroluminescence, 13, 16, 24, 27, 35, 152, 153, 174, 236
electromagnetic, 52, 125, 199, 212, 213, 214, 215, 218, 229, 230, 233, 242, 281
electromagnetic waves, 212, 230
electron, x, 3, 5, 10, 13, 15, 17, 23, 24, 25, 55, 58, 67, 68, 119, 120, 122, 124, 125, 126, 127, 129, 130, 131, 132, 146, 189, 195, 196, 197, 198, 199, 200, 201, 202, 203, 204, 205, 206, 207, 208, 234, 235, 238, 241, 242, 243, 244, 245, 249, 250, 251, 252, 254, 263
electron state, 122, 125, 146
electronic structure, 202
embossing, 27
emitters, ix, 7, 8, 38, 39, 117, 235, 244, 266, 271
employment, 5, 12
encapsulation, 20, 151, 158, 159
encoding, 88
end-users, 44, 78
energy consumption, vii, 1
energy efficiency, vii, 1
energy transfer, vii, 1, 3, 8, 13, 32, 37, 156, 161, 165, 166, 258
engineering, ix, x, 33, 117, 212, 231, 233, 235
environment, 77, 226, 254, 257
environments, 71
epitaxial growth, ix, 117
EPS, 80
equilibrium, 57, 130, 240, 247, 251, 254
ESD, 79
etching, 45, 175, 189, 226, 227, 228, 258, 260
ethanol, ix, 151, 158, 160, 164, 165, 166
europium, 155
evidence, 236, 237, 241
evolution, ix, 13, 195, 196, 197, 199, 261
excitation, ix, x, 53, 130, 151, 152, 153, 154, 155, 160, 161, 162, 163, 166, 195, 196, 197, 203, 205, 211, 237
exciton, vii, x, 1, 3, 8, 14, 20, 21, 25, 31, 34, 155, 195, 196, 197, 201, 202, 203, 205, 208
experimental condition, 55
exploitation, 71
exposure, 155, 156
extraction, x, 16, 18, 19, 21, 24, 28, 29, 30, 35, 36, 38, 40, 41, 51, 67, 191, 231, 232, 233, 234, 254, 256, 257, 258, 260, 262, 264, 265, 266

F

fabrication, 18, 21, 22, 27, 88, 114, 130, 146, 151, 152, 153, 154, 174, 211, 212, 227, 233, 258, 260, 263, 274
Fabrication, 40, 160, 229
Fermi level, 52, 53, 55, 120, 122, 125, 127, 241, 242, 251, 252
fiber, x, 80, 118, 120, 121, 125, 152, 195, 199, 202, 206, 229
fibers, x, 137, 195, 206
fidelity, 107
film thickness, 263, 265
films, 19, 24, 26, 29, 33, 34, 35, 37, 40, 152, 154, 157, 258, 267
filters, 115
filtration, 261
flexibility, 33, 95, 129, 196
fluctuations, 235, 236, 242, 243
fluorescence, 37
fluorine, 195
force, 248
Ford, 37
formation, 3, 26, 31, 34, 90, 98, 158, 160, 174, 178, 180, 181, 186, 202, 203, 221, 227, 241, 246
formula, 5, 119

France, 43
functional approach, 79
funding, 33

G

gain threshold, 211
gel, ix, 28, 151, 154, 157, 158, 159, 160, 162,
    163, 165, 166, 167
geometry, vii, 1, 27, 242, 246, 248, 260, 266
Georgia, 274
glasses, viii, 83, 84, 87, 89, 98, 108, 114, 115
glue, 68, 69
gratings, 258
grids, 23, 38, 248, 285
GRIN, 258
growth, ix, x, 43, 117, 119, 121, 125, 129, 173,
    211, 212, 231, 233, 235, 236, 243

H

Hamiltonian, 52, 197, 241
Hartree-Fock, 198
harvesting, 7
heat capacity, 48
heat transfer, 248, 254
height, 22, 134
hemisphere, 23
hexagonal lattice, 220
hexane, 160, 164
history, 13, 114, 232
host, vii, 1, 3, 4, 5, 8, 10, 11, 31, 32, 33, 34, 35
hot spots, 246
House, 78
human, 111, 115, 233, 234
hybrid, vii, x, 153, 157, 158, 159, 231, 233, 239,
    248, 251, 252
hydrogen, 44
hydrolysis, 157, 158, 159, 160
hypothesis, 55, 57, 58, 67, 161

I

ideal, viii, 83, 113, 118
identification, 56, 66
III-nitrides, x
illumination, vii, viii, 1, 2, 84, 104, 105, 107
image, viii, 22, 24, 27, 28, 31, 37, 38, 83, 84, 86,
    87, 89, 90, 91, 94, 95, 98, 100, 101, 102, 104,
    105, 106, 107, 109, 111, 112, 113, 115, 157,
    165, 175, 176, 189, 190
imaging systems, 115

imprinting, 196
impurities, ix, 173, 253
incidence, 20, 31, 254, 256, 257
India, 211
indium, 21, 23, 36, 37, 46, 235, 236, 260, 263
industry, 2
infrared spectroscopy, 118
insertion, 12
insulation, 263
integration, 78, 261, 266
interface, viii, 5, 7, 11, 15, 17, 24, 31, 33, 44, 51,
    63, 66, 68, 77, 80, 83, 119, 121, 126, 176, 181,
    185, 190, 252, 253, 254, 256, 257, 262
interface layers, 253
interference, 111, 281
Internal structure, 46, 47
inventions, 196
inversion, 51, 52, 55, 56, 77, 196, 197, 202, 203,
    204, 205, 206, 207, 208
ionicity, 118
ionization, 242
iridium, 34
irradiation, ix, 67, 68, 69, 80, 151, 152, 153, 154,
    155, 156, 157, 160, 162, 164, 166, 167, 174,
    175, 176, 177, 178, 179, 180, 185
islands, 236
isotropic media, 229
issues, 79, 84, 89, 105, 233, 260, 268
I-V curves, 49, 178, 183, 185, 186, 187

J

Japan, 34, 151, 167, 173, 175, 176, 177, 178,
    179, 181, 182, 183, 184, 185, 186, 187, 188,
    190, 191, 192, 276
joints, 44
Jordan, 38, 39

K

kinetic model, 239
kinetics, 43, 44

L

lanthanide, ix, 151, 152
laptop, viii, 84, 98
laser radiation, 208
lasers, ix, x, 117, 118, 125, 152, 195, 196, 211,
    228
lasing effect, 196, 197, 198, 199, 202
lattice parameters, 66, 77, 224

laws, 45, 59, 80

lead, x, 5, 7, 44, 76, 146, 211, 212, 213, 214, 217, 218, 221, 224, 226, 227, 228, 230, 246

leakage, x, 20, 44, 49, 59, 66, 68, 77, 107, 179, 190, 195, 206, 235, 240, 248

leisure, 268

lens, viii, 22, 45, 51, 83, 89, 90, 98, 100, 101, 102, 105, 111, 113, 258

lifetime, 10, 20, 36, 44, 50, 53, 55, 58, 63, 64, 70, 72, 74, 75, 76, 78, 80, 84, 118, 153, 236, 253

ligand, ix, 151, 152, 154, 155, 156, 158, 161, 165, 166

light beam, 101

light emitting diode, vii, 1, 2, 19, 34, 35, 36, 37, 38, 39, 40, 80, 84, 146, 173, 232, 246, 266

light scattering, viii, x, 25, 83, 195, 208

light-emitting diodes, vii, ix, 33, 34, 35, 36, 37, 38, 39, 40, 79, 81, 154, 173, 196

liquid phase, 158

lithography, 22, 27, 37, 40, 226, 260

localization, ix, 195, 197, 206, 235, 237, 248, 263

locus, 11, 13

low temperatures, 133, 237

lumen, 232, 234

luminescence, 51, 52, 55, 124, 152, 153, 187, 189, 196, 197, 232, 236, 237

luminescence efficiency, 153

Luo, 253, 269, 280, 281, 284

M

magnetic field, 214, 215, 216, 217

magnetic materials, 152, 159

magnitude, 127, 133, 236, 237, 239

majority, 241, 250, 251

management, 44, 48, 233, 253

manufacturing, viii, 43, 44, 83, 152, 212

mass, 53, 63, 119, 121, 129, 130, 238, 241, 242

materials, vii, x, 1, 11, 33, 35, 43, 44, 79, 80, 118, 127, 129, 151, 152, 154, 156, 159, 173, 190, 195, 211, 212, 213, 227, 229, 231, 233, 235, 239, 241, 254, 278

matrix, 2, 44, 98, 99, 118, 123, 125, 127, 158, 236, 266, 267

matter, iv, 53, 212

Maxwell equations, 260, 261, 285

measurement, 56, 57, 58, 59, 60, 61, 63, 78, 107, 109, 110, 111, 113, 158, 160, 180, 233

media, 214, 216, 229, 254, 262

median, 75

medical, 89, 98, 127, 152

memory, 147

mercury, 12

metal organic chemical vapor deposition, 46

metals, 174

meter, 99, 110, 111

methanol, 158, 160, 164

methodology, 45, 59, 73, 74, 76, 77, 78, 106, 212, 213, 217

microcavity, 24, 25, 36, 38, 39, 196, 230, 259

microelectronics, ix, 117

micro-lens, 22, 26

micrometer, 212

microscope, 23, 189, 190

microscopy, 234, 236

microspheres, 258

microstructure, 12, 27, 258

migration, 61, 253

miniaturization, 213

misfit dislocations, 43

mission, 59, 75, 152, 276

Missouri, 195

mixing, 4, 48, 160, 237

mobile phone, 2

MOCVD, 46, 236

modelling, x, 231, 232, 233, 267, 268

models, 45, 59, 77, 89, 98, 115, 239, 246, 248, 268, 279

modifications, 113

modules, vii, 43, 78

moisture, 234

mold, 22, 259

molecular beam, 46, 63, 80, 174, 273

molecular beam epitaxy, 46, 63, 80, 174, 273

molecular dynamics, 197, 198, 199

molecular structure, 153, 154, 155, 156, 159, 161, 163, 164, 165

molecular weight, 20

molecules, vii, 1, 3, 37, 212

momentum, 239, 242

monolayer, 258

Monte Carlo method, 239, 243

Moon, 114

Moses, 208

motivation, ix, 75, 117

MOVPE, 174, 229

N

nanocrystals, 258

nanodots, 14, 15, 35

nanofibers, 196

nanoimprint, 27, 40

nanolaser, 197

nanometer, 212

nanometers, 242

nanoparticles, 41
nanostructures, 118
Nanostructures, 271
nanowires, 19
neutral, 14, 57, 250, 251, 252, 254
neutrons, 67, 69
next generation, vii, 1
nitrides, x, 118, 231, 232, 233, 243, 254
nitrogen, 156, 157
NMR, 158
nodes, 48
non-linear optics, ix, 117
normal distribution, 74
Nuclear Magnetic Resonance, 158
numerical analysis, 225, 230

O

obstacles, 7
occlusion, 115
one dimension, 240
operations, 260
optical fiber, 80, 118, 120, 121, 137, 152, 260
optical gain, 52, 54, 55, 77
optical microscopy, 236
optical parameters, 45, 77
optical properties, 118, 125, 147, 152, 237, 256
optimization, x, 12, 21, 233, 268
optoelectronics, iv, vii, 80
oscillation, 199
overlap, 126, 129, 130, 131, 248, 258
oxidation, 31, 80, 152, 174, 175, 176, 178, 180
oxygen, ix, 151, 153, 156, 159, 166, 174

P

parallel, x, 22, 57, 77, 125, 195, 196, 199, 206, 208, 239, 257
participants, 98
periodicity, 29, 212, 217
permeability, 125, 215
permittivity, 119, 137, 215, 217, 239
phase shifts, 110
phonons, 55, 240
phosphorescence, 13, 22, 31, 34, 266
photodetector, 129, 187
photodetectors, ix, 117, 118, 127, 129
photographs, 99, 155
photolithography, 189, 226
photoluminescence, ix, 24, 26, 39, 40, 81, 151, 152, 178, 186, 221, 229, 235, 236
photonics, 26, 39

photons, 20, 29, 49, 58, 130, 135, 234, 242, 256, 257, 262, 266
photovoltaic cells, 154, 156, 157
physical mechanisms, x, 231, 233
physical phenomena, 239, 268
physical properties, x, 119, 129, 233
physics, 45, 55, 60, 78, 118, 146, 212, 229, 248, 281
pitch, 113
PL spectrum, 153, 154, 155, 160, 161, 162, 163
plane waves, 217
Plank constant, 238
platelets, 236
playing, 26
Poisson equation, 119, 129, 130, 131, 239, 240
polar, 235, 238, 244, 249, 250, 257, 265, 266, 273
polarity, 34, 238, 243, 244
polarization, x, 27, 102, 213, 229, 231, 233, 235, 236, 237, 238, 239, 241, 242, 243, 254, 257
polycarbonate, 32
polymer, ix, 19, 20, 25, 29, 33, 36, 39, 40, 44, 195, 196, 197, 198, 199, 200, 201, 202, 203, 204, 205, 206, 207, 208, 260
polymer chain, x, 195, 196, 197, 199, 200, 201, 202, 203, 205, 206, 207, 208
polymer chains, 199, 202, 203, 207, 208
polymer matrix, 44
polymers, 33, 35, 195, 196, 197
population, 14, 51, 52, 55, 56, 196, 197, 198, 202, 203, 204, 205, 206, 207
pop-up windows, 95
preparation, iv, 115
principles, 279
probability, 5, 6, 7, 15, 31, 52, 53, 129, 236, 257
probe, 237
process control, 44
propagation, x, 17, 66, 70, 78, 89, 196, 197, 206, 212, 213, 231, 257, 258, 260, 261, 262, 281
prototype, 98, 105
pumps, 198, 205, 206

Q

quantum confinement, ix, 117, 118, 120
quantum dot, ix, 14, 52, 53, 117, 146, 236, 266
quantum dots, ix, 14, 52, 53, 117, 146, 236, 266
quantum structure, 118, 129
quantum theory, 52, 66, 77
quantum well, ix, 12, 39, 46, 117, 118, 120, 121, 125, 129, 130, 132, 133, 136, 137, 146, 220, 236, 273, 276, 286

R

radiation, 20, 29, 35, 53, 56, 58, 77, 125, 199, 208, 233, 262, 265, 266, 267
Radiation, 274
radicals, 175, 178, 180
radio, 19
radius, 220, 221, 267
reaction temperature, 159
reactions, 158
reading, 108
reality, 113, 268
recombination, vii, x, 1, 4, 5, 6, 7, 13, 14, 15, 31, 50, 53, 55, 56, 57, 58, 60, 61, 62, 63, 64, 124, 130, 134, 189, 211, 234, 235, 236, 239, 240, 241, 243, 244, 245, 247, 249, 250, 251, 252, 258, 259, 260, 276
recommendations, iv
rectification, 179, 180, 187
recycling, 266, 267, 286
red shift, 126, 139
reflectivity, 228, 257
refraction index, 262
refractive index, 16, 18, 22, 23, 24, 29, 33, 36, 125, 136, 137, 144, 145, 146, 147, 215, 218, 221, 225, 254, 256, 258, 267
refractive index variation, 221
refractive indices, 29, 220, 254, 266, 267
relaxation, 53, 118, 120, 121, 125, 147, 198, 203, 205, 206, 219, 237, 258
relaxation process, 118, 121, 198, 203
reliability, vii, 44, 45, 60, 75, 76, 77, 78, 80, 81, 118, 268
rendition, 265
rent, 261
reproduction, 232
repulsion, 31, 198
requirements, vii, 43, 89, 98, 99, 101, 106, 265
researchers, 7, 151, 152, 153
resistance, 19, 44, 48, 49, 65, 74, 93, 178, 182, 186, 189, 190, 232, 235, 246, 252, 253, 254, 263, 265
resolution, viii, 2, 29, 57, 83, 84, 85, 89, 90, 91, 94, 95, 98, 105, 108, 111, 113, 175, 181, 236, 237
resonator, x, 195, 196, 206, 208, 259
resources, 239, 248
response, 20, 53, 55, 84, 92, 107, 110, 225, 226, 261
response time, 84, 107, 110
restrictions, 130
retail, 265
retardation, 98, 102

RIE, 189
room temperature, ix, 55, 57, 125, 131, 133, 152, 160, 173, 174, 179, 180, 187, 191, 192, 211, 221, 236
root, 93, 113
roughness, 16, 28, 226
routes, 5
Russia, 231

S

salts, 221
sapphire, 79, 235, 258, 263
saturation, 63, 127, 176, 182, 240
scalar field, 219
scattering, viii, x, 18, 25, 29, 83, 195, 197, 199, 202, 206, 208, 212, 213, 225, 229, 236, 258, 266, 267
scattering patterns, 267
Schrödinger equation, 119, 129
segregation, 236
semiconductor, viii, ix, x, 34, 43, 44, 49, 51, 52, 53, 56, 60, 61, 62, 66, 68, 69, 78, 79, 80, 117, 118, 125, 129, 174, 196, 204, 211, 212, 213, 218, 220, 224, 227, 232, 239, 246, 252, 254, 256, 257, 278
semiconductor lasers, 125
sensing, 127, 135, 228
sensitivity, viii, 43, 59, 70, 147, 234
sensors, 152, 157, 266
sequencing, 107
shame, 131
shape, 20, 37, 157, 166, 199, 224, 234, 242, 267
shortage, 196
showing, 22, 64, 69, 74, 89, 98, 118, 147
signals, 86, 107, 184
signs, 146
silane, ix, 151, 158, 159, 160, 161, 162, 164, 167
silica, ix, 18, 151, 153, 157, 158, 159, 160, 162, 164, 166, 167, 258
silicon, 152, 212, 243, 253, 277
silver, 26, 39
simulation, x, 38, 48, 69, 79, 99, 213, 221, 226, 229, 231, 233, 239, 254, 260, 261, 262, 266, 268
simulations, 23, 119, 213, 233, 248, 252, 261, 268, 277
single crystals, 196
SiO_2, 20, 23, 24, 40, 45, 158
SiO_2 films, 24
smoothing, 121
sodium, 158, 232
software, 89, 98, 99, 213, 233, 248

solar cells, 118, 173

sol-gel, ix, 28, 151, 153, 157, 158, 159, 160, 162, 163, 164, 166, 167

solid state, vii, x, 1, 2, 3, 32, 33, 34, 229

solidification, 158

solubility, 176

solution, ix, 19, 28, 29, 30, 33, 34, 36, 44, 55, 58, 84, 119, 151, 158, 159, 160, 164, 166, 196, 212, 213, 217, 218, 219, 229, 235, 240, 241, 242, 248, 252, 260, 261

solvents, 160, 163, 164, 166

solvothermal synthesis, vii, 159

SPCC, 26

specifications, 60, 70

spectroscopy, 39, 118, 237

speed of light, 125, 217

spin, 198, 242

St. Petersburg, 231

stability, ix, 31, 44, 51, 80, 113, 151, 152, 153, 154, 155, 156, 157, 158, 167, 216

Stark effect, 55, 235

state, vii, x, 1, 2, 3, 10, 13, 32, 33, 34, 37, 49, 50, 51, 53, 55, 66, 102, 108, 110, 118, 120, 121, 122, 125, 130, 135, 152, 155, 195, 196, 197, 199, 200, 229, 231, 261, 280

statistics, 238, 239

storage, 43, 45, 59, 60, 61, 69, 72, 76, 77, 81

stress, 44, 59, 60, 71, 81

structural defects, 80

style, 11, 12, 13, 35

subdomains, 248

substitution, 46

substrate, vii, 2, 15, 16, 17, 18, 20, 21, 22, 23, 24, 25, 26, 27, 28, 30, 31, 32, 35, 36, 38, 40, 46, 77, 79, 102, 160, 174, 175, 176, 179, 180, 185, 188, 189, 211, 212, 214, 220, 226, 227, 234, 235, 243, 254, 258, 259, 263, 266

substrates, 17, 18, 19, 20, 21, 28, 33, 36, 79, 227, 229, 235, 246, 281

succession, 57

Sun, 23, 31, 37, 38, 39, 41, 170, 208, 209, 273, 285, 286

superlattice, 235

suppression, 257

surface modification, 16

surface treatment, 258

susceptibility, 52, 135, 136

symmetry, 200, 205, 220, 238

synchronization, 86, 93, 102, 107, 109

synchronize, 105

synthesis, vii, 3, 158, 159

T

Taiwan, 1, 33, 83

target, 20, 37

techniques, vii, ix, 1, 3, 16, 18, 20, 27, 32, 71, 89, 93, 98, 117, 136, 174, 196

technology, vii, viii, ix, x, 1, 2, 32, 59, 69, 70, 73, 74, 78, 79, 81, 83, 84, 94, 95, 98, 114, 117, 129, 158, 212, 231, 233, 276

telecommunications, 78

temperature dependence, 61, 63, 66, 254

tensile strength, 66, 77

tension, 237

TEOS, ix, 151, 158, 160, 161, 162, 164

testing, vii, 5, 43

tetraethoxysilane, 158

thermal activation, 71

thermal activation energy, 71

thermal analysis, 254

thermal decomposition, 159

thermal degradation, 157

thermal expansion, 43

thermal oxidation, 80

thermal resistance, 44, 48, 49, 65, 74

thermal stability, 152, 156, 158

thermal treatment, 156, 161

thermalization, 254

thermodynamic equilibrium, 130

thin films, 33, 46

threshold level, x, 211

tics, 260

tin, 19, 21, 23, 36, 37, 258, 259, 260, 263

tin oxide, 21, 37, 258, 259, 263

TIR, 16, 18, 21, 31, 32, 254, 256

titanium, 151

total internal reflection, 15, 16, 254, 257

transformation, 69, 261

transition rate, 199

transmission, 26, 31, 40, 90, 91, 105, 118, 129, 133, 134, 135, 179, 180, 188, 189, 236, 256, 260, 262

Transmission Electron Microscopy (TEM), 66, 236

transparency, 19, 157, 235

transport, 14, 15, 36, 61, 67, 68, 115, 129, 178, 186, 189, 239, 241, 250, 251, 252

transportation, 4

treatment, 156, 161, 198, 260

trial, 121

triggers, 206

triphenylphosphine, 155

tunneling, 129, 178, 241

twist, 229

U

UK, 149, 276
uniform, 2, 65, 84, 105, 130, 166, 199, 219, 237, 244, 246, 248, 251, 254, 261, 262
updating, 91, 107, 112
USA, 37, 195
UV irradiation, 155
UV light, 153, 155, 160, 166
UV-irradiation, 156

V

vacancies, 235
vacuum, 152, 215, 217, 225, 239, 242
valence, 52, 53, 55, 129, 212, 235, 237, 238, 241, 242, 256
validation, 63, 242
vapor, ix, 20, 45, 151, 153, 156, 159, 166, 174
variables, 252
variations, 60, 72, 126, 137, 138, 142, 144, 185, 187
VCSEL, 211
vector, 125, 127, 217, 218, 242, 252
velocity, 50, 240, 242
vertical-cavity surface-emitting lasers, 228
vibration, 44, 158
videos, 115
vision, 104, 111, 115, 234
visualization, 81

W

Washington, 229
water, 153, 158, 159, 160, 164
wave vector, 127, 217, 218, 261
wavelengths, 120, 121, 122, 123, 124, 125, 126, 129, 136, 146, 224, 229, 236, 246, 250, 254
wear, viii, 84, 89
wells, ix, 12, 39, 46, 117, 118, 120, 121, 122, 125, 129, 130, 132, 133, 136, 146, 247
wide band gap, 118
windows, 95
wires, 21, 29, 32, 254
workers, 18, 31

X

X-axis, 74
XPS, 182

Y

yield, 3, 4, 10, 12, 111, 113
yttrium, 152, 267

Z

zinc, 118, 130
ZnO, 28, 233, 258, 271